Advances in Intelligent Systems and Computing

Volume 428

Series editor

Janusz Kacprzyk, Polish Academy of Sciences, Warsaw, Poland
e-mail: kacprzyk@ibspan.waw.pl

About this Series

The series "Advances in Intelligent Systems and Computing" contains publications on theory, applications, and design methods of Intelligent Systems and Intelligent Computing. Virtually all disciplines such as engineering, natural sciences, computer and information science, ICT, economics, business, e-commerce, environment, healthcare, life science are covered. The list of topics spans all the areas of modern intelligent systems and computing.

The publications within "Advances in Intelligent Systems and Computing" are primarily textbooks and proceedings of important conferences, symposia and congresses. They cover significant recent developments in the field, both of a foundational and applicable character. An important characteristic feature of the series is the short publication time and world-wide distribution. This permits a rapid and broad dissemination of research results.

Advisory Board

Chairman

Nikhil R. Pal, Indian Statistical Institute, Kolkata, India
e-mail: nikhil@isical.ac.in

Members

Rafael Bello, Universidad Central "Marta Abreu" de Las Villas, Santa Clara, Cuba
e-mail: rbellop@uclv.edu.cu

Emilio S. Corchado, University of Salamanca, Salamanca, Spain
e-mail: escorchado@usal.es

Hani Hagras, University of Essex, Colchester, UK
e-mail: hani@essex.ac.uk

László T. Kóczy, Széchenyi István University, Győr, Hungary
e-mail: koczy@sze.hu

Vladik Kreinovich, University of Texas at El Paso, El Paso, USA
e-mail: vladik@utep.edu

Chin-Teng Lin, National Chiao Tung University, Hsinchu, Taiwan
e-mail: ctlin@mail.nctu.edu.tw

Jie Lu, University of Technology, Sydney, Australia
e-mail: Jie.Lu@uts.edu.au

Patricia Melin, Tijuana Institute of Technology, Tijuana, Mexico
e-mail: epmelin@hafsamx.org

Nadia Nedjah, State University of Rio de Janeiro, Rio de Janeiro, Brazil
e-mail: nadia@eng.uerj.br

Ngoc Thanh Nguyen, Wroclaw University of Technology, Wroclaw, Poland
e-mail: Ngoc-Thanh.Nguyen@pwr.edu.pl

Jun Wang, The Chinese University of Hong Kong, Shatin, Hong Kong
e-mail: jwang@mae.cuhk.edu.hk

More information about this series at http://www.springer.com/series/11156

Erzsébet Merényi · Michael J. Mendenhall
Patrick O'Driscoll
Editors

Advances in Self-Organizing Maps and Learning Vector Quantization

Proceedings of the 11th International Workshop WSOM 2016, Houston, Texas, USA, January 6–8, 2016

 Springer

Editors
Erzsébet Merényi
Department of Statistics
Rice University
Houston, TX
USA

Patrick O'Driscoll
Applied Physics
Rice University
Houston, TX
USA

Michael J. Mendenhall
Department of Electrical and Computer
  Engineering
Air Force Institute of Technology
Wright-Patterson AFB, OH
USA

ISSN 2194-5357   ISSN 2194-5365 (electronic)
Advances in Intelligent Systems and Computing
ISBN 978-3-319-28517-7   ISBN 978-3-319-28518-4 (eBook)
DOI 10.1007/978-3-319-28518-4

Library of Congress Control Number: 2015959923

Printed on acid-free paper

This Springer imprint is published by SpringerNature
The registered company is Springer International Publishing AG Switzerland

Preface

This book contains the refereed contributions presented at the 11th Workshop on Self-Organizing Maps (WSOM 2016) held at Rice University, Houston, Texas (USA) during January 6–8, 2016. It brings together more than 90 researchers and practitioners from 15 countries in the field of self-organizing systems for data analysis, with a particular emphasis on self-organizing maps (SOMs) and learning vector quantization (LVQ). The book highlights key advances in these and closely related fields. WSOM 2016 is the 11th in a series of biennial international conferences started with WSOM'97 in Helsinki.

We would like to extend our thanks and gratitude to Prof. Teuvo Kohonen for serving as Honorary Chair of WSOM 2016. We are indebted to the WSOM Steering and Executive Committees for guidance and advice. We greatly appreciate the time, thoughtfulness, and participation of the WSOM 2016 plenary speakers: Prof. Marrie Cottrell (Université Paris 1 Panthéon-Sorbonne, France), Prof. Pablo Estévez (University of Chile and Millennium Institute of Astrophysics, Chile), and Prof. Risto Miikkulainen (University of Texas at Austin, USA). We would like to express our appreciation for all the hard and timely work performed by the Program Committee and additional reviewers. Above all, we would like to thank all of the authors whose contributions made this book a reality.

Special thanks to the local organizers (P. Huitron, Y. Adler, J. Taylor, and J. Vera-Garza, Rice University; J. Climer, Air Force Institute of Technology) for their participation. Their untiring work and attention to all details of WSOM 2016 has been invaluable. Last but not least, we gratefully acknowledge the Department of Statistics, Rice University, The School of Engineering, Rice University, and the Department of Electrical and Computer Engineering, Air Force Institute of Technology, for their support.

Houston, TX
January 2016

Erzsébet Merényi
Michael J. Mendenhall
Patrick O'Driscoll

Contents

Organization

WSOM 2016 was held during January 6–8, 2016 at Rice University, Houston, Texas. The conference was organized by faculty, staff, and students in the Department of Statistics, Rice University, Houston, Texas and in the Department of Electrical and Computer Engineering, Air Force Institute of Technology, Wright-Patterson AFB, Ohio.

Executive Committee

Honorary Chair:
Teuvo Kohonen, Academy of Finland, Finland

General Chair:
Erzsébet Merényi, Rice University, Houston, Texas (United States)

Publication Co-Chair:
Michael J. Mendenhall, Air Force Institute of Technology, Wright-Patterson AFB, Ohio (United States)

Local Chair:
Patrick O'Driscoll, Rice University, Houston, Texas (United States)

Steering Committee

Teuvo Kohonen, Academy of Finland, Finland
Marie Cottrell, Université Paris 1, Pantheón-Sorbonne, France
Pablo Estévez, University of Chile, Chile
Timo Honkela, Aalto University, Finland
Thomas Martinetz, University of Lübeck, Germany

Michel Verleysen, Université Catholique de Louvain, Belgium
Thomas Villmann, University of Applied Sciences Mittweida, Germany

Program Committee

Guilherme Barreto
Michael Biehl
Yoonsuck Choe
Marie Cottrell
Pablo Estévez
Jan Faigl
Barbara Hammer
Marika Kaden
Markus Koskela
John Lee
Paulo Lisboa
Thomas Martinetz
Risto Miikkulainen
Erkki Oja
Madalina Oltenau
Jaakko Peltonen
Gilbert Peterson
Andreas Rauber
Helge Ritter
Fabrice Rossi
Frank-Michael Schleif
Udo Seiffert
Sambu Seo
Kadim Taşdemir
Peter Tino
Alfred Ultsch
Marc van Hulle
Michel Verleysen
Nathalie Villa-Vialaneix
Thomas Villmann
Axel Wismüller
Hujun Yin

Additional Reviewers

Anas Abidin
Ajalmar Rêgo Da Rocha
Adora D'Souza
Jens Hocke
Amir Madany
José Everardo B. Maia
Luis Gustavo Mota Souza
Amouri Holanda Souza Jr.

Sponsoring Institutions

- Brown School of Engineering
  Rice University, Houston, Texas (United States)
- Department of Statistics
  Rice University, Houston, Texas (United States)
- Department of Electrical and Computer Engineering
  Air Force Institute of Technology
  Wright-Patterson AFB, Ohio (United States)
- Human Signatures Branch
  711 Human Performance Wing
  Wright-Patterson AFB, Ohio (United States)

Part I
Self-Organizing Map Learning, Visualization, and Quality Assessment

Theoretical and Applied Aspects
of the Self-Organizing Maps

**Marie Cottrell, Madalina Olteanu, Fabrice Rossi
and Nathalie Villa-Vialaneix**

Abstract The Self-Organizing Map (SOM) is widely used, easy to implement, has nice properties for data mining by providing both clustering and visual representation. It acts as an extension of the k-means algorithm that preserves as much as possible the topological structure of the data. However, since its conception, the mathematical study of the SOM remains difficult and has be done only in very special cases. In WSOM 2005, Jean-Claude Fort presented the state of the art, the main remaining difficulties and the mathematical tools that can be used to obtain theoretical results on the SOM outcomes. These tools are mainly Markov chains, the theory of Ordinary Differential Equations, the theory of stability, etc. This article presents theoretical advances made since then. In addition, it reviews some of the many SOM algorithm variants which were defined to overcome the theoretical difficulties and/or adapt the algorithm to the processing of complex data such as time series, missing values in the data, nominal data, textual data, etc.

Keywords SOM · Batch SOM · Relational SOM · Stability of SOM

1 Brief History of the SOM

Since its introduction by T. Kohonen in his seminal 1982 articles ([33, 34]), the self-organizing map (SOM) algorithm has encountered a very large success. This is due to its very simple definition, to the easiness of its practical development, to

M. Cottrell (✉) · M. Olteanu · F. Rossi
SAMM - Université Paris 1 Panthéon-Sorbonne, 90, rue de Tolbiac,
75013 Paris, France
e-mail: marie.cottrell@univ-paris1.fr

M. Olteanu
e-mail: madalina.olteanu@univ-paris1.fr

F. Rossi
e-mail: fabrice.rossi@univ-paris1.fr

N. Villa-Vialaneix (✉)
INRA, UR 0875 MIAT, BP 52627, 31326 Castanet Tolosan, France
e-mail: nathalie.villa@toulouse.inra.fr

© Springer International Publishing Switzerland 2016   3
E. Merényi et al. (eds.), *Advances in Self-Organizing Maps and Learning
Vector Quantization*, Advances in Intelligent Systems and Computing 428,
DOI 10.1007/978-3-319-28518-4_1

its clustering properties as well as its visualization ability. SOM appears to be a generalization of basic clustering algorithms and at the same time, provides nice visualization of multidimensional data.

The basic version of SOM is an on-line stochastic process which has been inspired by neuro-biological learning paradigms. Such paradigms had previously been used to model some sensory or cognitive processes where the learning is directed by the experience and the external inputs without supervision. For example, [49] illustrate the somatosensory mapping property of SOM. However, quickly in the eighties, SOM was not restricted to neuro-biology modeling and has been used in a huge number of applications (see e.g. [31, 46] for surveys), in very diverse fields as economy, sociology, text mining, process monitoring, etc.

Since then, several extensions of the algorithms have been proposed. For instance, for users who are not familiar with stochastic processes or for industrial applications, the variability of the equilibrium state was seen as a drawback because the learnt map is not always the same from one run to another. To address this issue, T. Kohonen introduces the batch SOM, ([37, 39]) which is deterministic and thus leads to reproducible results (for a given initialization). Also, the initial SOM (on-line or batch versions) was designed for real-valued multidimensional data, and it has been necessary to adapt its definition in order to deal with complex non vectorial data such as categorical data, abstract data, documents, similarity or dissimilarity indexes, as introduced in [30, 32, 35]. One can find in [36, 37, 40–42] extensive lists of references related to SOM. At this moment more than 10 000 papers have been published on SOM or using SOM.

In this paper, we review a large selection of the numerous variants of the SOM. One of the main focus of this survey is the question of *convergence* of the SOM algorithms, viewed as stochastic processes. This departs significantly from the classical learning theory setting. In this setting, exemplified by the pioneering results of [48], one generally assumes given an optimization problem whose solution is interesting: for instance, an optimal solution of the quantization problem associated to the k-means quality criterion. The optimization problem is studied with two points of view: the true problem which involves a mathematical expectation with respect to the (unknown) data distribution and its empirical counterpart where the expectation is approximated by an average on a finite sample. Then the question of convergence (or consistency) is whether the solution obtained on the finite sample converges to the true solution that would be obtained by solving the true problem.

We focus on a quite different problem. A specific stochastic algorithm such as the SOM one defines a series of intermediate configurations (or solutions). Does the series converge to something interesting? More precisely, as the algorithm maps the inputs (the data) to an output (the prototypes and their array), one can take this output as the result of the learning process and may ask the following questions, among others:

- How to be sure that the learning is over?
- Do the prototypes extract a pertinent information from the data set?

- Are the results stables?
- Are the prototypes well organized?

In fact, many of these questions are without a complete answer, but in the following, we review parts of the questions for which theoretical results are known and summarize the main remaining difficulties. Section 2 is devoted to the definition of SOM for numerical data and to the presentation of the general methods used for studying the algorithm. Some theoretical results are described in the next sections: Sect. 3 explains the one dimensional case while in Sect. 4, the results available for the multi-dimensional case are presented. The batch SOM is studied in Sect. 5. In Sect. 6, we present the variants proposed by Heskes to get an energy function associated to the algorithm. Section 7 is dedicated to non numerical data. Finally, in Sect. 8, we focus on the use of the stochasticity of SOM to improve the interpretation of the maps. The article ends with a very short and provisional conclusion.

2 SOM for Numerical Data

Originally, (in [33, 34]), the SOM algorithm was defined for vector numerical data which belong to a subset $\mathcal{X}$ of an Euclidean space (typically $\mathbb{R}^p$). Many results in this paper additionnaly require that the subset is bounded and convex. There are two different settings from the theoretical point of view:

- *continuous setting*: the input space $\mathcal{X}$ can be modeled by a probability distribution defined by a density function f;
- *discrete setting*: the data space $\mathcal{X}$ comprises N data points $x_1, \ldots, x_N$ in $\mathbb{R}^p$ (In this paper, by *discrete setting*, we mean a *finite* subset of the input space).

The theoretical properties are not exactly the same in both cases, so we shall later have to separate these two settings.

2.1 *Classical On-line SOM, Continuous or Discrete Setting*

In this section, let us consider that $\mathcal{X} \subset \mathbb{R}^p$ (*continuous or discrete setting*).

First we specify a regular lattice of K units (generally in a one- or two-dimensional array). Then on the set $\mathcal{K} = \{1, \ldots, K\}$, a neighborhood structure is induced by a neighborhood function h defined on $\mathcal{K} \times \mathcal{K}$. This function can be time dependent and, in this case, it will be denoted by $h(t)$. Usually, h is symmetrical and depends only on the distance between units k and l on the lattice (denoted by dist(k, l) in the following)). It is common to set $h_{kk} = 1$ and to have k_{kl} decrease with increasing distance between k and l. A very common choice is the step function, with value 1 if the distance between k and l is less than a specific radius (this radius can decrease with time), and 0 otherwise. Another very classical choice is

$$h_{kl}(t) = exp\left(-\frac{\text{dist}^2(k,l)}{2\sigma^2(t)}\right),$$

where $\sigma^2(t)$ can decrease over time to reduce the intensity and the scope of the neighborhood relations.

A prototype $m_k \in \mathbb{R}^p$ is attached to each unit k of the lattice. Prototypes are also called models, weight vectors, code-vectors, codebook vectors, centroids, etc. The goal of the SOM algorithm is to update these prototypes according to the presentation of the inputs in such a way that they represent the input space as accurately as possible (in a quantization point of view) while preserving the topology of the data by matching the regular lattice with the data structure. For each prototype m_k, the set of inputs closer to m_k than to any other one defines a cluster (also called a Voronoï cell) in the input space, denoted by C_k, and the neighborhood structure on the lattice induces a neighborhood structure on the clusters. In other words, after running the SOM process, close inputs should belong to the same cluster (as in any clustering algorithm) or to neighbor clusters.

From any initial values of the prototypes, $(m_1(0), \ldots, m_K(0))$, the SOM algorithm iterates the following steps:

1. At time t, if $m(t) = (m_1(t), \ldots, m_K(t))$ denotes the current state of the prototypes, a data point x is drawn according to the density f in $\mathcal{X}$ (*continuous setting*) or at random in the finite set $\mathcal{X}$ (*discrete setting*).
2. Then $c^t(x) \in \{1, \ldots, K\}$ is determined as the index of the *best matching unit*, that is

$$c^t(x) = \arg\min_{k \in \{1,\ldots,K\}} \|x - m_k(t)\|^2, \tag{1}$$

3. Finally, all prototypes are updated via

$$m_k(t+1) = m_k(t) + \epsilon(t)h_{kc^t(x)}(t)(x - m_k(t)), \tag{2}$$

where $\epsilon(t)$ is a learning rate (positive, less than 1, constant or decreasing).

Although this algorithm is very easy to define and to use, its main theoretical properties remain without complete proofs. Only some partial results are available, despite a large amount of works and empirical evidences. More precisely, $(m_k(t))_{k=1,\ldots,K}$ are K stochastic processes in $\mathbb{R}^p$ and when the number t of iterations of the algorithm grows, $m_k(t)$ could have different behaviors: oscillation, explosion to infinity, convergence in distribution to an equilibrium process, convergence in distribution or almost sure to a finite set of points in $\mathbb{R}^p$, etc.

This is the type of convergence that we will discuss in the sequel. In particular, the following questions will be addressed:

• Is the algorithm convergent in distribution or almost surely, when t tends to $+\infty$?
• What happens when ϵ is constant? when it decreases?
• If a limit state exists, is it stable?
• How to characterize the organization?

One can find in [9, 21] a summary of the main rigorous results with most references as well as the open problems without solutions until now.

2.2 Mathematical Tools Related to the Convergence of Stochastic Processes

The main methods that have been used to analyze the SOM convergence are summarized below.

- *The Markov Chain theory* for constant learning rate and neighboring function, which is useful to study the convergence and the limit states. If the algorithm converges in distribution, this limit distribution has to be an invariant measure for the Markov Chain. If it is possible to prove some strong organization, it has to be associated to an absorbing class;
- The *Ordinary Differential Equation method* (ODE), which is a classical method to study the stochastic processes.

If we write down the Eq. (2) for each $k \in \mathcal{K}$ in a vector form, we get

$$m(t + 1) = m(t) - \epsilon(t)\Phi(x, m(t)), \tag{3}$$

where Φ is a stochastic term. To study the behavior of such stochastic processes, it is often useful to study the solutions of the associated deterministic ordinary differential equation that describes the mean behavior of the process. This ODE is

$$\frac{dm}{dt} = -\phi(m), \tag{4}$$

where $\phi(m)$ is the expectation of $\Phi(., m)$ with respect to the probability distribution of the inputs x (*continuous setting*) or the arithmetic mean (*discrete setting*).

Here the kth—component of ϕ is

$$\phi_k(m) = \sum_{j=1}^{K} h_{kj} \int_{C_j} (x - m_k) f(x) dx, \tag{5}$$

for the continuous setting or

$$\phi_k(m) = \frac{1}{N} \sum_{j=1}^{K} h_{kj} \sum_{x_i \in C_j} (x_i - m_k), \tag{6}$$

that can be also written

$$\phi_k(m) = \frac{1}{N} \sum_{i=1}^{N} h_{kc(x_i)}(x_i - m_k), \tag{7}$$

for the discrete setting.

The possible limit states of the stochastic process in Eq. (2) would have to be solutions of the equation

$$\phi(m) = 0.$$

Then if the zeros of this function were the minima of a function (most often called *energy* function), it would be useful to apply the gradient descent methods.

- The Robbins-Monro algorithm theory which is used when the learning rate decreases under conditions

$$\sum_t \varepsilon(t) = +\infty \quad \text{and} \quad \sum_t \varepsilon(t)^2 < +\infty. \tag{8}$$

Unfortunately some remarks explain why the original SOM algorithm is difficult to study. Firstly, for dimension $p > 1$, a problem arises: it is not possible to define any absorbing class which could be an organized state. Secondly, although the process $m(t)$ can be written down as a classical stochastic process of Eq. (3), one knows since the papers [15, 16], that it does not correspond to an energy function, that is it is not a gradient descent algorithm in the *continuous setting*. Finally, it must be emphasized that no demonstration takes into account the variation of the neighborhood function. All the existing results are valid for a fixed size and intensity of the function h.

3   The One-Dimensional Case

A very particular setting is the one-dimensional case: the inputs belong to $\mathbb{R}$ and the lattice is a one-dimensional array (*a string*). Even though this case is of a poor practical utility, it is interesting because the theoretical analysis can be fully conducted.

3.1   *The Simplest One-Dimensional Case*

The simplest case was fully studied in the article [7]. The inputs are supposed to be uniformly distributed in [0, 1], the lattice is a one-dimensional array $\{1, 2, \ldots, K\}$, the learning rate ϵ is a constant smaller than $\frac{1}{2}$, the neighborhood function is a constant step function $h_{kl} = 0$ if $|k - l| > 1$ and 1 otherwise. In that case the process $m(t)$ is a

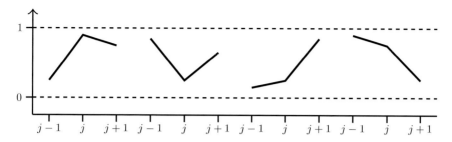

Fig. 1 Four examples of triplets of prototypes (m_{j-1}, m_j, m_{j+1}). For each j, $j-1$ and $j+1$ are its neighbors. The y-axis coordinates are the values of the prototypes that take values in $[0, 1]$. The first two triplets on the *left* are badly ordered. In the case under study, SOM will order them with a strictly positive probability. The last two triplets (on the *right*) are well ordered and SOM will never disorder them

homogeneous Markov Chain with continuous state space. The organization we look for is simply the ordering (ascending or descending) and so is easy to characterize. Let us describe the main steps of the proof.

1. There exists a decreasing functional: the number of badly ordered triplets (Fig. 1). But this is not sufficient to prove the convergence, it has to be strictly decreasing with a strictly positive probability.
2. The set of ordered dispositions is an absorbing class, composed of two classes which do not communicate: the increasing sequences class and the decreasing sequences class.
3. One shows that ordering (topology preservation in this special case) takes place after a finite time with a probability which is greater than a positive bound, and that the hitting time of the absorbing class is almost surely finite.
4. Then one shows that the Markov Chain has the Doeblin property: there exists an integer T, and a constant $c > 0$, such that, given that the process starts from any ordered state, and for all set E in $[0, 1]^n$, with positive measure, the probability to enter in E with less than T steps is greater than $c \, \mathrm{vol}(E)$.
5. This implies that the chain converges in distribution to a monotonous stationary distribution which depends on ϵ (which is a constant in that part).
6. If $\epsilon(t)$ tends towards 0 and satisfies the Robbins-Monro conditions (8), once the state is ordered, the Markov Chain almost surely converges towards a constant (monotonous) solution of an explicit linear system.

So in this very simple case, we could prove the convergence to a unique ordered solution such that

$$m_1(+\infty) < m_2(+\infty) < \cdots < m_K(+\infty),$$

or

$$m_1(+\infty) > m_2(+\infty) > \cdots > m_K(+\infty).$$

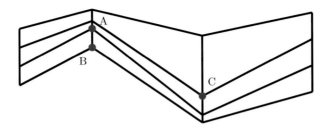

Fig. 2 This figure represents 2-dimensional prototypes (x and y-axes are not shown but are the standard *horizontal* and *vertical* axes) which are linked as their corresponding unit on the SOM grid. At this step of the algorithm, the x- and y- coordinates of the prototypes are well ordered. But contrarily to the one-dimensional case, this disposition can be disordered a with positive probability; in an 8-neighbors case, A is C's neighbor, but B is not a neighbor of C. If C is repeatedly the best matching unit, B is never updated, while A becomes closer and closer to C. Finally, the y coordinate of A becomes smaller than that of B and the disposition is disordered

Unfortunately, it is not possible to find absorbing classes when the dimension is larger than 1. For example, in dimension 2, with 8 neighbors, if the x- and y-coordinates are ordered, it is possible (with positive probability) to disorder the prototypes as illustrated in Fig. 2.

3.2   What We Know About the General One-Dimensional Case

We summarize in this section the essential results that apply to the general one dimension case (with constant neighborhood function and in the continuous setting). References and precise statements can be found in [9, 21]. Compared to the previous section, hypothesis on the data distribution and/or the neighborhood function are relaxed.

- The process $m(t)$ is almost surely convergent to a unique stable equilibrium point in a very general case: $\epsilon(t)$ is supposed to satisfy the conditions (8), there are hypotheses on the density f and on the neighborhood function h. Even though these hypotheses are not very restrictive, some important distributions, such as the χ^2 or the power distribution, do not fulfill them.
- For a constant ϵ, the ordering time is almost surely finite (and has a finite exponential moment).
- With the same hypotheses as before to ensure the existence and uniqueness of a stable equilibrium x^*, from any ordered state, for each constant ϵ, there exists an invariant probability measure π^ϵ. When ϵ tends to 0, this measure concentrates on the Dirac measure on x^*.

- With the same hypotheses as before to ensure the existence and uniqueness of a stable equilibrium x^*, from any ordered state, the algorithm converges to this equilibrium provided that $\epsilon(t)$ satisfies the conditions (8).

As the hypotheses are sufficiently general to be satisfied in most cases, one can say that the one-dimensional case is more or less well-known. However nothing is proved neither about the choice of a decreasing function for $\epsilon(t)$ to ensure simultaneously ordering and convergence, nor for the case of decreasing neighborhood function.

4 Multidimensional Case

When the data are p-dimensional, one has to distinguish two cases, the continuous setting and the discrete one.

4.1 Continuous Setting

In the p-dimensional case, we have only partial results proved by Sadeghi in ([52]). In this paper, the neighborhood function is supposed to have a finite range, the learning rate ϵ is a constant, the probability density function is positive on an interval (this excludes the discrete case). Then the algorithm weakly converges to a unique probability distribution which depends on ϵ.

Nothing is known about the possible topology preservation properties of this stationary distribution. This is a consequence of the difficulty of defining an absorbing organized state in a multi-dimensional setting. For example, two results of Flanagan and Fort-Pagès illustrate the complexity of the problem. These two apparently contradictory results hold. For $p = 2$, let us consider the set F^{++} of simultaneously ordered coordinates (respectively x and y coordinates). We then have:

- for a constant ϵ and very general hypotheses on the density f, the hitting time of F^{++} is finite with a positive probability ([17]),
- *but* in the 8-neighbor setting, the exit time is also finite with positive probability ([22]).

4.2 Discrete Setting

In this setting, the stochastic process $m(t)$ of Eqs. (2) and (4) derives from a potential function, which means that it is a gradient descent process associated to the energy. When the neighborhood function does not depend on time, [50] have proven that the stochastic process $m(t)$ of Eqs. (2) and (3) derives from a potential, that is it can be written

$$
\begin{aligned}
m_k(t+1) &= m_k(t) + \epsilon(t)h_{kc(x)}(t)(x - m_k(t)), \\
&= m_k(t) - \epsilon(t)\Phi_k(x, m(t)), \\
&= m_k(t) - \epsilon(t)\frac{\partial}{\partial m_k}E(x, m(t)),
\end{aligned}
$$

where $E(x, m)$ is a sample function of $E(m)$ with

$$
E(m) = \frac{1}{2N}\sum_{k=1}^{K}\sum_{j=1}^{K}h_{kj}\sum_{x_i \in C_j}\|m_k - x_i\|^2, \tag{9}
$$

or in a shorter expression

$$
E(m) = \frac{1}{2N}\sum_{i=1}^{N}\sum_{k=1}^{K}h_{kc(x_i)}\|m_k - x_i\|^2. \tag{10}
$$

In other words the stochastic process $m(t)$ is a stochastic gradient descent process associated to function $E(m)$. Three interesting remarks can be made:

1. The energy function is a generalization of the *distortion* function (or *intra classes variance* function) associated to the Simple Competitive Learning process (SCL, also known as the Vector Quantization Process/Algorithm), which is the stochastic version of the deterministic Forgy algorithm. The SCL process is nothing else than the SOM process where the neighborhood function is degenerated, i.e. when $h_{kl} = 1$ only for $k = l$ and $h_{kl} = 0$ elsewhere. In that case, E reduces to

$$
E(m) = \frac{1}{2N}\sum_{i=1}^{N}\|m_{c(x_i)} - x_i\|^2.
$$

   For that reason, E is called *extended intra-classes variance*.
2. The above result does not ensure the convergence of the process: in fact the gradient of the energy function is not continuous and the general hypotheses used to prove the convergence of the stochastic gradient descent processes are not valid. This comes from the fact that there are discontinuities when crossing the boundaries of the clusters associated to the prototypes, because the neighbors involved in the computation change from a side to another. However this energy gives an interesting insight on the process behavior.
3. In the 0-neighbor setting, the Vector Quantization algorithm converges, since there is no problem with the neighbors and the gradient is continuous. However there are a lot of local minima and the algorithm converges to one of these minima.

5 Deterministic Batch SOM

As the possible limit states of the stochastic process (2) would have to be solutions of the ODE equation

$$\phi(m) = 0,$$

it is natural to search how to directly get these solutions. The definition of the batch SOM algorithm can be found in [37, 39].

From Eq. (5), in the continuous setting, the equilibrium m^* must satisfy

$$\forall k \in \mathcal{K}, \quad \sum_{j=1}^{K} h_{kj} \int_{C_j} (x - m_k^*) f(x) dx.$$

Hence, for the continuous setting, the solution complies with

$$m_k^* = \frac{\sum_{j=1}^{K} h_{kj} \int_{C_j} x f(x) dx}{\sum_{j=1}^{K} h_{kj} \int_{C_j} f(x) dx}.$$

In the discrete setting, the analogous is

$$m_k^* = \frac{\sum_{j=1}^{K} h_{kj} \sum_{x_i \in C_j} x_i}{\sum_{j=1}^{K} h_{kj} |C_j|} = \frac{\sum_{i=1}^{N} h_{kc(x_i)} x_i}{\sum_{i=1}^{N} h_{kc(x_i)}}$$

Thus, the limit prototypes m_k^* have to be the weighted means of all the inputs which belong to the cluster C_k or to its neighboring clusters. The weights are given by the neighborhood function h.

Using this remark, it is possible to derive the definition of the batch algorithm.

$$m_k(t+1) = \frac{\sum_{j=1}^{K} h_{kj}(t) \int_{C_{j(m_k(t))}} x f(x) dx}{\sum_{j=1}^{K} h_{kj}(t) \int_{C_{j(m_k(t))}} f(x) dx}. \tag{11}$$

for the continuous setting, and

$$m_k(t+1) = \frac{\sum_{i=1}^{N} h_{kc^t(x_i)}(t) x_i}{\sum_{i=1}^{N} h_{kc^t(x_i)}(t)} \tag{12}$$

for the discrete case.

This algorithm is deterministic, and one of its advantages is that the limit states of the prototypes depend only on the initial choices. When the neighborhood is reduced to the unit itself, this batch algorithm for the SOM is nothing else than the classical Forgy algorithm ([18]) for clustering. Its theoretical basis is solid and a study of the

Table 1 Comparison summary

	On-line stochastic	Batch deterministic
No neighbor	VQ, SCL, k-means	Forgy, moving centers
With neighbors	SOM	Batch SOM

convergence can be found in [5]. One can prove ([19, 20]) that it is exactly a quasi-Newtonian algorithm associated to the extended distortion (energy) E (see Eq. (10)), when the probability to observe a x in the sample which is exactly positioned on the median hyperplanes (e.g. the boundaries of C_k) is equal to zero. This assumption is always true in the continuous setting but it is not relevant in the discrete setting since there is no guarantee that data points never belong to the boundaries which vary along the iterations.

The batch SOM algorithm is the extension of the Forgy algorithm with the introduction of the neighborhood between clusters, in the same way as the on-line SOM algorithm is for the Vector Quantization algorithm. It is not exactly a gradient descent algorithm, but it converges to a minimum of the energy E. Obviously there are many local minima. In conclusion, the relations between these clustering algorithms are summarized in Table 1.

6 Other Algorithms Related to SOM

As explained before, the on-line SOM is not a gradient algorithm in the continuous setting ([15, 16]). In the discrete setting, there exists an energy function, which is an extended intra-classes variance as in Eq. (10), but this function is not continuously differentiable. To overcome these problems, [27] proposes to slightly modify the on-line version of the SOM algorithm so it can be seen as a stochastic gradient descent on the same energy function. To do so, he introduces a new hard assignment of the winning unit and a soft version of this assignment.

6.1 Hard Assignment in the Heskes's Rule

In order to obtain an energy function for the on-line SOM algorithm, [27] modifies the rule for computing the best matching unit (BMU). In his setting, Eq. (1) becomes

$$c^t(x) = \arg\min_{k \in \{1,\dots,K\}} \sum_{j=1}^{K} h_{kj}(t)\|x - m_k(t)\|^2 \qquad (13)$$

Table 2 Smoothness of the energy function

	Discrete setting	Continuous setting
Kohonen rule for computing BMU	Energy: discontinuous (but finite on V)	Energy: continuous
	Gradient: discontinuous (infinite on V)	Gradient: discontinuous
Heskes rule for computing BMU	Energy: continuous	Energy: continuous
	Gradient: discontinuous (finite on V)	Gradient: continuous

The energy function considered here is

$$E(m) = \frac{1}{2} \sum_{j=1}^{K} \sum_{k=1}^{K} h_{kj}(t) \int_{x \in C_j(m)} \|x - m_k(t)\|^2 f(x) dx, \qquad (14)$$

where $C_j(m)$ is the cluster (Voronoï cell) associated to the j-th prototype. The regularity properties of the energy function and of its gradient are summarized in Table 2, as discussed in [27].

6.2 Soft Topographic Mapping (STM)

The original SOM algorithm is based on a hard winner assignment. Generalizations based on soft assignments were derived in [24, 27]. First, let us remark that the energy function in the discrete case can also be written as

$$E(m, c) = \frac{1}{2} \sum_{k=1}^{K} \sum_{i=1}^{N} c_{ik} \sum_{j=1}^{K} h_{kj}(t) \|m_j(t) - x_i\|^2$$

where c_{ik} is equal to 1 if x_i belongs to cluster k and zero otherwise. This crisp assignment may be smoothed by considering $c_{ik} \geq 0$ such that $\sum_{k=1}^{K} c_{ik} = 1$. The soft assignments may be viewed as the probabilities of input x_i to belong to class k.

Since the optimization of the energy function with gradient descent-like algorithms would get stuck into local minima, the problem is transformed into a deterministic annealing scheme. The energy function is smoothed by adding an entropy term and transforming it into a "free energy" cost function, parameterized by a parameter β:

$$F(m, c, \beta) = E(m, c) - \frac{1}{\beta} S(c),$$

where $S(c)$ is the entropy term associated to the full energy. For low values of β, only one global minimum remains and may be easily determined by gradient descent or

EM schemes. For $\beta \to +\infty$, the free energy has exactly the same expression as the original energy function.

When using deterministic annealing, one begins by computing the minimum of the free energy at low values of β and then attempts to compute the minimum for higher values of β (β may be chosen to grow exponentially), until the global minimum of the free energy for $\beta \to +\infty$ is equal to the global minimum of the original energy function.

For a fixed value of β, the minimization of the free energy leads to iterating over two steps given by Eqs. (15) and (16), in batch version, and very similar to the original SOM (the neighborhood function h is not varied during the optimization process):

$$\mathbb{P}(x_i \in C_k) = \frac{\exp(-\beta e_{ik})}{\sum_{j=1}^{K} \exp(-\beta e_{ij})}, \tag{15}$$

where $e_{ik} = \frac{1}{2} \sum_{j=1}^{K} h_{jk}(t) \|x_i - m_j(t)\|^2$ and

$$m_k(t) = \frac{\sum_{i=1}^{N} x_i \sum_{j=1}^{K} h_{jk}(t) \mathbb{P}(x_i \in C_j)}{\sum_{i=1}^{N} \sum_{j=1}^{K} h_{jk}(t) \mathbb{P}(x_i \in C_j)} \tag{16}$$

The updated prototypes are written as weighted averages over the data vectors. For $\beta \to +\infty$, the classical batch SOM is retrieved.

6.3 Probabilistic Views on the SOM

Several attempts have been made in order to recast the SOM algorithm (and its variants) into a probabilistic framework, namely the general idea of mixture models (see e.g. [45]). The central idea of those approaches is to constrain a mixture of Gaussian distributions in a way that mimic the SOM grid. Due to the heuristic nature of the SOM, the resulting models depart quite significantly from the SOM algorithms and/or from standard mixture models. We describe below three important variants. Other variants are listed in e.g. [53].

6.3.1 SOM and Regularized EM

One of the first attempts in this direction can be found in [28]. Based on his work on energy functions for the SOM, Heskes shows in this paper that the batch SOM can be seen as a form of regularized Expectation Maximization (EM) algorithm.[1]

---

[1] EM is the standard algorithm for mixture models.

As mentioned above, the starting point of this analysis consists in introducing an isotropic Gaussian mixture with K components. The multivariate Gaussian distributions share a single precision parameter β, with the covariance matrix $\frac{1}{\beta}\mathbf{I}$, and are centered on the prototypes.

However, up to some constant terms, the opposite of the log likelihood of such a mixture corresponds to the k-means quantization error. And therefore, maximizing the likelihood does not provide any topology preservation. Thus Heskes introduces a regularization term which penalizes prototypes that do not respect the prior structure (the term does not depend directly on the data points), see [28] for details. Then Heskes shows that applying the EM principle to the obtained regularized (log) likelihood leads to an algorithm that resembles the batch SOM one.

This interpretation has very interesting consequences, explored in the paper. It is easy for instance to leverage the probabilistic framework to handle missing values in a principled (non heuristic) way. It is also easy to use other mixtures e.g. for non numerical data (such as count data). However, the regularization itself is rather ad hoc (it cannot be easily interpreted as a prior distribution on the parameters, for instance). In addition, the final algorithm is significantly different from the batch SOM. Indeed, as in the case of the STM, crisp assignments are replaced by probabilistic ones (the crispness of the assignments is controlled by the precision parameter β). In addition, as in STM, the neighborhood function is fixed (as it is the core of the regularization term). To our knowledge, the practical consequences of those differences have not been studied in detail on real world data. While one can argue that β can be increased progressively and at the same time, one can modify the neighborhood function during the EM algorithm, this might also have consequences that remain untested.

6.3.2 SOM and Variational EM

Another take at this probabilistic interpretation can be found in [53]. As in [28] the first step consists in assuming a standard mixture model (e.g. a K components Gaussian isotropic mixture for multivariate data). Then the paper leverages the variational principle (see e.g. [29]).

In summary, the variational principle is based on introducing an arbitrary distribution q on the latent (hidden) variables Z of the problem under study. In a standard mixture model, the hidden variables are the assignment ones, which map each data point to a component of the mixture (a cluster in the standard clustering language). One can show that the integrated log likelihood of a mixture model with Θ as parameters, $\log p(X|\Theta)$, is equal to the sum of three components: the complete likelihood (knowing both the data points X and the hidden variables Z) integrated over the hidden variables with respect to q, $\mathbb{E}_q \log p(X, Z|\Theta)$, the entropy of q, $H(q)$, and the Kullback-Leibler divergence, $KL(q|p(Z|X, \Theta))$, between q and the posterior distribution of the hidden variables knowing the data points $p(Z|X, \Theta)$. This equality allows one to derive the EM algorithm when the posterior distribution of the hidden variables knowing the data points can be calculated exactly. The variational approach consists in replacing this distribution by a simpler one when it cannot be calculated.

In standard mixture models (such as the multivariate Gaussian mixture), the variational approach is not useful as the posterior distribution of the hidden variables can be calculated. However [53] propose nevertheless to use the variational approach as a way to enforce regularity in the mixture model. Rather than allowing $p(Z|X, \Theta)$ to take an arbitrary form, they constrain it to a subset of probability distributions on the hidden variables that fulfill topological constraint corresponding to the prior structure of the SOM. See [53] for details.

This solution shares most of the advantages of the older proposal in [28], with the added value of being based on a more general principle that can be applied to any mixture model (in practice, [28] makes sense only for the exponential family). In addition, [53] study the effects of shrinking the neighborhood function during training and conclude that it improves the quality of the solutions. Notice that, in [53], the shared precision of the Gaussian distributions (β) is not a meta-parameter as in [28] but a regular parameter that is learned from the data.

6.3.3 The Generative Topographic Mapping

The Generative Topographic Mapping (GTM, [2]) is frequently presented as a probabilistic version of the SOM. It is rather a mixture model inspired by the SOM rather than an adaptation. Indeed the aim of the GTM designers was not to recover a learning algorithm close to a SOM variant, but rather to introduce a mixture model that enforce topology preservation.

The GTM is based on uniform prior distribution on a fixed grid which is mapped via an explicit smooth nonlinear mapping to the data space (with some added isotropic Gaussian noise). It can be seen as a constrained Gaussian mixture, but with yet another point of view compared to [28, 53]. In [28], the constraint is enforced by a regularization term on the data space distribution while in [53] the constraint is induced at the latent variable level (via approximating $p(Z|X, \Theta)$ by a smooth distribution). In the GTM the constraint is induced on the data space distribution because it is computed via a smooth mapping. In other words, the centers of the Gaussian distributions are not freely chosen but rather obtained by mapping a fixed grid to the data space via the nonlinear mapping.

The nonlinear mapping is in principle arbitrary and can therefore implement various type of regularity (i.e. topology constraints). The use of Gaussian kernels lead to constraints that are quite similar to the SOM constraints. Notice that those Gaussian kernels are not to be confused with the isotropic Gaussian distributions used in the data space (the same confusion could arise in [53] where Gaussian kernels can be used to specify the constraints on $p(Z|X, \Theta)$).

Once the model has been specify (by choosing the nonlinear mapping), its parameters are estimated via an EM algorithm. The obtained algorithm is quite different from the SOM (see [28] for details), at least in its natural formulation. However the detailed analysis contained in [28] shows that the GTM can be reformulated in a way that is close to the batch SOM with probabilistic assignments (as in e.g. the STM). Once again, however, this is not exactly the same algorithm. In practice, the results on real world data can be quite different. Also, as all the probabilistic variants discussed in this section, the GTM benefits from the probabilistic setting that enables principled missing data analysis as well as easy extensions to the exponential family of distributions in order to deal with non numerical data.

7 Non Numerical Data

When the data are not numerical, the SOM algorithm has to be adapted. See for example [11, 30, 32, 35, 38, 40–43], where some of these adaptations are presented. Here we deal with categorical data collected in surveys and with abstract data which are known only by a dissimilarity matrix or a kernel matrix.

7.1 Contingency Table or Complete Disjunctive Table

Surveys collect answers of the surveyed individuals who have to choose an answer to several questions among a finite set of possible answers. The data can consist in

- a simple contingency table, where there are only two questions, and where the entries are the numbers of individuals who choose a given pair of categories,
- a Burt table, that is a full contingency table between all the pairs of categories of all the questions,
- a complete disjunctive table that contains the answers of all the individuals, coded in 0/1 against dummy variables which represent all the categories of all the questions.

In all these settings, the data consist in a positive integer-valued matrix, which can be seen as a large "contingency table". In classical data analysis, one uses Multiple Correspondence Analysis (MCA) that are designed to deal with these tables. MCA is nothing else than two simultaneous weighted Principal Component Analysis (PCA) of the table and of its transposed, using the χ^2 distance instead of the Euclidean distance. To use a SOM algorithm with such tables, it is therefore sufficient to apply a transformation to the data, in order to take into account the χ^2 distance and the weighting, in the same way that it is defined to use Multiple Correspondence Analysis. After transforming the data, two coupled SOM using the rows and the columns of the transformed table can thus be trained. In the final map, related categories belong to the same cluster or to neighboring clusters. The reader interested by a detailed explanation of the algorithm can refer to [3]. More details and real-world examples can also be found in [8, 10]. Notice that the transformed tables are numerical data tables, and so there is no particular theoretical results to comment on. All the results that we presented for numerical data still hold.

7.2 Dissimilarity Data

In some cases, complex data such as graphs (social networks) or sequences (DNA sequences) are described through relational measures of resemblance or dissemblance, such as kernels or dissimilarity matrices. For these general situations, several extensions of the original algorithm, both in on-line and batch versions, were proposed during the last two decades. A detailed review of these algorithms is available in [51].

More precisely, these extensions consider the case where the data are valued in an arbitrary space $\mathcal{X}$, which is not necessarily Euclidean. The observations are described either by a pairwise dissimilarity $\mathbf{D} = (\delta(x_i, x_j))_{i,j=1,\dots,N}$, or by a kernel matrix $\mathbf{K} = (K(x_i, x_j))_{i,j=1,\dots,N}$.[2] The kernel matrix $\mathbf{K}$ naturally induces an Euclidean distance matrix, but the dissimilarity matrix $\mathbf{D}$ may not necessarily be transformed into a kernel matrix.

The first class of algorithms designed for handling relational data is based on the median principle (*median SOM*): prototypes are forced to be equal to an observation, or to a fixed number of observations. Hence, optimal prototypes are computed by searching through $(x_i)_{i=1,\dots,N}$, instead of $\mathcal{X}$, as in [13, 14, 32, 43]. The original steps of the algorithm are thus transformed in a discrete optimization scheme, which is performed in batch mode:

1. Equation (1) is replaced by the affectation of *all* data to their best matching units: $c(x_i) = \arg\min_{k=1,\dots,K} \delta(x_i, m_k(t))$;
2. Equation (2) is replaced by the update of all prototypes within the dataset $(x_i)_{i=1,\dots,N}$: $m_k(t) = \arg\min_{x_i \,:\, i=1,\dots,N} \sum_{j=1}^{N} h_{c(x_j)k}(t)\delta(x_i, x_j)$.

Since the algorithm explores a finite set when updating the prototypes, it is necessarily convergent to a local minimum of the energy function. However, this class of algorithms exhibits strong limitations, mainly due to the restriction of the prototypes to the dataset, in particular, a large computational cost (despite efficient implementations such as in [6]) and no interpolation effect which yields to a deterioration of the quality of the map organization.

The *second class* of algorithms, *kernel SOM and relational/dissimilarity SOM*, rely on expressing prototypes as convex combinations of the input data. Although these convex combinations do not usually have sense in $\mathcal{X}$ (consider, for instance, that input data are various texts), a convenient embedding in an Euclidean or a pseudo-Euclidean space gives a sound theoretical framework and gives sense to linear combinations of inputs.

For *kernel SOM*, it is enough to use the kernel trick as given by [1] which prove that there exists a Hilbert space $\mathcal{H}$, also called feature space, and a mapping $\psi : \mathcal{X} \to \mathcal{H}$,

---

[2]A kernel is a particular case of symmetric similarity such that K is a symmetric matrix, semi-definite positive with $K(x_i, x_i) = 0$ and satisfies the following positive constraint

$$\forall M > 0, \ \forall (x_i)_{i=1,\dots,M} \in \mathcal{X}, \ \forall (\alpha_i)_{i=1,\dots,M}, \quad \sum_{i,j} \alpha_i \alpha_j K(x_i, x_j) \geq 0.$$

called feature map, such that $K(x, x') = \langle \psi(x), \psi(x') \rangle_{\mathcal{H}}$. In the case where data are described by a *symmetric dissimilarity measure*, they may be embedded in a pseudo-Euclidean space $\psi : x \in \mathcal{X} \rightarrow \psi(x) = (\psi^+(x), \psi^-(x)) \in \mathcal{E}$, as suggested in [23]. $\mathcal{E}$ may be written as the direct decomposition of two Euclidean spaces, $\mathcal{E}_+$ and $\mathcal{E}_-$, with a non-degenerate and indefinite inner product defined as

$$\langle \psi(x), \psi(y) \rangle_{\mathcal{E}} = \langle \psi^+(x), \psi^+(u) \rangle_{\mathcal{E}^+} - \langle \psi^-(x), \psi^-(u) \rangle_{\mathcal{E}^-}$$

The distance naturally induced by the pseudo-Euclidean inner product is not necessarily positive.

For both kernel and relational/dissimilarity SOM, the input data are embedded in $\mathcal{H}$ or $\mathcal{E}$ and prototypes are expressed as convex combinations of the images of the data by the feature maps. For example, in the kernel case,

$$m_k(t) = \sum_{i=1}^{N} \gamma_{ki}^t \psi(x_i) \text{, with } \gamma_{ki}^t \geq 0 \text{ and } \sum_i \gamma_{ki}^t = 1.$$

The above writing of the prototypes allows the computation of the distance from an input x_i to a prototype $m_k(t)$ in terms of the coefficients γ_{ki}^t and the kernel/dissimilarity matrix only. For kernel SOM, one has

$$\|\psi(x_i) - m_k(t)\|^2 = \left(\gamma_k^t\right)^T \mathbf{K}\gamma_k^t - 2\mathbf{K}_i\gamma_k^t + \mathbf{K}_{ii} \text{ ,}$$

where $\mathbf{K}_i$ is the ith row of $\mathbf{K}$ and $\left(\gamma_k^t\right)^T = \left(\gamma_{k,1}^t, \ldots, \gamma_{k,N}^t\right)$. For relational/dissimilarity SOM, one obtains a similar expression

$$\|\psi(x_i) - m_k(t)\|^2 = \mathbf{D}_i\gamma_k^t - \frac{1}{2}\left(\gamma_k^t\right)^T \mathbf{D}\gamma_k^t \text{ .}$$

The first step of the algorithm, finding the best matching unit of an observation, as introduced in Eq. (1), can thus be directly generalized to kernels and dissimilarities, both for on-line and batch settings.

In the batch framework, the updates of the prototypes are identical to the original algorithm (see Eq. (12)), by simply noting that only the coefficients of the x_i's (or of their images by the feature maps) are updated:

$$m_k(t+1) = \sum_{i=1}^{N} \frac{h_{kc^t(x_i)}(t)}{\sum_{j=1}^{N} h_{kc^t(x_j)}(t)} \psi(x_i) \Leftrightarrow \gamma_{ki}^{t+1} = \frac{h_{kc^t(x_i)}(t)}{\sum_{j=1}^{N} h_{kc^t(x_j)}(t)} \qquad (17)$$

This step is the same, both for batch kernel SOM, [54], and for batch relational SOM, [26].

In the on-line framework, updating the prototypes is similar to the original algorithm, as in Eq. (2). Here also, the update rule concerns the coefficients γ_{ki}^t only, and the linear combination of them remains convex:

$$\gamma_k^{t+1} = \gamma_k^t + \varepsilon(t) h_{kc^t(x_i)}(t) \left(\mathbf{1}_i - \gamma_k^t\right),\tag{18}$$

where x_i is the current observation and $\mathbf{1}_i$ is a vector in $\mathbb{R}^N$, with a single non-null coefficient, equal to 1, on the i-th position. As previously, this step is identical for on-line kernel SOM, [44], and for on-line relational SOM, [47].

In the case where the dissimilarity is the squared distance induced by the kernel, kernel SOM and relational SOM are strictly equivalent. Moreover, in this case, they are also fully equivalent to the original SOM algorithm for numerical data in the feature (implicit) Euclidean space induced by the dissimilarity or the kernel, as long as the prototypes are initialized in the convex hull of the input data. The latter assertion induces that the theoretical limitations of the original algorithm also exist for the general kernel/relational versions. Furthermore, these may worsen for the relational version since the non-positivity of the dissimilarity measure adds numerical instability when using a gradient-descent like scheme for updating the prototypes.

The *third class* of algorithms uses the *soft topographic maps* setting introduced in Sect. 6.2. Indeed, in the algorithm described in Eqs. (15) and (16), the soft assignments depend on the distances between input data and prototypes only, while prototypes update consists in making an update if the coefficients of the input data. Using a mean-field approach and similarly to the previous framework for kernel and dissimilarity/relational SOM, [24] obtain the extensions of soft topographic mapping (STM) algorithm for kernels and dissimilarities. The updates for the prototype coefficients are then expressed as

$$\gamma_{ki}(t+1) = \frac{\sum_{j=1}^{K} h_{jk}(t)\mathbb{P}(x_i \in A_j)}{\sum_{l=1}^{N} \sum_{j=1}^{K} h_{jk}(t)\mathbb{P}(x_l \in A_j)},\tag{19}$$

where $m_k(t) = \sum_{i=1}^{N} \gamma_{ki}^t \psi(x_i)$ and ψ is the feature map.

8 Stochasticity of the Kohonen Maps for the On-line Algorithm

Starting from a given initialization and a given size of the map, different runs of the on-line stochastic SOM algorithm provide different resulting maps. On the contrary, the batch version of the algorithm is a deterministic algorithm with always provides the same results for a given initialization. For this reason, the batch SOM algorithm is often preferred over the stochastic one because its results are reproducible. However, this hides the fact that all the pairs of observations which are associated in a given cluster do not have the same significance. More precisely, interpreting a SOM result, we can use the fact that close input data belong to close clusters, i.e. their best matching units are identical or adjacent. But if two given observations are classified in the same or in neighboring units of the map, then they may not be close in the input space. This drawback comes from the fact that there is not perfect matching between a multidimensional space and a one- or two-dimensional map.

More precisely, given a pair of observations (data), $\{x_i, x_j\}$, three cases can be distinguished, depending on the way their respective mapping on the map can be described:

- *significant association*: the pair is classified in the same cluster or in neighboring clusters because x_i and x_j are close in the input space. The observations are said to attract each other;
- *significant non-association*: the pair is never classified in neighboring clusters and x_i and x_j are remote in the input space. The observations are said to repulse each other;
- *fickle pair*: the pair is sometimes classified in the same cluster or in neighboring clusters but x_i and x_j are not close in the input space: their proximity on the map is due to randomness.

The stochasticity of the on-line SOM results can be used to precisely qualify every pairs of observations by performing several runs of the algorithm. The question is addressed in a bootstrap framework in [12] and used for text mining applications in [3, 4]. The idea is simple: since the on-line SOM algorithm is stochastic, its repetitive use may allow to identify the pairs of data in each case.

More precisely, if L is the number of different and independent runs of the on-line SOM algorithm and if $Y_{i,j}$ denotes the number of times x_i and x_j are neighbors on the resulting map in the L runs, a *stability index* can be defined: for the pair (x_i, x_j), this index is equal to:

$$\mathcal{M}_{i,j} = \frac{Y_{i,j}}{L}.$$

Using an approximation of the binomial distribution that would hold if the data were neighbors by chance in a pure random way, and a test level of 5%, for a K-units map, the following quantities are introduced

$$A = \frac{9}{K} \text{ and } B = 1.96\sqrt{\frac{9}{KL}\left(1 - \frac{9}{K}\right)}. \tag{20}$$

These values give the following decision rule to qualify every pair $\{x_i, x_j\}$:

- if $\mathcal{M}_{i,j} > A + B$, the association between the two observations is significantly frequent;
- if $A - B \leq \mathcal{M}_{i,j} \leq A + B$, the association between the two observations is due to randomness. $\{x_i, x_j\}$ is called a fickle pair;
- if $\mathcal{M}_{i,j} < A - B$, the non-association between the two observations is significantly frequent.

In [12], the method is used in order to qualify the stability and the reliability of the global Kohonen map, while both other papers ([3, 4]) study the fickle data pairs for themselves. In these last works, the authors also introduce the notion of *fickle word* which is defined as an observation which belongs to a huge number of *fickle pairs* by choosing a threshold.

These fickle pairs and fickle words can be useful in various way: first, fickle pairs can be used to obtain more robust maps, by distinguishing stable neighboring and non neighboring pairs from fickle pairs. Also, once identified, fickle words can be removed from further studies and representations: for instance, Factorial Analysis visualization is improved. In a text mining setting, [4] have shown that a graph of co-occurrences between words can be simplified by removing fickle words and [3] have used the fickle words for interpretation: they have shown that the fickle words form a lexicon shared between the studied texts.

9   Conclusion

We have reviewed some of the variants of the SOM, for numerical and non numerical data, in their stochastic (on-line) and batch versions. Even if a lot of theoretical properties are not rigorously proven, the SOM algorithms are very useful tools for data analysis in different contexts. Since the Heskes's variants of SOM have a more solid theoretical background, SOM can appear as an easy-to-develop approximation of these well-founded algorithms. This remark should ease the concern that one might have about it.

On a practical point of view, SOM is used as a statistical tool which has to be combined with other techniques, for the purpose of visualization, of vector quantization acceleration, graph construction, etc. Moreover, in a big data context, SOM-derived algorithms seem to have a great future ahead since the computational complexity of SOM is low (proportional to the number of data). In addition, it is always possible to train the model with a sample randomly extracted from the database and then to continue the training in order to adapt the prototypes and the map to the whole database. As most of the stochastic algorithm, SOM is particularly well suited for stream data (see [25] which proposed a "patch SOM" to handle this kind of data). Finally, it would also be interesting to have a look at the robust associations revealed by SOM, to improve the representation and the interpretation of too verbose and complex information.

References

1. Aronszajn, N.: Theory of reproducing kernels. Trans. Am. Math. Soc. **68**(3), 337–404 (1950)
2. Bishop, C.M., Svensén, M., Williams, C.K.I.: GTM: the generative topographic mapping. Neural Comput. **10**(1), 215–234 (1998)
3. Bourgeois, N., Cottrell, M., Deruelle, B., Lamassé, S., Letrémy, P.: How to improve robustness in kohonen maps and display additional information in factorial analysis: application to text minin. Neurocomputing **147**, 120–135 (2015a)

4. Bourgeois, N., Cottrell, M., Lamassé, S., Olteanu, M.: Search for meaning through the study of co-occurrences in texts. In: Rojas, I., Joya, G., Catala, A. (eds.) Advances in Computational Intelligence, Proceedings of IWANN 2015, Part II. LNCS, vol. 9095, pp. 578–591. Springer, Switzerland (2015b)
5. Cheng, Y.: Convergence and ordering of Kohonen's batch map. Neural Comput. **9**, 1667–1676 (1997)
6. Conan-Guez, B., Rossi, F., El Golli, A.: Fast algorithm and implementation of dissimilarity self-organizing maps. Neural Netw. **19**(6–7), 855–863 (2006)
7. Cottrell, M., Fort, J.C.: Étude d'un processus d'auto-organisation. Annales de l'IHP, section B **23**(1), 1–20 (1987)
8. Cottrell, M., Letrémy, P.: How to use the Kohonen algorithm to simultaneously analyse individuals in a survey. Neurocomputing **63**, 193–207 (2005)
9. Cottrell, M., Fort, J.C., Pagès, G.: Theoretical aspects of the SOM algorithm. Neurocomputing **21**, 119–138 (1998)
10. Cottrell, M., Ibbou, S., Letrémy, P.: Som-based algorithms for qualitative variables. Neural Netw. **17**, 1149–1167 (2004)
11. Cottrell, M., Olteanu, M., Rossi, F., Rynkiewicz, J., Villa-Vialaneix, N.: Neural networks for complex data. Künstliche Intelligenz **26**(2), 1–8 (2012). doi:10.1007/s13218-012-0207-2
12. de Bodt, E., Cottrell, M., Verleisen, M.: Statistical tools to assess the reliability of self-organizing maps. Neural Netw. **15**(8–9), 967–978 (2002). doi:10.1016/S0893-6080(02)00071-0
13. El Golli, A., Conan-Guez, B., Rossi, F.: Self organizing map and symbolic data. J. Symb. Data Anal. **2**(1), 11 (2004a)
14. El Golli, A., Conan-Guez, B., Rossi, F.: A self organizing map for dissimilarity data. In Banks, D., House, L., McMorris, F.R., Arabie, P., Gaul, W. (eds.) Classification, Clustering, and Data Mining Applications (Proceedings of IFCS 2004), pp. 61–68. IFCS, Chicago, Illinois (USA), Springer, 7 (2004b)
15. Erwin, E., Obermayer, K., Schulten, K.: Self-organizing maps: ordering, convergence properties and energy functions. Biol. Cybern. **67**(1), 47–55 (1992a)
16. Erwin, E., Obermayer, K., Schulten, K.: Self-organizing maps: stationnary states, metastability and convergence rate. Biol. Cybern. **67**(1), 35–45 (1992b)
17. Flanagan, J.A.: Self-organisation in Kohonen's som. Neural Netw. **6**(7), 1185–1197 (1996)
18. Forgy, E.W.: Cluster analysis of multivariate data: efficiency versus interpretability of classifications. Biometrics **21**, 768–769 (1965)
19. Fort, J.-C., Cottrell, M., Letrémy, P.: Stochastic on-line algorithm versus batch algorithm for quantization and self organizing maps. In: Neural Networks for Signal Processing XI, 2001, Proceedings of the 2001 IEEE Signal Processing Société Workshop, pp. 43–52. IEEE, North Falmouth, MA, USA (2001)
20. Fort, J.-C., Letrémy, P., Cottrell, M.: Advantages and drawbacks of the batch Kohonen algorithm. In Verleysen, M. (ed.) European Symposium on Artificial Neural Networks, Computational Intelligence and Machine Learning (ESANN 2002), pp. 223–230. d-side publications, Bruges, Belgium (2002)
21. Fort, J.C.: SOM's mathematics. Neural Netw. **19**(6–7), 812–816 (2006)
22. Fort, J.C., Pagès, G.: About the kohonen algorithm: strong or weak self-organisation. Neural Netw. **9**(5), 773–785 (1995)
23. Goldfarb, L.: A unified approach to pattern recognition. Pattern Recognit. **17**(5), 575–582 (1984). doi:10.1016/0031-3203(84)90056-6
24. Graepel, T., Burger, M., Obermayer, K.: Self-organizing maps: generalizations and new optimization techniques. Neurocomputing **21**, 173–190 (1998)
25. Hammer, B., Hasenfuss, A.: Topographic mapping of large dissimilarity data sets. Neural Comput. **22**(9), 2229–2284 (2010)
26. Hammer, B., Hasenfuss, A., Rossi, F., Strickert, M.: Topographic processing of relational data. In: Group, Bielefeld University Neuroinformatics (ed.) Proceedings of the 6th Workshop on Self-Organizing Maps (WSOM 07). Bielefeld, Germany, September 2007

27. Heskes, T.: Energy functions for self-organizing maps. In: Oja, E., Kaski, S. (eds.) Kohonen Maps, pp. 303–315. Elsevier, Amsterdam (1999). http://www.snn.ru.nl/reports/Heskes.wsom. ps.gz
28. Heskes, Tom: Self-organizing maps, vector quantization, and mixture modeling. IEEE Trans. Neural Netw. 12(6), 1299–1305 (2001)
29. Jordan, M.I., Ghahramani, Z., Jaakkola, T.S., Saul, L.K.: An introduction to variational methods for graphical models. Mach. Learn. 37(2), 183–233 (1999)
30. Kaski, S., Honkela, T., Lagus, K., Kohonen, T.: Websom—self-organizing maps of document collections. Neurocomputing 21(1), 101–117 (1998a)
31. Kaski, S., Jari, K., Kohonen, T.: Bibliography of self-organizing map (SOM) papers: 1981–1997. Neural Comput. Surv. 1(3&4), 1–176 (1998b)
32. Kohohen, T., Somervuo, P.J.: Self-organizing maps of symbol strings. Neurocomputing 21, 19–30 (1998)
33. Kohonen, T.: Self-organized formation of topologically correct feature maps. Biol. Cybern. 43, 59–69 (1982a)
34. Kohonen, T.: Analysis of a simple self-organizing process. Biol. Cybern. 44, 135–140 (1982b)
35. Kohonen, T.: Median strings. Pattern Recognit. Lett. 3, 309–313 (1985)
36. Kohonen, T.: Self-organization and Associative Memory. Springer, Berlin (1989)
37. Kohonen, T.: Self-Organizing Maps. Springer Series in Information Science, vol. 30. Springer, Berlin (1995)
38. Kohonen, T.: Self-organizing maps of symbol strings, Technical Report a42, Laboratory of computer and information science, Helsinki University of technoligy, Finland (1996)
39. Kohonen, T.: Comparison of som point densities based on different criteria. Neural Comput. 11, 2081–2095 (1999)
40. Kohonen, T.: Self-Organizing Maps, vol. 30, 3rd edn. Springer, Berlin (2001)
41. Kohonen, T.: Essentials of self-organizing map. Neural Netw. 37, 52–65 (2013)
42. Kohonen, T.: MATLAB Implementations and Applications of the Self-Organizing Map. Unigrafia Oy, Helsinki (2014)
43. Kohonen, T., Somervuo, P.J.: How to make large self-organizing maps for nonvectorial data. Neural Netw. 15(8), 945–952 (2002)
44. Mac Donald, D., Fyfe, C.: The kernel self organising map. In: Proceedings of 4th International Conference on knowledge-based Intelligence Engineering Systems and Applied Technologies, pp. 317–320 (2000)
45. McLachlan, G., Peel. D.: Finite mixture models. Wiley (2004)
46. Oja, M., Kaski, S., Kohonen, T.: Bibliography of self-organizing map (SOM) papers: 1998–2001 addendum. Neural Comput. Surv. 3, 1–156 (2003)
47. Olteanu, M., Villa-Vialaneix, N.: On-line relational and multiple relational SOM. Neurocomputing 147, 15–30 (2015). doi:10.1016/j.neucom.2013.11.047
48. Pollard, D., et al.: Strong consistency of k-means clustering. Ann. Stat. 9(1), 135–140 (1981)
49. Ritter, H., Schulten, K.: On the stationary state of Kohonen's self-organizing sensory mapping. Biol. Cybern. 54, 99–106 (1986)
50. Ritter, H., Martinetz, T., Schulten, K.: Neural Computation and Self-Organizing Maps, an Introduction. Addison-Wesley (1992)
51. Rossi, F.: How many dissimilarity/kernel self organizing map variants do we need? In Villmann, T., Schleif, F.M., Kaden, M., Lange, M. (eds.) Advances in Self-Organizing Maps and Learning Vector Quantization (Proceedings of WSOM 2014), Advances in Intelligent Systems and Computing, vol. 295, pages 3–23. Mittweida, Germany, Springer, Berlin (2014). doi:10.1007/978-3-319-07695-9_1
52. Sadeghi, A.: Convergence in distribution of the multi-dimensional Kohonen algorithm. J. of Appl. Probab. 38(1), 136–151 (2001)
53. Verbeek, J.J., Vlassi, N., Kröse, B.J.A.: Self-organizing mixture models. Neurocomputing 63, 99–123 (2005)
54. Villa, N., Rossi, F.: A comparison between dissimilarity SOM and kernel SOM for clustering the vertices of a graph. In: 6th International Workshop on Self-Organizing Maps (WSOM: 2007) Bielefield, Germany, 2007. Bielefield University, Neuroinformatics Group (2007). ISBN 978-3-00-022473-7. doi:10.2390/biecoll-wsom2007-139

Aggregating Self-Organizing Maps with Topology Preservation

Jérôme Mariette and Nathalie Villa-Vialaneix

Abstract In the online version of Self-Organizing Maps, the results obtained from different instances of the algorithm can be rather different. In this paper, we explore a novel approach which aggregates several results of the SOM algorithm to increase their quality and reduce the variability of the results. This approach uses the variability of the algorithm that is due to different initialization states. We use simulations to show that our result is efficient to improve the performance of a single SOM algorithm and to decrease the variability of the final solution. Comparison with existing methods for bagging SOMs also show competitive results.

Keywords Self-Organizing Maps · Aggregation · Topology preservation

1 Introduction

Self-Organizing Maps (SOM), [1] have been shown to be powerful methods for analyzing high dimensional and complex data (see, for instance, [2] for applications of the method to many different areas). However, the method suffers from its lack of good convergence properties. In its original version, the theoretical convergence of the algorithm has only be proved in very limited cases [3] and even in the modified version in which the training of the SOM is expressed as an energy minimization problem [4], different runs of the algorithm give different results, that can be very dependent on the initialization. This problem is even more critical when the data set to be analyzed is complex or high dimensional.

This paper addresses the issue of aggregating several results of the SOM algorithm, all obtained on the same data set. Several attempts to combine SOMs while preserving their topological properties have been proposed in the literature [5–9]. In this paper,

J. Mariette (✉) · N. Villa-Vialaneix
INRA, UR 0875 MIA-T, BP 52627 , 31326 Castanet Tolosan Cedex, France
e-mail: jerome.mariette@toulouse.inra.fr

N. Villa-Vialaneix
e-mail: nathalie.villa@toulouse.inra.fr

© Springer International Publishing Switzerland 2016
E. Merényi et al. (eds.), *Advances in Self-Organizing Maps and Learning Vector Quantization*, Advances in Intelligent Systems and Computing 428,
DOI 10.1007/978-3-319-28518-4_2

we present a novel method to combine several SOMs while preserving their topology. The proposed method combines several ideas taken from the different methods and allows to explore initialization states. It is both simple and efficient. We present a full comparison of the different options to aggregate the results of different SOMs and discuss the most relevant choices. Finally, we show that our approach is a competitive alternative to the existing methods on real data applications.

The remainder of the paper is organized as follows: in Sect. 2, an overview of aggregation methods for SOMs is presented. In Sect. 3, the proposed method is described. Finally, Sect. 4 presents experimental results and comparisons.

2   An Overview of Aggregation Methods for SOMs

Suppose that B results of the SOM algorithm are given for the items $(x_i)_{i=1,\ldots,n}$, $(\mathcal{M}^b)_{b=1,\ldots,B}$. Each of these results, $\mathcal{M}^b$ is well defined by its set of prototypes $(p_u^b)_{u=1,\ldots,U}$ and comes with an associated clustering function $\phi^b : x \in \mathbb{R}^d \rightarrow \arg\min_{u=1,\ldots,U} \|x - p_u^b\|^2$. For the b-th SOM, the clusters will be denoted by $(\mathcal{C}_u^b)_{u=1,\ldots,U}$, where $\mathcal{C}_u^b = \{x_i : \phi^b(x_i) = u\}$. The purpose is to build a *fused* or a *merged* map, $\mathcal{M}^*$, with prototypes $(p_u^*)_{u=1,\ldots,U}$ and a clustering function ϕ^* which improves and summarizes the B maps into a unique consensual map. Note that all SOMs have been trained from the same data $(x_i)_{i=1,\ldots,n}$ or from a subset (e.g., a bootstrap sample) of this data set. They can also have been trained from different descriptors of the observations (e.g., from different sets of variables observed on the same items): in this case, the fused map thus corresponds to a map integrating the different descriptors. However, for the sake of simplicity, we will restrict our description and simulation to the first case (same observations, or eventually, bootstrap samples from the same observations and same descriptors).

As already explained in [5] in the context of a one-dimensional grid, there is no ground truth for cluster labelling in the unsupervised framework. A first strategy to overcome this issue is to perform a re-labelling of the clusters based on the clustering only: [6] merge together the clusters of different maps with a majority vote scheme. A "fused" prototype is defined as the centroid of the grouped cluster prototypes over $b = 1, \ldots, B$ and a topology is deduced posterior to the definition of the clusters. Another approach that uses the different maps in an indirect way is described in [10]: in this paper, we proposed to use a subset of $(x_i)_i$, using the most representative observations of the set of B maps, to train a final SOM from a simpler and more robust data set. This method is well suited to handle very large data sets. However, both approaches do not necessary produce a map with a topology similar to the B merged SOMs and make use of only a small part of the information provided by the B learned SOMs.

Several attempts to explicitly take advantage of the prior (common) structure of the maps have been proposed in the literature. A first method consists in constraining the B SOMs to be as similar as possible by a common initialization. This initialization can be derived, for instance, from a PCA of $(x_i)_i$. Then, the different maps are fused by averaging the prototypes of the clusters situated at the same position the B SOMs [7] or by using a majority vote scheme to classify the observations [5]. Alternatively, [5, 8] also propose to make the B SOMs similar by initializing the b-th SOM with the final prototypes of the previous one. Baruque and Corchado [8] improves this approach by weighting the averaging of the prototypes by a cluster quality index. Similarly, [11] uses a similar strategy to handle streaming or large data sets, splitting the data into several patches that are sequentially processed by a different SOM algorithm initialized with the result of the previous one. However, these methods do not allow to explore the possibilities of different initializations, which can be an issue in SOM. Moreover, a sequential initialization of the B SOMs prevents from training them in parallel, which can be an important issue if B is large: using a large B is advised for stabilizing the result of the algorithm.

Another approach to preserve the topology property of the map is to align the different maps on one of them, which serves as a reference for the topology: in [12], the map is chosen arbitrarily, and the other maps are fused sequentially to this first one, averaging the prototypes $(p_u^b)_u$ of the current map to the closest prototypes of the current fused map $(p_u^*)_u$. To leverage the problem of the choice of the map that is used to align the other maps, [9] proposes to choose a reference map that is the best one according to a given clustering quality criterion. However, this method makes the result strongly dependent on the choice of the first map because only its topology is used, whereas the topologies of the next maps are not utilized as such.

3 Description of the Optimal Transformation Method

It is well known that the quality of the SOM strongly depends on its initialization. Given different maps obtained from different (random) initializations, we propose to find the "best" transformation that can be used to obtain two comparable results between two distinct maps. The optimal one-to-one transformation between prototypes in general might be difficult to define so we restrict ourselves to transformations that strictly preserve the topology of the map, i.e. the set of linear isometric transformations (rotation and/or symmetry). To do so, only square maps with m rows and columns are considered (i.e., using the notations introduced in the previous section, $U = m^2$): in these maps, the clusters are supposed to be positioned on a 2D grid at coordinates $\{(k_1, k_2)\}_{k_1, k_2=1,...,m}$.

Then, $\mathcal{T}$ denotes the set of all transformations, $T : \mathbb{R}^2 \to \mathbb{R}^2$, that let the map globally invariant: more precisely, $\mathcal{T}$ is composed of the set of rotations $\{r_\theta\}_{\theta \in \{0, \pi/2, \pi, 3\pi/2\}}$ and of the transformations $\{r_\theta \circ s\}_\theta$, with s the symmetry with respect to the axis passing by the points $\left(\frac{m+1}{2}, 0\right)$ and $\left(\frac{m+1}{2}, m\right)$. For a given map $\mathcal{M}$ with prototypes $(p_u)_u$ and a given $T \in \mathcal{T}$, the transformed map $T(\mathcal{M})$ is the map in which the unit u, with coordinates (k_1^u, k_2^u) in $\mathbb{N}^2$, has a prototype denoted by p_u^T which is the prototype $p_{u'}$ of the original map, u' being the unit located at $T^{-1}(k_1^u, k_2^u)$.

When comparing two maps, the mean of the squared distances (in $\mathbb{R}^d$) between the prototypes of the two maps that are located at the same position is calculated.

For two maps $\mathcal{M}$ and $\mathcal{M}'$, with respective prototypes $(p_u)_u$ and $(p'_u)_u$, we define a distance between two maps as the distance between their respective prototypes positionned at the same coordinates:

$$D\left(\mathcal{M}, \mathcal{M}'\right) = \frac{1}{m^2} \sum_{u=1}^{m^2} \|p_u - p'_u\|^2. \tag{1}$$

The best transformation between the current fused map and the next map to be fused is chosen according to this distance. The two maps are then fused using the optimal transformation before they are merged, as described in Algorithm 1. The optimal

---

Algorithm 1 Optimal transformation

Initialization $\mathcal{M}^{*,1} \leftarrow \mathcal{M}^1$
for $b : 2 \rightarrow B$ **do**
   Optimal transformation
$$T_b^* := \arg\min_{T \in \mathcal{T}} D\left(\mathcal{M}^{*,b-1}, T(\mathcal{M}^b)\right)$$

   Fusion between $\mathcal{M}^{*,b-1}$ and $T_b^*(\mathcal{M}^b)$. **Provides:** $\mathcal{M}^{*,b}$
end for
Return $\mathcal{M}^* := \mathcal{M}^{*,B}$

---

transformation is found by computing the distance between the maps to be fused, $T_b^*(\mathcal{M}^b)$, and a reference map, which can be the first of the list, $\mathcal{M}^1$, for instance.[1] The fusion between the map is performed as suggested in [7] by averaging the prototypes located at the same position:

$$\forall u = 1, \ldots, m^2, \qquad p_u^* := \frac{1}{B} \sum_{b=1}^{B} p_u^{b,T}. \tag{2}$$

In the method described in the previous section, all maps are fused in an arbitrary order. However, as pointed out in [9], the maps may have very different qualities and may also be very different: merging a very peculiar map with a poor quality might lead to deterioration of the results instead of improving them. In this section, two strategies are presented to leverage this problem.

The first one uses a measure of quality of the maps and first rank the maps from the one with the best quality to the one with the worse quality: $\mathcal{M}^{(1)}, \ldots, \mathcal{M}^{(B)}$. Standard quality measures for SOM can be used to perform this ranking [13]: (i) the *quantization error* (QE), $\sum_{u=1}^{m^2} \sum_{i: x_i \in \mathcal{C}_u^*} \|x_i - p_u^*\|^2$, which is a clustering quality measure, disregarding the map topology; (ii) the *topographic error* (TE) which is the

---

[1] The current fused map, $\mathcal{M}^{*,b-1}$ has also been used as a reference map, with no difference in the final result. Using $\mathcal{M}^1$ is thus a better strategy, because optimal transformation can be computed in parallel.

simplest of the topographic preservation measure: it counts the ratio of second best matching units that are in the direct neighborhood on the map of the best matching units for every $(x_i)_i$. However, for small maps and relatively simple problems, this measure has a small variability and can lead to many equally ranked maps.

Therefore, another approach is introduced to make a trade-off, while ranking the maps, between clustering and topographic qualities: the average rank of the maps is computed as:

$$r^b = \frac{r^b_{\text{quanti}} + r^b_{\text{topo}}}{2} \tag{3}$$

where r^b_{quanti} is the rank of the map $\mathcal{M}^b$ according to its quantization error (the best map is ranked first) and similarly for r^b_{topo} with the topographic error and the maps were finally ranked by increasing order of $(r^b)_b$.

Taking advantage of this ordering of the maps, the previous method can be modified using two different strategies:

1. the *similarity strategy*: following an idea similar to [9], the maps are merged by similarity: the merging process is initialized with the best map: $\mathcal{M}^{*,1} \leftarrow \mathcal{M}^{(1)}$. Then, this map is merged only with the maps that resemble this reference map. To do so, a simple ascending hierarchical clustering is performed between the maps $(T_b^*(\mathcal{M}^b))_{b=1,\dots,B}$, with $(T_b^*)_b$ obtained by comparison with the reference map $\mathcal{M}^1$. This clustering is based on the distance introduced in (1) and the hierarchical tree is cut using the method described in [14]. Finally, the maps in the same cluster as $\mathcal{M}^{(1)}$ are fused to $\mathcal{M}^{*,1}$;

2. the *ordering strategy*: an alternative approach is performed sequentially by merging the maps by increasing rank $\mathcal{M}^{(1)}$, $\mathcal{M}^{(2)}$, …The merging process is stopped at $\mathcal{M}^{(B')}$ with $B' \leq B$ (and usually $B' < B$) when the quality of the fused map $\mathcal{M}^{*,B'}$ would not increase anymore by merging it with $\mathcal{M}^{(B'+1)}$ (actually, two strategies are investigated: stopping when the quality measure is not increasing or stopping when the quality measure has not increased for the last $5\%B$ fused maps).

4   Simulations

Methodology. In all the simulations, $B = 100$ maps are generated using the standard SOM. The optimal B has not been investigated in this paper and the number of fused maps was simply taken large enough so that the fusion makes sense. All maps were built with approximately $m = \sqrt{\frac{n}{10}}$ units and $5 \times n$ iterations of the stochastic algorithm and equipped with a Gaussian neighborhood controlled with the Euclidean distance between units on the grid. The size of the neighborhood was progressively decreased during the training. All simulations have been performed using the R package **SOMbrero**.[2] The 100 maps are then fused using one of the

---

[2]http://cran.r-project.org/web/packages/sombrero, version 1.0.

strategies described below and the performance of the different methods are finally assessed using various quality criteria for the resulting maps $\mathcal{M}^*$: (i) two criteria already mentioned in Sect. 3 that are standard to measure the quality of the SOM: (i) QE and TE; (ii) a criterion which uses the ground truth, when available (i.e., an a priori group for the observations), the normalized mutual information (NMI) [15] between the unit of the map and the a priori group. This criterion quantifies the resemblance between the a priori group and the clustering provided by the SOM (it is comprised between 0 and 1, a value of 1 indicating a perfect matching between the two classifications). Note that this criterion must be interpreted with care because if the a priori groups are split between several units of the map, each of these units being composed of one group only (which is expected for SOM results), the criterion can be lower than when the groups are split between less units which are all composed of several groups (which would be a less expected result). Thus, this criterion has to be interpreted only together with the QE and the TE values.

The performance of the method is also assessed in term of *stability*. It is expected that several runs of one aggregating method give similar (thus stable) results. This stability is estimated in terms of: (i) the distance between two final maps obtained from two different runs of the same method. If $\mathcal{M}^*$ and $\widetilde{\mathcal{M}}^*$ are two maps, the quantity $D(\mathcal{M}^*, T^*(\widetilde{\mathcal{M}}^*))$, where D is defined as in (1) and $T^* := \arg\min_{T \in \mathcal{T}} D(\mathcal{M}^*, T(\widetilde{\mathcal{M}}^*))$, is computed. This gives an estimation of the dissemblance between two maps from the prototype (hence the topological) perspective. If calculated over 250 different final maps, this quantity helps to quantify the stability of the final prototypes provided by a given aggregation method; (ii) the NMI between the final classes of two final maps obtained from two different runs of the same method. This gives an estimation of the dissemblance from the clustering perspective for a given aggregation method.

250 fusions for each method are performed using the methodology described above. This permits to compute average quality as stability criteria as well as to have an overview of the distribution of these criteria when the method is repeated.

Compared methods. The comparisons performed in this section aim at comparing our approach to existing ones (which are described in Sect. 2) as well as to investigate several options of the method (as discussed in Sect. 3).

First, our method, which merges several maps obtained from several initialization states, is compared to the standard bagging approach, in which several maps are trained from bootstrap samples from the similar initialization states. More precisely, bootstrap strategies are:

- the method denoted by **B-Rand**, which uses a common random initialization to learn $B = 100$ maps from 100 bootstrap samples coming from the original data set. Then, the prototypes that are positioned at the same coordinates, are averaged to obtain the final map $\mathcal{M}^*$ (as suggested in [7]);
- the method denoted by **B-PCA**, which uses a common PCA initialization to learn $B = 100$ maps from 100 bootstrap samples coming from the original data set (as suggested by [5]). The PCA initialization consists of initializing the prototypes by regularly positioning them along the coordinates of the projection of the data set

on the first two axis of the PCA. Then, the prototypes that are positioned at the same coordinates, are averaged to obtain the final map $\mathcal{M}^*$;

- the method denoted by **B-Seq**, which uses a sequential initialization of the $B = 100$ maps: the first map is initialized randomly and trained with a bootstrap sample and the b-th map is initialized with the final prototypes of the $(b - 1)$-th map and trained with another bootstrap sample. Finally, the final map $\mathcal{M}^*$, is obtained by averaging the prototypes of the $B = 100$ maps, that are positioned at the same coordinates, as suggested in [8].

These strategies are compared with our method and its bootstrap version, respectively denoted by **RoSyF** (for "Rotation and Symmetry Fusion") and **B-RoSyF**. **RoSyF** learns $B = 100$ maps, each from a different random initial state and using the whole data set $(x_i)_{i=1,...,n}$ and **B-RoSyF** learns $B = 100$ maps from 100 bootstrap samples coming from the original data set.

Finally, we also compare **RoSyF** with the approach consisting in selecting only one map from the B maps, the map supposed to be the best for instance. More precisely, using the $B = 100$ maps generated during the training of the **RoSyF** method, we selected one of the $B = 100$ maps (i) randomly (this method is denoted by **Best-R**), (ii) with the smallest QE (this method is denoted by **Best-QE** or (iii) with the smallest TE (this method is denoted by **Best-TE**).

Datasets and results. This section compares the results obtained on two datasets coming from the UCI Machine Learning Repository[3] as available in the R package **mlbench**.[4] More precisely, the data "Glass" ($n = 214$, $d = 10$ and 7 a priori groups) [16] and the data "Vowel" ($n = 990$, $d = 10$ and 11 a priori groups) [17] are used. The SOM parameters are set to $m = 5$ and 1 000 iterations for "Glass" and $m = 10$ with 5 000 iterations for "Vowel". The different strategies, and especially the relevance of using different initial states instead of different bootstrap samples with the same initialization, is evaluated. The results are provided in Table 1.

First, note that for almost all quality criteria and datasets, **RoSyF** obtain better results than the methods based on different bootstrap samples (all differences are significant according to Wilcoxon test, risk 5 %). **B-RoSyF** slightly deteriorates **RoSyF** performances. Cottrell et al. [18, 19] reported that the SOM algorithm is highly insensitive to initialization if run on the same data set as compared to what is obtained if bootstrap samples are used. However, it seems that the quality of the aggregated map is much better when different initial states are used on the same data set rather than different bootstrap samples with a common initial state, whatever this initial state is. Second, the TE obtained by **RoSyF** is always the lowest, just after the one obtained by **Best-TE** (which always selects the map with the lowest TE) but with a better QE and a better NMI. Again, all these differences are significant according to Wilcoxon tests (risk: 5 %). On a clustering quality point of view, **RoSyF** is the method that obtains the second lowest quantization error, just after **Best-QE** which is designed to select the map with the lowest QE. Also, from a classification point of view, its performance is also very good: in average, **RoSyF** ranks first for the NMI criterion.

---

[3]http://archive.ics.uci.edu/ml.

[4]http://cran.r-project.org/web/packages/mlbench.

Table 1 Method performance comparison (mean and standard deviation of different quality criteria; QE has been multiplied by 100)

	B-Rand	B-PCA	B-Seq	B-RoSyF	RoSyF	Best-R	Best-QE	Best-TE
"Glass"								
mean QE	855.10	855.93	854.97	609.84	597.81	595.09	**560.69**	593.68
sd QE	10.30	9.43	9.24	23.10	9.82	15.52	**5.45**	13.96
mean TE (%)	11.95	12.42	11.77	0.01	0.01	0.10	0.04	**0.00**
sd TE (%)	6.09	6.53	6.45	0.04	0.07	0.24	0.17	**0.00**
mean NMI (%)	15.80	15.77	16.00	**18.92**	17.86	15.64	16.37	15.87
sd NMI (%)	3.38	3.15	3.30	2.09	**1.38**	2.20	2.03	2.21
"Vowel"								
mean QE	847.57	847.73	847.91	550.78	545.88	547.44	**531.30**	548.23
sd QE	11.82	10.88	11.63	5.18	**1.01**	7.10	2.39	6.72
mean TE (%)	5.89	6.06	5.80	0.07	0.07	0.19	0.20	**0.00**
sd TE (%)	3.62	3.46	3.37	0.10	0.08	0.14	0.14	**0.00**
mean NMI (%)	7.11	6.76	7.03	9.47	9.57	**9.64**	9.53	9.53
sd NMI (%)	1.44	1.37	1.49	0.12	**0.11**	0.66	0.54	0.72

Also note that all quality criteria have a low variability: the standard deviations is almost always the lowest: **RoSyF** is the method which has the best coefficient of variation (mean divided by the standard deviation) for all quality criteria.

Table 2 (and Fig. 1 for the dataset "Vowel") provides a comparison of the stability criteria. For this data set, **RoSyF** has the best stability, either in term of prototype stability (even though **B-PCA** and **B-Seq** also have a good prototype stability) and even more in term of class stability. These differences are significant according to Wilcoxon tests (risk: 5 %). The results indicate that the method is indeed appropriate to improve the quality of the final map but also that it is very stable and gives very similar results if used several times, with different initializations of the prototypes and different training of the merged maps.

The relevance of stopping the merging process before all the maps have been fused has also been evaluated.[5] This comparison shows that there is only a small benefit in stopping the merging process before all maps have been used: most strategies lead to an highly deteriorated TE. Only stopping the training process when TE increases (**TE-Inc**) or based on the similarity strategy described in Sect. 3 are valid approaches in terms of quality criteria. However, a stability analysis shows that all these strategies strongly deteriorate the stability of the final map: merging all maps is the approach that provides the best stability, either in term of prototype comparison than in term of class comparison, except for **TE-Inc** which provides a slightly more stable clustering

---

[5]For the sake of paper length, detailed results are not reported but only described.

Table 2 Method stability comparison (mean and standard deviation of different stability criteria; D has been multiplied by 10 000)

	B-Rand	B-PCA	B-Seq	B-RoSyF	RoSyF	Best-R	Best-QE	Best-TE
"Glass"								
mean D	70.85	67.22	**67.06**	149.65	67.07	2047.14	1302.27	1581.49
sd D	38.62	32.32	**31.24**	335.14	310.74	1557.08	1170.39	1186.28
mean NMI (%)	64.77	65.60	65.88	83.54	**87.47**	49.15	54.41	49.86
sd NMI (%)	6.37	6.32	6.23	5.83	**5.11**	10.81	9.57	10.26
"Vowel"								
mean D	59.89	61.33	59.21	15.30	**11.07**	681.87	535.32	716.81
sd D	31.19	33.33	31.42	5.77	**3.87**	275.23	185.06	343.41
mean NMI (%)	57.32	56.83	57.70	90.83	**92.39**	72.53	74.94	72.11
sd NMI (%)	5.32	5.21	5.20	1.59	**1.33**	3.29	2.66	3.37

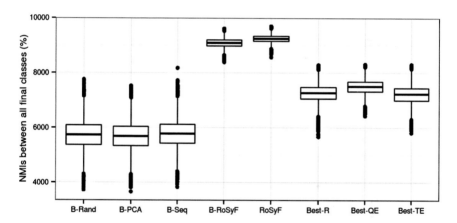

Fig. 1 Normalized mutual information (NMI) between pairs of clusterings obtained from the 250 final maps generated by the different approaches

but very different prototypes. All these strategies use only few maps (less than 10 maps in average), except again **TE-Inc** which uses 89.4 maps in average for the "Glass" dataset and is thus very close to the maximum number of available maps (100). Actually, additional simulations (not shown for the sake of paper length) merging more than 100 maps proved that the stability increases with the number of fused maps (up to a certain number which was for our dataset between 500 to 1000 maps). A trade-off has thus to be found between computational time required to generate a large number of maps and stability of the results. This question is still under study.

5 Conclusion

Although most work on SOM ensembles are based on bootstrapping techniques, this paper presents an approach allowing to explore different initial states for the map. The method improves the stability of the fused map, both in term of prototypes and in terms of clustering. We are currently investigating how to choose an optimal number B of maps to fuse as well as weighting schemes based on various quality criteria: this approach is already promising to improve the results, especially the stability of the final map.

References

1. Kohonen, T.: Self-Organizing Maps, vol. 30, 3rd edn. Springer, Berlin (2001)
2. Kohonen, T.: MATLAB Implementations and Applications of the Self-Organizing Map. Unigrafia Oy, Helsinki (2014)
3. Cottrell, M., Fort, J., Pagès, G.: Theoretical aspects of the SOM algorithm. Neurocomputing **21**, 119–138 (1998)
4. Heskes, T.: Energy functions for self-organizing maps. In: Oja, E., Kaski, S. (eds.) Kohonen Maps, pp. 303–315. Elsevier, Amsterdam (1999)
5. Petrakieva, L., Fyfe, C.: Bagging and bumping self organising maps. Comput. Inf. Syst. J. **9**, 69–77 (2003)
6. Saavedra, C., Salas, R., Moreno, S., Allende, H.: Fusion of self organizing maps. In: Proceedings of the 9th International Work-Conference on Artificial Neural Networks (IWANN 2007) (2007)
7. Vrusias, B., Vomvoridis, L., Gillam, L.: Distributing SOM ensemble training using grid middleware. In: Proceedings of IEEE International Joint Conference on Neural Networks (IJCNN 2007), pp. 2712–2717 (2007)
8. Baruque, B., Corchado, E.: Fusion Methods for Unsupervised Learning Ensembles. Studies in Computational Intelligence, vol. 322. Springer, Berlin (2011)
9. Pasa, L., Costa, J.: Guerra de Medeiros, M.: Fusion of Kohonen maps ranked by cluster validity indexes. In: Polycarpou, M., de Carvalho, A., Pan, J., Woźniak, M., Quintian, H., Corchado, E. (eds.) Proceedings of the 9th International Conference on Hybrid Artificial Intelligence Systems (HAIS 2014), vol. 8480, pp. 654–665. Salamanca, Spain, Springer International Publishing Switzerland (2014)
10. Mariette, J., Olteanu, M., Boelaert, J., Villa-Vialaneix, N.: Bagged kernel som. In: Proceedings of WSOM, Mittweida, Germany (2014) Forthcoming
11. Hammer, B., Hasenfuss, A.: Topographic mapping of large dissimilarity data sets. Neural Comput. **22**(9), 2229–2284 (2010)
12. Georgakis, A., Li, H., Gordan, M.: An ensemble of som networks for document organization and retrieval akrr (2005). In: Proceedings of International Conference on Adaptive Knowledge Representation and Reasoning (AKRR 2005) (2005)
13. Polzlbauer, G.: Survey and comparison of quality measures for self-organizing maps. In: Paralic, J., Polzlbauer, G., Rauber, A. (eds.) Proceedings of the Fifth Workshop on Data Analysis (WDA'04), pp. 67–82. Vysoke Tatry, Slovakia, Elfa Academic Press, Sliezsky dom (2004)
14. Langfelder, P., Zhang, B., Horvath, S.: Defining clusters from a hierarchical cluster tree: the dynamic tree cut package for R. Bioinformatics **24**(5), 719–720 (2008)
15. Danon, L., Diaz-Guilera, A., Duch, J., Arenas, A.: Comparing community structure identification. J. Stat. Mech. P09008 (2005)

16. Towell, G., Shavlik, J.: Interpretation of artificial neural networks: mapping knowledge-based neural networks into rules. Proceedings of Advances in Neural Information Processing Systems 4 (1992)
17. Niranjan, M., Fallside, F.: Neural networks and radial basis functions in classifying static speech patterns. Comput. Speech Lang. 4(3), 275–289 (1990)
18. Cottrell, M., de Bodt, E., Verleisen, M.: A statistical tool to assess the reliability of self-organizing maps. In: Allinson, N., Yin, H., Allinson, J., Slack, J. (eds.) Advances in Self-Organizing Maps (Proceedings of WSOM 2001), pp. 7–14. Lincoln, UK, Springer (2001)
19. de Bodt, E., Cottrell, M., Verleisen, M.: Statistical tools to assess the reliability of self-organizing maps. Neural Netw. 15(8–9), 967–978 (2002)

ESOM Visualizations for Quality Assessment in Clustering

Alfred Ultsch, Martin Behnisch and Jörn Lötsch

Abstract Classical clustering algorithms as well as intrinsic evaluation criteria impose predefined structures onto a data set. If the structures do not fit the data, the clustering will fail and the evaluation criteria will lead to erroneous conclusions. Recently, the abstract U-matrix has been defined for emergent self-organizing maps (ESOM). In this work the abstract forms of the P- and the U* are defined in analogy to the P- and the U*-matrix on ESOM. The abstract U*-matrix can be used for AU*-clustering of data by taking account of density and distance structures. For AU*-clustering the structures seen on the ESOM serve as a supervising quality measure. In this way it can be determined whether an AU*-clustering represents important structures inherent to the high dimensional data. Importantly, AU*-clustering does not impose a geometric cluster shape, which may not fit the underlying data structure, onto the data set. The approach is demonstrated on benchmark data as well as real world data from spatial science.

Keywords Self-organizing maps · U-matrix

A. Ultsch (✉)
DataBionics Research Group, University of Marburg, Hans-Meerwein-Straße,
35032 Marburg, Germany
e-mail: ultsch@Mathematik.Uni-Marburg.de

M. Behnisch
Leibniz Institute of Ecological Urban and Regional Development (IOER),
Weberplatz 1, 01217 Dresden, Germany

J. Lötsch
Institute of Clinical Pharmacology, Goethe - University, Theodor-Stern-Kai 7,
60590 Frankfurt Am Main, Germany

J. Lötsch
Fraunhofer Institute of Molecular Biology and Applied Ecology - Project Group
Translational Medicine and Pharmacology (IME-TMP), Theodor-Stern-Kai 7,
60590 Frankfurt Am Main, Germany

© Springer International Publishing Switzerland 2016
E. Merényi et al. (eds.), *Advances in Self-Organizing Maps and Learning Vector Quantization*, Advances in Intelligent Systems and Computing 428,
DOI 10.1007/978-3-319-28518-4_3

39

1 Introduction

It is known that classical clustering algorithms can frequently fail to produce a correct clustering even on data with a clearly defined cluster structure and for which the correct number of clusters is provided as input. This can be demonstrated, for example, on the "Lsun" data set (Fig. 1) from the Fundamental Clustering Problems Suite (FCPS) published as benchmark problems for clustering algorithms [1].

Lsun consists of three clearly separated sets of points on an x-y plane in the form of two elongated rectangular sets forming the letter L and a circular shaped set of points forming the "sun" (Fig. 1, left panel). Popular clustering algorithms such as k-means, Ward, complete- and average linkage all fail to cluster this data set correctly. Figure 1 shows the result of a k-means respectively Ward clustering with the correct number of clusters (i.e. 3) as input (Fig. 1, middle and right panels). The reason for this not uncommon phenomenon of incorrect clustering is that these algorithms imply a geometrical model for the cluster structure. That is, k-means clustering produces a spherical cluster shape, while Ward hierarchical clustering produces a hyperelliptic shape. If this implicit assumption on cluster shape does not fit the underlying data structure, the clustering will fail.

Emergent self-organizing feature maps (ESOM) [2] using the U-matrix [3] represent a topology-preserving mapping of high-dimensional data points $x_i \epsilon R^D$ onto a two-dimensional grid of neurons. In a 3D-display of the U-matrix (e.g. see Fig. 2 in [4]), valleys, ridges and basins indicate a distance-based cluster structure in the data set. Figure 2 (left panel) shows the U-matrix for the Lsun data. The P-matrix on the ESOM enables the visualization of density structures within the data. Both measures, i.e. densities and distances, are combined in the U*-matrix [3] (Figs. 2 and 3). In this way it is possible to discover cluster structures in a data set that are both density- and distance-based. However, ESOM is simply a method to project data from the D-dimensional data space into the plane or the three dimensional landscapes of the

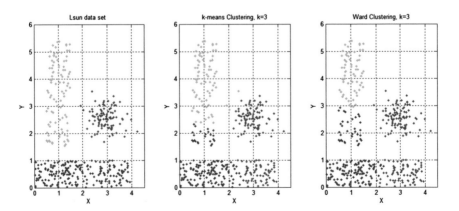

Fig. 1 Lsun data set and some clustering examples

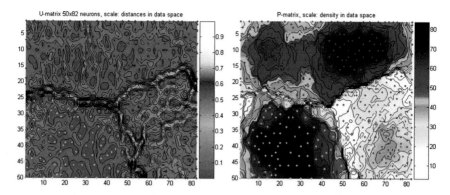

Fig. 2 U- and P-matrix of the Lsun data set

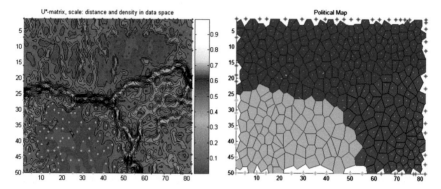

Fig. 3 U*-matrix and "Political Map" for Lsun

U-, P- and U*matrix (ESOM matrices). If cluster structures are revealed through the ESOM matrices, a clustering algorithm is required that can reproduce the structures.

The recently introduced "Abstract U-matrix" (AU-matrix) [5] formally explains the structures seen in the U-matrix. In this work, the abstract P (AP-matrix) and abstract U* (AU*-matrix) are defined. Classical clustering algorithms can be used on the AU*-matrix. The validity of this type of clustering can be assessed by comparing results with the structures seen on ESOM matrices in the form of "Political Maps". The approach is demonstrated on the Lsun data set and on a real-world data set from spatial science research.

2   Methods

The ESOM displays the U-matrix on top of an SOM on the output grid arranged in r rows and c columns using a large ($r^*c > 4000$) number of neurons. Large U-heights in the U-matrix indicate a wide gap in the data space whereas low U-heights indicate that the points are close to one another. In a 3D display of the U-matrix, valleys, ridges and basins indicate distance-based cluster structures in the data. The P-matrix [3] displays the point density $p(x) = |\{data\ points\ x_i|\ d(x_i, x) <= r\}|$ estimated as the number of data points in a sphere of radius r around x at each grid point on the ESOM's output grid. The U*-matrix combines distance structures (U-matrix) and density structures (P-matrix) into a single matrix (U*-matrix) [3].

The combination can be formalized as pointwise matrix multiplication: $U^* = U^*F(P)$, where $F(P)$ is a matrix of factors $f(p)$ that are determined through a linear function f on the densities $p = p(x)$ of the P-matrix. The function f is calculated such that $f(p) = 1$ if the density p is equal to the median and $f(p) = 0$ if p is equal to the 95-percentile ($p95$) of the densities in the P-matrix. For $p(x) > p95 : f(p) = 0$, which indicates that x is well within a cluster and results in zero heights in the U* matrix. The P-matrix allows the identification of density-based clusters in data sets. The U*-matrix shows a consistent picture of density and distance structures in the data.

The abstract U-matrix (AU-matrix) is a three-dimensional structure with the Voronoi cells of the best-matching units (BMUs) of the data as floor and the data distances corresponding to adjacent Voronoi cells as walls [5]. The AU-matrix can be calculated as the product of the adjacency matrix Del of the Delaunay graph of the best-matching units (BMU) with the matrix of distances D between the data points, i.e. $AU = Del * D$. In analogy to the P-matrix, the abstract P-matrix is defined as follows: Let $Del(i, j)$ be an edge in Del. This implies that the Voronoi cells of data points x_i and x_j are adjacent. The point (midpoint) $m_{i,j} = mean(x_i, x_j)$ is the point in data space corresponding to $AU(i, j)$. The abstract P-matrix (AP-matrix) contains the densities of all these midpoints: $AP(i, j) = p(mi, j)$. The Abstract U*matrix (AU*-matrix) is calculated in the same way as the U*-matrix (see above). It defines a distance between the data points that takes into account (i) the topology preserving projection of the SOM, (ii) the U-matrix structure and (iii) the density structure of the data. The "Political Map" of an ESOM is a coloring of the Voronoi cells of the BMUs, with different colors for each cluster. Figure 3 (right panel) shows a Political Map for a Ward clustering of the AU*-matrix. A correct clustering using the AU* distances (AU*-clustering) coincides with the structures seen on the ESOM-matrices. Thus, AU*-clustering is a clustering of the data whose results can be visually inspected and supervised using the ESOM-matrices and, in particular, using "Political Maps". This concurs with the structures seen in the other ESOM matrices and enables the validation or invalidation of the data clustering.

3   Relationship to Other Approaches

The Abstract U-matrix (AU-matrix), as well as the extensions presented here (AP-matrix, AU*-matrix), are concepts which help to understand what an empirical U-Matrix, respectively P-Matrix and U*-Matrix, shows which is constructed by the learning algorithm of an SOM on a data set. The concepts presented here are designed for emergent SOMs (ESOM). These have the property of using SOM which have a very large number of neurons, even substantially more neurons than data points. From our perspective, the number of neurons can be thought of as the pixel resolution of a digital photo camera: the more pixels (neurons) the better the image resolution, i.e. the representation of high dimensional data space. It is clear that time and costs for data processing increase with the number of neurons. However, two factors serve to reduce this burden: improved learning algorithms for the SOMs and Moore's law, which famously states that computing power doubles every two years.

A different approach to Kohonen maps is the so-called k-means-SOM, which uses only few units to represent (clusters of) data. For example, Cottrell and de Bodt use 4×4 units to represent the 150 data points in the Iris data set [6]. In contrast to these approaches, ESOMs represent more of the high dimensional space in their neurons than just the BMUs of the data points. BMUs on ESOM only have more than one data point as attractors if they are practically identical in data space. The connectivity matrix CONN [7–9] assumes non-zero density of data points within the attractor field, i.e. the number of data points projected onto one BMU. The number of data points in these Voronoi cells represents a frequency count. However, this is *not* a valid density measure, since the volumes of the Voronoi cells of different BMUs may be quite different.

A single wall of AU matrix represents the true distance information between two points in data space. A valid density information at the midpoints between BMU and second BMU (notation taken from [7–9]) is calculated for the AP-matrix, since the same volumes, i.e. spheres of a predefined radius, are used. The AU*-matrix therefore represents the true distance information between two points weighted by the true density at the midpoint. The representation is such that high densities shorten the distance and low densities stretch this distance. Using transitive closure for these weighted distances allows classical clustering algorithms (AU*-clustering) to actually perform distance- and density-based clustering, taking into account the complex topology of partially entwined clusters within the data.

As the walls of the AU*-matrix are "paper-thin" there is hardly any way to actually display the AU*-matrix directly. However, an empirical given U*-matrix can and should be adjusted, scaled and normalized to fit best the properties of the AU*-matrix. Such a normalized U*-matrix can then be understood as a visualization of the abstract AU*-matrix.

4 AU*-clustering of the Benchmark Data Set

A top view of the U-matrix using a geographical analogy for color-coding of distances separates the two classes visually as a ridge between valleys (Fig. 2 left panel). This allows the identification of the number of clusters. The P-matrix (Fig. 2 right panel) shows particularly low data densities at those neurons where high values in the

U-matrix are observed. This confirms that the parameter for the density calculation, i.e. the radius of the Parzen window (sphere), is correctly chosen. Furthermore, it shows that the density in the red class (sun) is considerably lower than in the two L-classes in Lsun.

The U*-matrix shown in the left panel of Fig. 3 displays enhanced ridges between the prospective clusters and indicates the cluster centers. The results of the AU*-clustering using Ward clustering on the AU*-matrix are shown as the "Political Map" in Fig. 3. Clustering accuracy using AU*-clustering of the Lsun data was 100% as compared with the true classification shown in Fig. 1 (left panel).

5   AU*-clustering Applied to FCPS Data Sets

AU*-clustering (AU*C) is the application of a classical clustering algorithm using the AU* distances taken from the Abstract AU*-matrix. Here AU*C-clustering was applied to the data sets in the Fundamental Clustering Problems Suite (FCPS) [10]. FCP was accessed on September 15th, 2015, and downloaded from http://www.uni-marburg.de/fb12/datenbionik/downloads/FCPS.

FCPS offers a variety of clustering problems that any algorithm should be able to handle when facing real world data [10], and thus serves as an elementary benchmark for clustering algorithms. FCPS consists of data sets with known a priori classifications that are to be reproduced by the algorithm. All data sets are intentionally created to be simple, enabling visualization in two or three dimensions. Each data set represents a certain problem that is solved by known clustering algorithms with varying degrees of success. This is done in order to reveal the benefits and shortcomings of the algorithms in question. Standard clustering methods, e.g. single-linkage, ward und k-means, are not able to solve the FCPS problems satisfactorily [10].

Here the accuracy of data clustering, i.e. agreement of U*C on FCPS with the a priori classification, was as follows:

Data Set	Accuracy (%)
Atom	100.00
Chainlink	100.00
EngyTime	95.00
Hepta	100.00
Lsun	100.00
Target	100.00
Tetra	99.00
TwoDiamonds	100.00
WingNut	100.00
GolfBall	100.00

6 AU*-clustering Applied to Spatial Science Data

The AU*-clustering was applied to a data set describing the dynamics of land consumption in all of Germany's municipalities (n = 11, 441; data valid as of 31.12.2010). The data set captures changes in land consumption in the years 2000 to 2010. Land consumption dynamics (LCD) are described along four dimensions: changes in land usage, changes in population density, changes in trade tax revenues and changes in municipal populations. The rededication of open space into settlement and transportation areas has long been the subject of debate. In many related works, clustering has been employed as a popular method intended to answer specific research questions such as: "How many forms of land consumption exist in Germany?" Most recent approaches have used a Ward or k-means clustering [11, 12]. However, many of these approaches have not validated the clustering. As mentioned above, k-means and Ward clustering algorithms are limited to finding clusters of specific shape, e.g. spherical or ellipsoid respectively for a predefined number of clusters.

The LCD data was ESOM projected onto a grid of $50 \times 160 = 8000$ neurons. Figure 4 shows the U*-matrix of this projection. An AU*-clustering of the data resulted in eight different clusters. Figure 5 shows the political map of this clustering. A comparison with the U* Matrix of the same data set shows excellent coincidence of the observed structures. The ESOM matrices in Figs. 4 and 5 are toroid, i.e. the borders top-bottom and left-right connect to one another [3]. The identified clusters could be related to previously unknown structures of spatial effects in land consumption in German municipalities. For example, one of the clusters indicates that an increase in trade tax per inhabitant was unexpectedly associated with a loss in open spaces and also in population. This points to possible problems in municipal development. Another cluster could be characterized as comprising communities

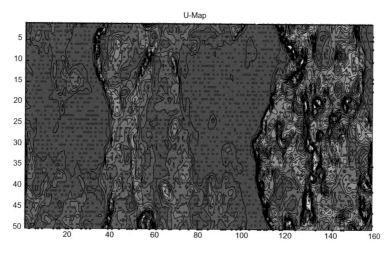

Fig. 4 U*-matrix of the LCD-data set

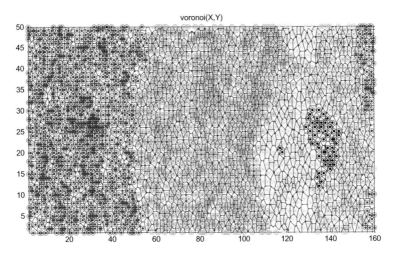

Fig. 5 Political Map of an AU*-clustering of the LCD data set

undergoing the highest change in land consumption within one decade. This could be observed particularly in periurban rural areas. Such results help in the development and optimization of planning programs for sustainable land development. Moreover, the results can be used to help establish a monitoring framework and as the basis for support systems for spatial decision-making. Thus, AU*-clustering offers a deeper multidimensional description of the characteristics of municipal land consumption for cooperating spatial experts.

7 Discussion

Clustering algorithms belong to the class of unsupervised algorithms in Machine Learning. As no desired or "correct" results are available, the results of the algorithm cannot be directly evaluated with respect to their correctness, i.e. no extrinsic evaluation is readily possible. Intrinsic evaluation measures for clustering methods try to capture numeric features of distances with respect to the assumed clusters. They rely on the assignment of low values to the distances within a cluster and of large values to the distances between clusters. However, these measures also implicitly define the geometrical structure of an optimal cluster. For example, the popular silhouette coefficient [13] compares the average distance to elements within the same cluster with the average distance to elements in other clusters. This defines the sphere as the optimal cluster shape. As a consequence, silhouette coefficients do not favor the best cluster structure but rather the cluster structure found by a k-means clustering. Therefore, intrinsic evaluation measures do not allow for the conclusion that some clustering algorithms are better than others as they rely on the existence of the structure imposed by either algorithm. If the data set in fact contains a differing structure, they will neither provide the correct clustering nor allow the quality of the results to be determined. The Ward and k-means results for the Lsun data set demonstrate this effect (Fig. 1).

ESOM are based on the topology-preserving projection of the data onto the output plane by the underlying SOM. The structures seen on the ESOM matrices therefore allow visual (in-)validation of the cluster structures in the data. Such structures may be defined by distances (U-matrix), densities (P-matrix) or a combination of both (U*-matrix). The abstract form of these three matrices can be used to understand the perceived structures. In this paper, it is proposed that they may be used for clustering (AU*-clustering). The result of a clustering using the AU*-matrix can be compared to the structures seen in the U*-matrix using "Political Maps". This means that if the clustering reproduces the observed structures, it correctly represents (topologically) the structural features of a data set. The algorithm does not impose a model of cluster structure onto the data set. In the data on land consumption dynamics, the AU*-clustering approach produced a map showing eight different types of dynamics. It could be validated with regard to the ESOM matrices constructed for this data set. The resulting clusters were meaningful for the experts in spatial development and planning.

8   Conclusions

Clustering belongs to unsupervised machine learning algorithms for which no "correct" results exist a priori. Classical clustering algorithms and intrinsic evaluation measures of cluster quality impose a predefined structure onto a data set, which can lead to mis-clustering if the imposed structures do not fit the data. By contrast, the here presented professionally constructed ESOM represents a topologically correct projection of the data. The U-Matrix allows visual inspection of distance structures while the P-matrix enables assessment of density structures in the data, and the U*-matrix combines both. In this work the abstract form of these matrices was used for data clustering (AU*-clustering) where the structures seen in the ESOM matrices proofed as a valid quality measure. It can therefore be concluded that this clustering represents important structures in the data without requiring an implicit predefinition of cluster shape or number.

References

1. Ultsch, A.: Clustering with SOM: U*C. Workshop on Self-Organizing Maps, pp. 75–82. Paris (2005)
2. Kohonen, T.: Self-organized formation of topologically correct feature maps. Biol. Cybern. **43**, 59–69 (1982)

3. Ultsch, A.: Maps for Visualization of High-Dimensional Data Spaces. WSOM, pp. 225–230, Kyushu, Japan (2003)
4. Lötsch, J., Ultsch, A.: A machine-learned knowledge discovery method for associating complex phenotypes with complex genotypes. application to pain. J. Biomed. Inf. **46**, 921–928 (2013)
5. Lötsch, J., Ultsch, A.: Exploiting the structures of the U-matrix. In: Villmann, T., Schleif, F.-M., Kaden, M., Lange, M. (eds.) Adv. Intell. Syst. Comput., vol. 295, pp. 248–257. Springer, Heidelberg (2014)
6. Cottrell, M, de Bodt, E.: A Kohonen Map Representation to Avoid Misleading Interpretations, ESANN'96, pp. 103–110. DeFacto, Bruges, Belgium (1996)
7. Tademir, K., Mernyi, E.: Exploiting data topology in visualization and clustering of self-organizing Maps. IEEE Trans. Neural Netw. **20**(4), 549–562 (2009)
8. Merényi, E., Tasdemir, K., Zhang, L.: Learning highly structured manifolds: harnessing the power of SOMs. In: Biehl, M., Hammer, B., Verleysen, M., Villmann, T. (eds.) Similarity Based Clustering. Lecture Notes in Computer Science, LNAI 5400, pp. 138–168. Springer, Berlin (2009)
9. Tasdemir, K., Merényi, E.: A validity index for prototype based clustering of data sets with complex structures. IEEE Trans. Syst. Man Cybern. Part B. 02/2011 **41**(4), 1039–1053 (2011)
10. Ultsch, A.: Clustering with SOM: U*C. In: Proceedings Workshop on Self-Organizing Maps WSOM, pp. 75–82. Paris, France (2005)
11. Kroll, F., Haase, D.A.: Does demographic change affect land use patterns? a case study from Germany. Land Use Policy **27**, 726–737 (2010)
12. Hietel, E., Waldhardt, R., Otte, A.: Analysing land-cover changes in relation to environmental variables in Hesse Germany. Landsc. Ecol. **19**(5), 473–489 (2004). Springer, New York
13. Rousseeuw, P.J.: Silhouettes: a graphical aid to the interpretation and validation of cluster analysis. Comput. Appl. Math. **20**, 53–65 (1987)

SOM Quality Measures: An Efficient Statistical Approach

Lutz Hamel

Abstract We are interested in practical tools for the quantitative evaluation of self-organizing maps (SOMs). Recently it has been argued that any quality measure for SOMs needs to evaluate the embedding or coverage of a map as well as its topological quality. Over the years many different quality measures for self-organizing maps have been proposed. However, many of these only measure one aspect of a SOM or are computationally very expensive or both. Here we present a novel, computationally efficient statistical approach to the evaluation of SOMs. Our approach measures both the embedding and the topological quality of a SOM.

1 Introduction

We are interested in practical tools for the quantitative evaluation of trained self-organizing maps (SOM) [10]. Here we present an efficient statistical approach to the evaluation of SOM quality. A nice overview of common SOM quality measures appears in [14]. The measures described there report on either the quality of map embedding in the data input space, sometimes called coverage (e.g. quantization error [10]) or on the topological quality of the map (e.g. topographic error [9]). Another measure not mentioned in the above overview is the neighborhood preservation [3] which similarly to the topographic error strictly measures the topological quality of a map.

More recently it has been argued that any SOM quality measure needs to report on both the embedding of the map in the input data space as well as the topological quality of a map [2]. To this we would like to add that any practical SOM quality measure also has to be computationally efficient. Most quality measures fail these requirements: they either only measure one aspect of a SOM or they are computationally very expensive or both. Here we propose a statistical approach that measures

L. Hamel (✉)
Department of Computer Science and Statistics, University of Rhode
Island, RI 02881 Kingston, USA
e-mail: hamel@cs.uri.edu; lutz.hamel@gmail.com

© Springer International Publishing Switzerland 2016   49
E. Merényi et al. (eds.), *Advances in Self-Organizing Maps and Learning
Vector Quantization*, Advances in Intelligent Systems and Computing 428,
DOI 10.1007/978-3-319-28518-4_4

both the embedding and the topological quality of a map and is computationally efficient even for large training data sets and/or maps. Our proposed measure computes the quality of a SOM as a pair of numbers: (1) the embedding accuracy, (2) the estimated topographic accuracy. The embedding accuracy is a quality measure we first explored in [6] as a convergence criterion and we reexamine it here in this new context. The estimated topographic accuracy is a novel statistical approach to the topological quality of a map. Besides developing our statistical approach here we also provide a preliminary validation.

The remainder of this paper is structured as follows. Section 2 examines our notion of embedding summarizing major results. We develop the estimated topographic accuracy in Sect. 3. Our implementation is briefly discussed in Sect. 4. We provide the results of our preliminary validation in Sect. 5. Section 6 provides conclusions and points to further work.

2   Map Embedding Accuracy

Yin and Allinson have shown that under some mild assumptions the neurons of a large enough self-organizing map will converge on the probability distribution of the training data given infinite time [19]. This is the motivation for our map embedding accuracy:

> A SOM is completely embedded if its neurons appear to be drawn from the same distribution as the training instances.

This was the basic insight of our original SOM convergence criterion [6]. Here we briefly summarize and adjust our terminology with respect to embedding.

Our view of embedding naturally leads to a two-sample test [12]. Here we view the training data as one sample from some probability space $\mathbf{X}$ having the probability density function $p(x)$ and we treat the neurons of the SOM as another sample. We then test to see whether or not the two samples appear to be drawn from the same probability space. If we operate under the simplifying assumption that each of the d features of the input space $\mathbf{X} \subset \mathbb{R}^d$ are normally distributed and independent of each other, we can test each of the features separately. This assumption leads to a fast algorithm for identifying SOM embedding: We define a feature as embedded if the variance and the mean of that feature appear to be drawn from the same distribution for both the training data and the neurons. If all the features are embedded then we say that the map is completely embedded.

The following is the formula for the $(1 - \alpha) * 100\%$ confidence interval for the ratio of the variances from two random samples [12],

$$\frac{s_1^2}{s_2^2} \cdot \frac{1}{f_{\frac{\alpha}{2}, n_1-1, n_2-1}} < \frac{\sigma_1^2}{\sigma_2^2} < \frac{s_1^2}{s_2^2} \cdot f_{\frac{\alpha}{2}, n_1-1, n_2-1}, \tag{1}$$

where s_1^2 and s_2^2 are the values of the variance from two random samples of sizes n_1 and n_2 respectively, and where $f_{\frac{\alpha}{2}, n_1-1, n_2-1}$ is an F distribution with $n_1 - 1$ and $n_2 - 1$ degrees of freedom. To test for SOM embedding, we let s_1^2 be the variance of a feature in the training data and we let s_2^2 be the variance of that feature in the neurons of the map. Furthermore, n_1 is the number of training samples and n_2 is the number of neurons in the SOM. The variance of a particular feature of both training data and neurons appears to be drawn from the same probability space if 1 lies in the confidence interval denoted by Eq. (1): the ratio of the underlying variance as modeled by input space and the neuron space, respectively, is approximately equal to one, $\sigma_1^2/\sigma_2^2 \approx 1$, up to the confidence interval.

In the case where $\bar{x}_1$ and $\bar{x}_2$ are the values of the means from two random samples of size n_1 and n_2, and the variances of these samples are σ_1^2 and σ_2^2 respectively, the following formula provides $(1 - \alpha) * 100\%$ confidence interval for the difference between the means [12],

$$\mu_1 - \mu_2 > (\bar{x}_1 - \bar{x}_2) - z_{\frac{\alpha}{2}} \cdot \sqrt{\frac{\sigma_1^2}{n_1} + \frac{\sigma_2^2}{n_2}}, \tag{2}$$

$$\mu_1 - \mu_2 < (\bar{x}_1 - \bar{x}_2) + z_{\frac{\alpha}{2}} \cdot \sqrt{\frac{\sigma_1^2}{n_1} + \frac{\sigma_2^2}{n_2}}. \tag{3}$$

The mean of a particular feature for both training data and neurons appears to be drawn from the same probability space if 0 lies in the confidence interval denoted by Eqs. (2) and (3). Here $z_{\frac{\alpha}{2}}$ is the appropriate z score for the chosen confidence interval.

We say that a feature is embedded if the above criteria for both the mean and variance of that feature are fulfilled. We can now define the *map embedding accuracy* for d features,

$$ea = \frac{1}{d} \sum_{i=1}^{d} \rho_i, \tag{4}$$

where

$$\rho_i = \begin{cases} 1 & \text{if feature } i \text{ is embedded,} \\ 0 & \text{otherwise.} \end{cases}$$

The map embedding accuracy is the fraction of the number of features which are actually embedded (i.e. those features whose mean and variance were adequately modeled by the neurons in the SOM). With a map embedding accuracy of 1 a map is fully embedded. In order to enhance the map embedding accuracy in our implementation [7], we multiply each embedding term ρ_i by the significance of the corresponding feature i which is a Bayesian estimate of that feature's relative importance [5].

The computational complexity of our map embedding accuracy is,

$$O((n + m) \times d) \tag{5}$$

with n the number of training examples, m the number of neurons, and d the number of features. For most cases we have that $d \ll n$ and $d \ll m$, therefore we can say our algorithm is quasi-linear in the sum of the number of training examples and number of neurons. This means that computing the map embedding accuracy is efficient for most cases.

In essence our map embedding accuracy measures the same thing as the quantization error: the effective representation of the training data by the neurons of a map. There is one big difference; our map embedding accuracy indicates when a map is completely embedded, that is, it indicates when statistically there is no difference between the population of training points and the population of neurons. No such criterion exists for the quantization error. The ramification is that the map embedding accuracy can be used as a measure across different sized maps where the quantization error cannot [14]. A more in-depth statistical analysis of our map embedding accuracy can be found in [13].

3 Estimated Topographic Accuracy

Many different approaches to measuring the topological quality of a map exist, e.g. [11, 18]. But perhaps the simplest measure of the topological quality of a map is the *topographic error* [9] defined as:

$$te = \frac{1}{n} \sum_{i=1}^{n} err(\boldsymbol{x}_i) \tag{6}$$

with

$$err(\boldsymbol{x}_i) = \begin{cases} 1 & \text{if } bmu(\boldsymbol{x}_i) \text{ and } 2bmu(\boldsymbol{x}_i) \text{ are not neighbors,} \\ 0 & \text{otherwise.} \end{cases}$$

for training data $\{\boldsymbol{x}_1, \ldots, \boldsymbol{x}_n\}$ where $bmu(\boldsymbol{x}_i)$ and $2bmu(\boldsymbol{x}_i)$ are the best matching unit and the second-best matching unit for training vector $\boldsymbol{x}_i$ on the map, respectively. We define the *topographic accuracy* of a map as,

$$ta = 1 - te. \tag{7}$$

Computing the topographic accuracy can be very expensive, especially for large training data sets and/or maps. If we let n be the size of the training data, m the number of neurons of the map, and d the number of features of the training data, then the complexity of computing the topographic accuracy is,

$$O(n \times m \times d). \tag{8}$$

One way to ameliorate the situation is to sample the training data and use this sample S to estimate the topographic accuracy. If we let s be the size of the sample then the *estimated topographic accuracy* is,

$$ta' = 1 - \frac{1}{s}\sum_{i=1}^{s} err(\boldsymbol{x}_i) \tag{9}$$

with $\boldsymbol{x}_i \in S$ and complexity $O(s \times m \times d)$. As we will see later in the paper we can get accurate values for ta' with very small samples. Therefore we can assume $s \ll m$. Also, in most cases we have $d \ll m$. Therefore, the complexity of ta' becomes quasi-linear in the number of neurons of the map which again represents a very efficient algorithm to compute the estimated topographic accuracy.

In addition to computing the value for the estimated topographic accuracy we use the bootstrap [4] to compute values for an appropriate confidence interval in order to give us further insight into the estimated topographic accuracy in relation to the actual value for the topographic accuracy whose value should fall within the bootstrapped confidence interval.

It is easy to see from (9) that for topological faithful maps the estimated topographic accuracy should be close to 1. We then say that the map is *fully organized*.

4   Implementation

We maintain an R package called `popsom` [7] in the CRAN repository [15]. The functionality discussed in this paper has been implemented in that package and is available as of package version 3.0.[1] Here is a sample session using our package:

```
 1: > library(popsom)
 2: > data(iris)
 3: > df <- subset(iris,select=-Species)
 4: > labels <- subset(iris,select=Species)
 5: > m <- map.build(df, labels, xdim=15, ydim=10, train=1000)
 6: > q <- map.quality(m)
 7: > cat(sprintf("embedding: %3.2f\n",q$embedding))
 8: embedding: 0.81
 9: > acc <- q$accuracy$acc
10: > lo <- q$accuracy$lo
11: > hi <- q$accuracy$hi
12: > cat(sprintf("accuracy: %3.2f (%3.2f-%3.2f)\n",acc,lo,hi))
13: accuracy: 0.94 (0.86-1.00)
14: >
```

---

[1] The 3.0 version should be available on CRAN by August 2015.

The first four lines deal with loading the package and the data and then preparing the data for building maps. On the fifth line we build a map with dimensions 15×10 using 1000 training iterations. On line six we compute the map quality. This computes a value with multiple components which we print out separately on the following lines. The embedding accuracy is 0.81 and the estimated topographic accuracy is 0.94. The bootstrapped 95 % confidence interval for the estimated topographic accuracy is 0.86–1.00. One way to interpret this interval is that there is a 95 % probability that the topographic accuracy computed on the whole training data lies within the interval 0.86–1.00.

5 Preliminary Validation

For our preliminary validation we use the same experiments as in [14]; namely we use the Iris data set [1] (4 independent variables, 150 instances, 3 classes) and the Epil data set [16] (8 independent variables, 236 instances, 2 classes). We build SOMs with the following sizes for the Iris data set:

- small Iris map: 5×3 (15 nodes)
- medium Iris map: 11×6 (66 nodes)
- large Iris map: 23×11 (253 nodes)

and SOMs of the following sizes for the Epil dataset:

- small Epil map: 5×4 (20 nodes)
- medium Epil map: 10×8 (80 nodes)
- large Epil map: 22×15 (330 nodes)

Map quality does depend largely on two factors: the map size and the number of training iterations applied to a map. Therefore, the big difference between our study and the original study is that we not only track map sizes but also the number of training iterations applied to each map. This allows us to observe the respective quality measures with regards to map sizes and training iterations. Table 1 shows our results for the Iris data set. Here we have the following abbreviations:

- *iter*: training iterations
- *qerr*: the quantization error defined as

$$qerr = \frac{1}{n} \sum_{i=1}^{n} ||bmu(\boldsymbol{x}_i) - \boldsymbol{x}_i||^2, \qquad (10)$$

where $||bmu(\boldsymbol{x}_i) - \boldsymbol{x}_i||$ represents the Euclidean distance between point $\boldsymbol{x}_i$ and its best matching unit $bmu(\boldsymbol{x}_i)$ on the map
- *ea*: embedding accuracy as defined by (4)
- *ta*: topographic accuracy as defined by (7)
- *ta'*: estimated topographic accuracy as defined by (9)
- *(lo-hi)*: bootstrap estimate of the 95 % confidence interval of *ta'*

Table 1 Results for the Iris data set

iter	qerr	ea	ta	ta$'$	(lo-hi)
*** 5 × 3 ***					
1	43.95	0.81	0.69	0.74	(0.64–0.86)
10	16.10	0.13	0.83	0.82	(0.70–0.92)
100	5.14	0.68	0.91	0.92	(0.84–0.98)
1000	3.29	1.00	0.95	0.94	(0.88–1.00)
10000	3.36	1.00	1.00	1.00	(1.00–1.00)
*** 11 × 6 ***					
1	28.36	0.96	0.09	0.06	(0.00–0.14)
10	20.01	0.28	0.47	0.44	(0.28–0.58)
100	4.10	0.00	0.95	0.88	(0.82–0.96)
1000	1.27	0.96	0.99	1.00	(1.00–1.00)
10000	1.24	1.00	0.99	1.00	(1.00–1.00)
*** 23 × 11 ***					
1	36.67	0.81	0.00	0.00	(0.00–0.00)
10	18.12	0.81	0.17	0.14	(0.06–0.22)
100	3.29	0.00	0.82	0.76	(0.64–0.88)
1000	0.59	0.81	0.98	1.00	(1.00–1.00)
10000	0.46	1.00	1.00	1.00	(1.00–1.00)

We can observe that the quantization error decreases for the most part for all map sizes as the number of training iterations applied to the maps increases. One of the big issues with the quantization error as a quality measure is to determine when it is sufficiently small for the map to be considered to be a good map. That is, with the quantization error there is no indication when a map is completely embedded. Reducing the quantization error to zero is usually not the solution as then the map will likely overfit the data as is usual with statistical models whose training error was reduced to zero. Notice that the quantization error is non-zero for fully embedded and fully organized maps.

Both the embedding accuracy (*ea*) and topographic accuracy (*ta*) increase with the number of training iterations applied to a map until both reach 1 indicating that the map is fully embedded and completely organized, respectively. There is phenomenon where the random initialization of an untrained map can look like a fully embedded map except that it is completely unorganized according to the topographic accuracy.

We can observe that the estimated topographic accuracy (*ta$'$*) is a good estimate for the topographic accuracy (*ta*) as it usually falls within a couple of 1/100's of the actual value.

Finally, the topographic accuracy value *ta* falls within the bootstrap estimate of the 95 % interval except for the cases where the map is completely unorganized or the map is fully organized. In these boundary cases the 95 % confidence interval does not fully predict the value of *ta*. In all the computations we use a sample size of 50 to both compute the value of *ta'* and to compute the bootstrap estimate of the

Table 2 Results for the Epil data set

iter	qerr	ea	ta	ta$'$	(lo-hi)
*** 5 × 4 ***					
1	21.06	0.91	0.37	0.34	(0.24–0.48)
10	12.08	0.30	0.54	0.50	(0.36–0.66)
100	5.50	0.23	0.92	0.90	(0.82–0.98)
1000	2.53	0.98	1.00	1.00	(1.00–1.00)
10000	2.01	0.91	1.00	1.00	(1.00–1.00)
100000	2.17	0.91	1.00	1.00	(1.00–1.00)
*** 10 × 8 ***					
1	20.67	0.00	0.23	0.10	(0.02–0.18)
10	18.49	0.00	0.06	0.04	(0.00–0.10)
100	4.27	0.30	0.90	0.88	(0.78–0.96)
1000	1.02	0.91	0.98	1.00	(1.00–1.00)
10000	0.82	0.91	0.98	0.98	(0.92–1.00)
100000	0.93	0.91	0.97	0.98	(0.94–1.00)
*** 22 × 15 ***					
1	17.76	0.00	0.00	0.00	(0.00–0.00)
10	16.99	0.30	0.06	0.02	(0.00–0.06)
100	8.52	0.30	0.62	0.62	(0.48–0.74)
1000	0.45	0.53	0.93	0.98	(0.94–1.00)
10000	0.27	0.68	1.00	1.00	(1.00–1.00)
100000	0.33	0.99	0.98	1.00	(1.00–1.00)

confidence interval. We take a look at the effects of the sample size on the value of ta' and the bootstrap estimate in the next section.

Table 2 shows the results of our experiments for the Epil data set. We can make observations very similar to the observations we made on the Iris data set: The quantization error decreases with training, both *ea* and *ta* increase with training until they both reach 1, ta' is a fairly accurate estimate of *ta*, and the bootstrap estimate of the range of the actual value *ta* is correct except for the boundary cases. However, the Epil data set seems to be inherently more complex than the Iris data set because even with 100,000 iterations the embedding accuracy never quite reaches 1 even for the small map.

It is interesting to see that in most cases the topographic accuracy converges on 1 much faster than the embedding accuracy, that is, in those cases *ta* indicates that a map is fully organized without being fully embedded. Also, as we observed earlier, an untrained map can appear to be fully embedded without being fully organized. Therefore, both quality measures are necessary to fully evaluate the goodness of a map and of course we prefer maps where both indices are close to 1. In our implementation we could have created some sort of linear combination of both indices in order to come up with a single quality index. However, we prefer the additional information separate embedding and topographic accuracies purvey.

Table 3 Effects of the sample size on the estimated topographic accuracy

k	ta	ta'	(lo-hi)
*** Iris ***			
15	0.95	1.00	(1.00–1.00)
50	0.95	0.96	(0.90–1.00)
100	0.95	0.94	(0.89–0.98)
150	0.95	0.95	(0.91–0.98)
*** Epil ***			
25	0.97	1.00	(1.00–1.00)
100	0.97	0.96	(0.92–0.99)
200	0.97	0.97	(0.94–0.99)
236	0.97	0.97	(0.94–0.99)

5.1 Sample Size and Estimated Topographic Accuracy

In order to see the effect the sample size has on the estimated topographic accuracy and the corresponding bootstrap estimate of the confidence interval we trained the respective medium sized maps for both the Iris and the Epil data set using 1000 iterations. We then computed the topographic accuracy ta (7), the estimated topographic accuracy ta' (9), and the bootstrap estimate of the 95 % confidence interval using sample sizes k that roughly corresponded to 10, 30, 60, and 100 % of the training data. Table 3 shows the results. What is surprising that even with very small samples we obtain accurate estimates of the topographic accuracy. On the other hand, the bootstrap estimate of the confidence interval improves with larger sample sizes.

With a sample size that corresponds to 100 % of the data the interpretation of the confidence interval slightly shifts. Here we see that the precise value of the topographic accuracy and in turn the value of the topographic error is data depend. The confidence interval at 100 % of the training data tells us that if we were to select another set of data points from the same distribution as the training data in order to compute the topographic accuracy we would expect a value within the given interval.

6 Conclusions and Further Work

We are interested in practical tools for the quantitative evaluation of self-organizing maps. Here we presented a novel statistical approach to the evaluation of SOMs which directly measures the embedding accuracy or coverage of a map and its topographic accuracy. Both quality indices can be computed in quasi-linear time for most cases making them computationally very efficient. We have provided an implementation of our quality measure in form of an R package.

Our preliminary validation seems to show that in essence our embedding accuracy measures the same thing as the quantization error: the effective representation of the training data by the neurons of a map. However, the embedding accuracy has the advantage that it indicates when a map is fully embedded, i.e., statistically there will be no improvement to the map with further training. Our preliminary validation also seems to show that our estimated topographic accuracy is very accurate with respect to the topographic accuracy computed on the whole training data set even when using very small samples.

In terms of a more rigorous validation we would like to test our quality measures against standard test suites such as FCPS [17] and on large real-world data sets. Finally, in order to dispense with our normality and independence assumptions of our data we consider switching to a multi-variate, non-parametric Kolmogorov-Smirnov goodness of fit test [8]. Experiments with the univariate Kolmogorov-Smirnov test seem promising.

Acknowledgments The author would like to thank Gavino Puggioni for suggesting the non-parametric goodness of fit tests.

References

1. UCI machine learning repository: Iris data set. http://archive.ics.uci.edu/ml/datasets/Iris, (Feb 2012)
2. Beaton, D., Valova, I., MacLean, D.: Cqoco: a measure for comparative quality of coverage and organization for self-organizing maps. Neurocomputing **73**(10), 2147–2159 (2010)
3. De Bodt, E., Cottrell, M., Verleysen, M.: Statistical tools to assess the reliability of self-organizing maps. Neural Netw. **15**(8–9), 967978 (2002)
4. Efron, B., Tibshirani, R.J.: An Introduction to the Bootstrap. CRC Press (1994)
5. Hamel, L., Brown, C.W.: Bayesian probability approach to feature significance for infrared spectra of bacteria. Appl. Spectrosc. **66**(1), 48–59 (2012)
6. Hamel, L., Ott, B.: A population based convergence criterion for self-organizing maps. In: Proceedings of the 2012 International Conference on Data Mining, Las Vegas, Nevada (July 2012)
7. Hamel, L., Ott, B., Breard, G.: popsom: Self-Organizing Maps With Population Based Convergence Criterion. http://CRAN.R-project.org/package=popsom (2015), r package version 3.0
8. Justel, A., Peña, D., Zamar, R.: A multivariate Kolmogorov-Smirnov test of goodness of fit. Stat. Probab. Lett **35**(3), 251–259 (1997)
9. Kiviluoto, K.: Topology preservation in self-organizing maps. In: IEEE International Conference on Neural Networks, pp. 294–299. IEEE (1996)
10. Kohonen, T.: Self-organizing maps. Springer series in information sciences. Springer, Berlin (2001)
11. Merényi, E., Tasdemir, K., Zhang, L.: Learning highly structured manifolds: harnessing the power of SOMs. In: Similarity-based clustering, pp. 138–168. Springer (2009)
12. Miller, I., Miller, M.: John E. Freund's Mathematical Statistics with Applications, 7th Edn. Prentice Hall (2003)
13. Ott, B.H.: A convergence criterion for self-organizing maps. Master's thesis, University of Rhode Island (2012)

14. Pölzlbauer, G.: Survey and comparison of quality measures for self-organizing maps. In: Proceedings of the Fifth Workshop on Data Analysis (WDA-04), pp. 67–82. Elfa Academic Press (2004)
15. Team, R.C.: R: A Language and Environment for Statistical Computing. R Foundation for Statistical Computing, Vienna, Austria (2013). http://www.R-project.org/. ISBN:3-900051-07-0
16. Thall, P.F., Vail, S.C.: Some covariance models for longitudinal count data with overdispersion. Biometrics 657–671 (1990)
17. Ultsch, A.: Clustering with SOM: U* C. In: Proceedings of Workshop on Self-Organizing Maps, pp. 75–82. Paris, France (2005)
18. Villmann, T., Der, R., Herrmann, M., Martinetz, T.M.: Topology preservation in self-organizing feature maps: exact definition and measurement. IEEE Trans. Neural Netw. **8**(2), 256–266 (1997)
19. Yin, H., Allinson, N.M.: On the distribution and convergence of feature space in self-organizing maps. Neural Comput. **7**(6), 1178–1187 (1995)

SOM Training Optimization Using Triangle Inequality

Denny, William Gozali and Ruli Manurung

Abstract Triangle inequality optimization is one of several strategies on the k-means algorithm that can reduce the search space in finding the nearest prototype vector. This optimization can also be applied towards Self-Organizing Maps training, particularly during finding the best matching unit in the batch training approach. This paper investigates various implementations of this optimization and measures the efficiency gained on various datasets, dimensions, maps, cluster size and density. Our experiments on synthetic and real life datasets show that the number of comparisons can be reduced to 24 % and the running time can also reduced to between 63 and 87 %.

Keywords Self-Organizing Map · Optimization · Implementation · Triangle inequality

1 Introduction

Clustering is an exploratory data analysis technique that aims to discover the underlying structures in data. A cluster is a set of similar observations of entities, but these observations are dissimilar to observations of entities in other clusters. Self-Organizing Maps (SOM) are suitable for such exploratory data analysis as SOMs perform vector quantization, projection of high dimensional data to low dimensional maps, and provide various visualizations.

This paper aims to optimize SOM training. This optimization is useful when training large maps, especially training an Emergent SOM with at least a few thousands nodes, typically above 4000 nodes [9]. One strategy to make the algorithm run

Denny (✉) · W. Gozali · R. Manurung
Faculty of Computer Science, University of Indonesia, Depok, Indonesia
e-mail: denny@cs.ui.ac.id

W. Gozali
e-mail: willam.gozali@ui.ac.id

R. Manurung
e-mail: maruli@cs.ui.ac.id

© Springer International Publishing Switzerland 2016
E. Merényi et al. (eds.), *Advances in Self-Organizing Maps and Learning
Vector Quantization*, Advances in Intelligent Systems and Computing 428,
DOI 10.1007/978-3-319-28518-4_5

61

faster is to reduce the search space in finding the Best Matching Unit (BMU). In the unoptimized version, or exhaustive full search (EFS), finding the BMU for a data vector is done by iterating through all possible prototype vectors, which takes $\mathcal{O}(k)$. One approach to reduce the search space is by using efficient data structures that support nearest neighbor queries, such as the kd-tree. However, the performance for kd-tree nearest neighbor searching is exponential in the dimensionality of the data.

One interesting method to reduce the BMU search space is presented in [2], which exploits the property of distance metrics that satisfy the triangle inequality. The approach was implemented for the k-means algorithm, and was shown to be efficient even when the dimensionality of the data is high. The result produced using this optimization is exactly the same as when using EFS, meaning no approximation occurs.

As k-means and SOM have similarities in finding the BMU, we are interested in investigating the efficiency of triangle inequality optimization when used for SOM. This paper investigates various implementation strategies for triangle inequality optimization on SOM training algorithms in Java.

The next section discusses SOM training and its optimization. Section 3 discusses various implementations of triangle inequality optimizations on SOM training. Section 4 then evaluates the proposed optimizations using synthetic and real life datasets. Sections 5 and 6 present and discuss the results. Conclusions and future works are provided in Sect. 7.

2   SOM Training and Optimizations

There are two approaches to training a SOM: sequential training and batch training [4]. In the sequential training algorithm, the map is trained iteratively by taking training data vectors one by one from a training data vector sequence, finding the BMU for the selected training data vector on the map, and updating the BMU and its neighbours closer toward the data vector. This process of finding the BMU and updating the prototype vectors is repeated until a predefined number of training iterations or epochs is completed. The proposed optimization is not suitable for this approach as the prototype vectors change after finding one BMU.

Unlike the sequential training algorithm, in the batch training algorithm the whole dataset, instead of a single data vector, is presented to the map before updating the values of the prototype vectors. In the batch training algorithm, the values of new prototype vectors $\mathbf{m}_j(t+1)$ are weighted averages of the training data vectors $\mathbf{x}_i$, where the weight is the neighbourhood kernel value $h_{b_i,j}$ centred on the BMU b_i [4]. The proposed optimization is suitable for this approach, since the prototype vectors are updated at once.

Finding the BMU in the batch training algorithm can be performed in parallel as the prototype vectors of the map $\mathcal{M}$ are updated after finding BMUs of all data vectors. Therefore, it can utilize a multi-processor environment to speed up the training process. Parallelizing SOM training can be performed by partitioning the dataset or partitioning the map [7, 8].

To reduce the search complexity of Kohonen's SOM, Koikkalainen and Oja [5] proposed a tree-structured topological feature map (TSTFM) that reduces the computational complexity to find BMU from $\mathcal{O}(|\mathcal{M}|)$ to $\mathcal{O}(log\,|\mathcal{M}|)$. This efficiency is achieved by restructuring the prototype vectors into pyramid-structured prototype vectors where each node has 2×2 subnodes. With pyramidal structure, TSTFM is not flexible in terms of the ratio of each side length. Cheung and Constantinides [1] compared various fast nearest neighbour search algorithms for SOM and VQ. Their experiments were performed on a small map size (100 prototype vectors) using a non-parallel execution environment. They argued that triangle inequality was not considered because the cost of preprocessing outweighed the gain in search. This is true when the triangle inequality is applied on the sequential training algorithm, but not for the batch training algorithm as discussed in this paper. Laha et al. [6] also proposed optimization in BMU searching by exploiting topological order property. To handle map folding, they use EFS when folding is detected. Furthermore, they also use another SOM as the second layer to partition the prototype vectors of the first layer SOM. Therefore, the results of this optimization is an approximation, which may lead to inferior results compared to an EFS.

3 SOM Training Algorithm with Triangle Inequality Optimization

Using Lemma 1, the search space when finding the BMU for a data vector $\mathbf{x}_i$ can be reduced. This is done by picking any prototype vector $\mathbf{m}$, and checking another prototype vector starting from the closest one to $\mathbf{m}$. When a prototype vector $\mathbf{m}_l$ is encountered, where $d\,(\mathbf{m}, \mathbf{m}_l) \geq 2 \cdot d\,(\mathbf{x}_i, \mathbf{m})$, then no further BMU searching is necessary for $\mathbf{x}_i$. The rest of the prototype vectors cannot be the BMU for $\mathbf{x}_i$, as their distance to $\mathbf{x}_i$ cannot be smaller than $d\,(\mathbf{x}_i, \mathbf{m})$. Figure 1 shows the graphical representation in two dimensional space.

Lemma 1 *Given three vectors,* $\mathbf{a}$, $\mathbf{b}$, *and* $\mathbf{c}$, *if* $d\,(\mathbf{b}, \mathbf{c}) \geq 2 \cdot d\,(\mathbf{a}, \mathbf{b})$, *then* $d\,(\mathbf{a}, \mathbf{c}) \geq d\,(\mathbf{a}, \mathbf{b})$.

The next problem is choosing the ideal prototype vector $\mathbf{m}$ which minimizes the search space. Intuitively, smaller $2 \cdot d\,(\mathbf{x}_i, \mathbf{m})$ leads to a smaller number of prototype vectors to be checked. The BMU for $\mathbf{x}_i$ in the previous epoch $(pBMU\,(i))$ is a suitable candidate for $\mathbf{m}$.

Fig. 1 Prototype vector $\mathbf{m}_i$ where $i = \{4, 5, 6\}$ cannot be BMU for $\mathbf{x}_i$, as $d\,(\mathbf{m}, \mathbf{m}_i) \geq 2 \cdot d\,(\mathbf{x}_i, \mathbf{m})$. Thus, these prototype vectors can be ignored in finding BMU

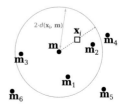

Based on this main idea, we derived three variants of triangle inequality optimization: static (STI), dynamic (DTI), and lazy (LTI). The first two are already presented in [2], and the third one is our proposed variant. We also propose a memory optimization for the static variant.

3.1 Static Triangle Inequality (STI)

This is a straightforward implementation from the main idea. At the start of an epoch, a list of prototype vectors ordered by distance to a certain prototype vector is needed. This will be used for iterating the prototype vector in radial fashion, starting from the closest one to the centroid of the hyper-sphere. Let $C\,(i, j)$ be the index of the jth closest prototype vector from $\mathbf{m}_i$. This list can be built by computing all pair distances between prototype vectors, and sort it using an efficient sorting method such as Quicksort. It takes an overall $\mathcal{O}\left(k^2 \log k\right)$ for computing every possible $C\,(i, j)$.

To find the BMU for data vector $\mathbf{x}_i$, the candidates are the prototype vectors inside hyper-sphere with centroid $pBMU\,(i)$ with radius $2 \cdot d\left(\mathbf{x}_i, \mathbf{m}_{pBMU(i)}\right)$. Using $C\,(pBMU\,(i), j)$, iterate the prototype vector starting from $j = 1$, until a prototype vector outside the hyper-sphere is encountered, as shown in Algorithm 1. This variant can be implemented within a multithreading environment without problem, as finding the BMU for any $\mathbf{x}_i$ and $\mathbf{x}_j$ is mutually independent.

Algorithm 1 Finding BMU for $\mathbf{x}_i$, using STI approach.

1: **function** FINDBMU(i)
2:    $curBMU \leftarrow pBMU\,(i)$
3:    $centroid \leftarrow pBMU\,(i)$
4:    $j \leftarrow 1$
5:    **while** $(j \leq k - 1) \wedge \left(d\left(\mathbf{m}_{C(centroid,j)}, \mathbf{m}_{centroid}\right) < 2 \cdot d\,(\mathbf{x}_i, \mathbf{m}_{centroid})\right)$ **do**
6:       **if** $d\left(\mathbf{x}_i, \mathbf{m}_{C(centroid,j)}\right) < d\,(\mathbf{x}_i, \mathbf{m}_{curBMU})$ **then**
7:          $curBMU \leftarrow C\,(centroid, j)$
8:       $j \leftarrow j + 1$
9:    **return** $curBMU$

While finding the BMU, the value of any prototype vector must not change. Otherwise, $C\,(i, j)$ has to be recomputed again. This is the reason why triangle inequality optimization applies only for batch training in SOM.

This variant can be optimized for less memory usage, from $\mathcal{O}\left(k^2\right)$ into $\mathcal{O}\,(N + k)$. Our proposed approach is computing $C\,(i, j)$ on the fly. First, group all data vectors according to its $pBMU$ using bucket sort. The result is buckets $\mathcal{B}_i$, which is a list of data vector indexes j, where $pBMU\,(j) = i$.

$$\mathcal{B}_i := \left\{ j \;\middle|\; \mathbf{x}_j \in \mathcal{D} \wedge pBMU\,(j) = i \right\},$$

Fig. 2 While looking for the BMU, $\mathbf{m}_2$ is found closer to $\mathbf{x}_i$. Changing the hyper-sphere's centroid from $\mathbf{m}$ (*left*) to $\mathbf{m}_2$ (*right*) shrinks the radius

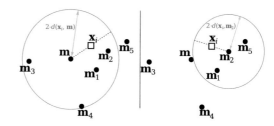

Storing all $\mathcal{B}_i$ takes $\mathcal{O}\,(N + k)$ space. Subsequently, process all data vectors in $\mathcal{B}_1$ to find its BMU. To do this, $C\,(1, j)$ for all j is required. However, $C\,(i, j)$ where $i \neq 1$ is not needed. Therefore we can just compute $C\,(1, j)$ and it takes $\mathcal{O}\,(k)$ space. After finding BMUs for all data vectors in $\mathcal{B}_1$, then repeat for $\mathcal{B}_2$, $\mathcal{B}_3$, and so on. The overall memory complexity needed is $\mathcal{O}\,(N + k)$. While using buckets reduces the memory required, the time complexity remains the same.

3.2 Dynamic Triangle Inequality (DTI)

While looking for the BMU, the prototype vector closer to $\mathbf{x}_i$ may be encountered. If the hyper-sphere's centroid is changed into this prototype vector, the radius for the hyper-sphere can be shrunk as illustrated in Fig. 2. A smaller radius means a smaller search space. Algorithm 2 shows the implementation for this approach. Similar with the STI, DTI can also be implemented for a multithreading environment.

While the hyper-sphere's radius become smaller, some of the prototype vectors may be checked twice. For example, $\mathbf{m}_1$ in Fig. 2 is checked twice when the centroid is $\mathbf{m}$ and $\mathbf{m}_2$. Furthermore, several prototype vectors that were not inside the previous hyper-sphere may be inside the smaller hyper-sphere, as shown by $\mathbf{m}_5$ in Fig. 2.

Algorithm 2 Finding BMU for $\mathbf{x}_i$, using DTI approach.

1: **function** FINDBMU(i)
2:    $curBMU \leftarrow pBMU\,(i)$
3:    $centroid \leftarrow pBMU\,(i)$
4:    $j \leftarrow 1$
5:    **while** $(j \leq k - 1) \wedge \left(d\left(\mathbf{m}_{C(centroid, j)}, \mathbf{m}_{centroid}\right) < 2 \cdot d\,(\mathbf{x}_i, \mathbf{m}_{centroid})\right)$ **do**
6:        **if** $d\left(\mathbf{x}_i, \mathbf{m}_{C(centroid, j)}\right) < d\,(\mathbf{x}_i, \mathbf{m}_{curBMU})$ **then**
7:            $curBMU \leftarrow C\,(centroid, j)$
8:            $centroid \leftarrow C\,(centroid, j)$
9:            $j \leftarrow 1$
10:        **else**
11:            $j \leftarrow j + 1$
12:    **return** $curBMU$

3.3   Lazy Triangle Inequality (LTI)

For a prototype vector $\mathbf{m}_i$, sometimes the information needed is only $C(i, 1)$, $C(i, 2)$, $C(i, 3), \ldots, C(i, p)$, where $p < k - 1$. Moreover, when convergence is almost reached, p can be much smaller than $k - 1$. This property motivates another proposed variant utilizing a partial sort, as only some of the closest prototype vectors to $\mathbf{m}_i$ are needed.

The suitable partial sort algorithm is Heap Sort. At the beginning of an epoch, initialize k heap data structures $\mathcal{H}_1, \mathcal{H}_2, \mathcal{H}_3, \ldots, \mathcal{H}_k$. These heaps store paired values, a distance and index, with smaller distances on top:

$$\mathcal{H}_i := \left\{ \left(d\left(\mathbf{m}_i, \mathbf{m}_j\right), j\right) \middle| 1 \le j \le k, j \ne i \right\}$$

For the BMU searching phase, the value of $C(i, j)$ is computed on demand. If a value of $C(i, j)$ is required and it has not been computed, then find its value from $\mathcal{H}_i$ by popping it. If the c entries for all $C(i, j)$ is required, then building all required $C(i, j)$ takes amortized $\mathcal{O}(c \log k)$, rather than fixed $\mathcal{O}(k^2 \log k)$ as shown in previous variants. Notice that this variant only modifies how $C(i, j)$ is computed. In this research, LTI adapts the DTI for BMU searching, where the hyper-sphere's radius shrinks over time.

4   Dataset and Evaluation Metrics

Our experiments involve 25 synthetic datasets and a real life dataset. The synthetic datasets are generated using Mersenne Twister pseudorandom number generator, with normal distribution. These datasets can be divided into four groups according to their evaluation metrics:

various These approaches are evaluated using synthetic datasets with various dimensions (2 or 3), cluster size (uniform or various), and cluster density (uniform or various). Therefore, there are eight datasets generated in this group. For all datasets, $N = 100,000$.

dim To test the performance in high dimension, datasets with the same characteristics but varying in dimension are generated. There are 16 datasets with dimension $2, 3, \ldots 9$, and $10, 20, \ldots, 80$ dimension with $N = 100,000$.

huge To test the performance in a huge dataset, a dataset with $N = 1,000,000$ and $dim = 25$ is generated.

real To test the performance in a real life dataset, a dataset from a telecommunication company with $N = 130,589$ and $dim = 6$ is used. This dataset describes wireless data usage from a population in a month.

Each dataset is normalized using range normalization and tested using EFS and all three variants of triangle inequality optimization. The map size k used for **various** and **huge** are 8×6, 16×12, and 32×24. The **dim** and **real** are tested using map with size 32×24. Rough training were used for each test, with linear initialization and 20 epoch. Experiments were performed in a quad core machine, CPU speed 2.5 GHz, and 4 GB memory.

5 Experimental Result

To ensure the correctness of the implementation of these optimizations, the map produced by EFS and optimized search were compared. Given the same initial map, training dataset, and training parameters, the batch training will produce the same map.

The overall result is positive, as there is improvement in training time and significant reduction in the number of comparisons. As the number of comparisons from the LTI is the same as for the DTI, these numbers are not shown.

Test Group **various**. Experiments on all datasets in this group show similar results. Figure 3 shows the averaged result from all datasets in this group. The number of comparisons between data vector and prototype vector is reduced further when grid size is increased. By using triangle inequality optimization, the number of comparisons can be reduced down to 15 % of the EFS. However, the DTI is not significantly different to the STI.

The result for training time is similar, as reducing the number of comparisons will reduce execution time. However, LTI takes much longer training time. This could be caused by the locking mechanism in multithreading environment. Further experiments using single thread implementation shows that LTI is also not significantly better than the other methods. By profiling the execution of the STI and DTI approaches, it is showed that the precomputation phase that calculate and sort the pairwise distance does not take significant time. For $k = 32 \times 24$ and $N = 100,000$, the precomputation contributes 10.3 % in the BMU searching phase. However, LTI could be useful when k is close to N. As k gets larger, the precomputation phase takes more time.

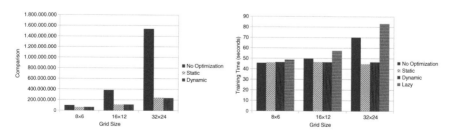

Fig. 3 Averaged number of comparisons (*left*) and training time (*right*) for each dataset on test group **various**

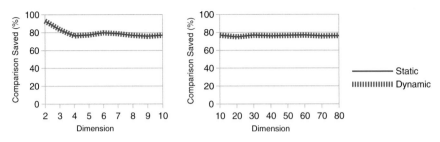

Fig. 4 The percentage of comparison on test group **dim** compared to EFS. Both approaches achieve the same level of efficiency

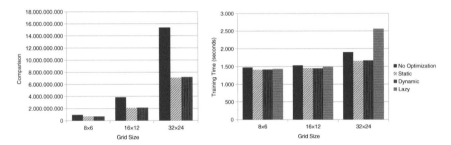

Fig. 5 Average number of comparisons (*left*) and training time (*right*) on test group **huge**

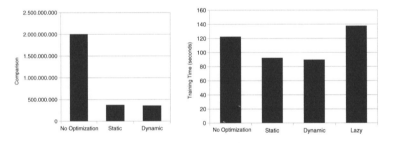

Fig. 6 Averaged number of comparisons (*left*) and training time (*right*) on test group **real**

Test Group **dim**. Figure 4 shows that the triangle inequality optimization saves a lot of comparison. As the dimension gets larger, the efficiency decreases and is finally stable around 80 %.

Test Group **huge**. Figure 5 indicates that the result is similar with test group **various**. In comparison, the savings is up to 53 %, while training time saved reaches 12 %. This result is not as good as test group **various**, which is to be expected as the dataset's dimensionality is much higher (25 vs. 2 or 3).

Test Group **real**. As shown in Fig. 6, the optimization also works well on a real-life dataset. Compared to the EFS, comparison needed is much smaller and training was 25 % faster.

6 Analysis and Discussion

From each test group, there is no significant difference between the STI and DTI variants. While the DTI variant keeps shrinking the hyper-sphere, prototype vectors that re-enter it does not make this variant much more efficient.

The efficiency of triangle inequality optimization fell for higher dimensions, and finally stagnates. This phenomenon can be explained by observing the nature of distance functions which satisfy the triangle inequality. Let $\mathcal{D}$ be a set of vectors with dim dimensions, uniformly distributed. Let $\mathbf{p}$ be one of the vectors in $\mathcal{D}$, $\mathbf{r}$ a random vector in $\mathcal{D}$, and $pdf(x)$ the probability density function so that $d(\mathbf{p}, \mathbf{r}) = x$. For higher dimensions, $pdf(x)$ will be concentrated in a single value as shown in Fig. 7.

With triangle inequality optimization, prototype vector $\mathbf{r}$ is omitted in finding BMU if $d(\mathbf{p}, \mathbf{r}) \geq 2 \cdot d(\mathbf{p}, \mathbf{q})$, where $\mathbf{q}$ is a data vector. This implies that the number of prototype vectors omitted is proportional to the probability that a random vector $\mathbf{r}$ satisfies $d(\mathbf{p}, \mathbf{r}) \geq 2 \cdot d(\mathbf{p}, \mathbf{q})$. This probability is equal to:

$$P_{omit} = \int_{2 \cdot d(\mathbf{p},\mathbf{q})}^{\infty} pdf(x)\, dx$$

This probability can be viewed as the filled area under the curve, starting from $2 \cdot d(\mathbf{p}, \mathbf{q})$ towards infinity as shown in Fig. 7. As the dimensionality gets higher, the area gets smaller. This is the reason why the efficiency of triangle inequality optimization fell as the dimension gets higher. While theoretically this optimization will not be significantly efficient when dim is very large, empirical results show that it still saves about 75 % comparison when $dim = 80$.

Another interesting fact is that the number of comparisons per epoch strictly reduces. Figure 8 shows the average comparison percentage per epoch, compared with EFS. This could be an indicator when convergence has been reached, especially when the number of comparisons does not change significantly.

Heskes [3] proposed an energy function for SOM by redefining the method to choose the BMU. In Heskes SOM, the BMU is chosen by finding the minimum of the average distance between data vector $\mathbf{x}_i$ to prototype vector $\mathbf{m}_j$ and also to its neighbours according to the neighbourhood function, as:

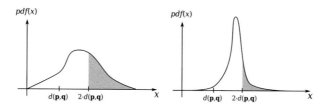

Fig. 7 Shaded region shows P_{omit}. The region gets smaller when the dimensionality of the data is higher (low dimension on the *left* and high dimension on the *right*)

Fig. 8 The percentage of
comparison per epoch for
test group **various**
compared to EFS. Both
approaches achieve the same
level of efficiency

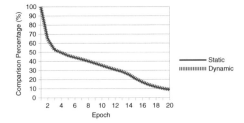

$$BMU\ (\mathbf{x}_i, \mathcal{M}) = \arg \min_{1 \leq j \leq |\mathcal{M}|} \left\{ \sum_k^{|\mathcal{M}|} h_{k,j}(t) \cdot d(\mathbf{x}_i, \mathbf{m}_k(t)) \right\},$$

where $h_{k,j}(t)$ is the neighbourhood function between map units k and j, and t is
the epoch. Triangle inequality optimization might not be suitable for Heskes winner
determination as the distance is calculated to several prototype vectors.

7 Conclusions and Future Works

Based on our experiments in implementing triangle inequality optimization, time
reduction is quite significant on high dimensional datasets even for various cluster
sizes and densities. The reduction is more significant for larger maps.

This optimization can be implemented using both static and dynamic approaches.
The reduction between these approaches are not significantly different. We have
extended the dynamic approach using lazy triangle inequality. However, this implementation is not faster.

Since triangle inequality optimization has not been implemented in many open
source SOM implementations, this optimization can be applied in these implementations. Implementation using vantage point tree and other metric trees as data structures to speed up BMU searching can be investigated.

References

1. Cheung, E., Constantinides, A.: Fast nearest neighbour search algorithms for self-organising
   map and vector quantisation. In: Conference Record of the Asilomar Conference on Signals,
   Systems and Computers, vol. 2, pp. 946–950 (1993)
2. Elkan, C.: Using the triangle inequality to accelerate k-means. In: Twentieth International Conference on Machine Learning (2003)
3. Heskes, T.: Energy functions for Self-Organizing Maps. In: Oja, E., Kaski, S. (eds.) Kohonen
   Maps, pp. 303–315. Elsevier, Amsterdam, The Netherlands (1999)
4. Kohonen, T.: Self-Organizing Maps. Springer Series in Information Sciences, vol. 30, 3rd edn.
   Springer, Berlin, Heidelberg (2001)

5. Koikkalainen, P., Oja, E.: Self-organizing hierarchical feature maps. In: Proceedings IJCNN-90, vol. 2, pp. 279–285. IEEE Service Center (1990)
6. Laha, A., Chanda, B., Pal, N.: Fast codebook searching in a SOM-based vector quantizer for image compression. Signal Image Video Process. **2**(1), 39–49 (2008)
7. Ozdzynski, P., Lin, A., Liljeholm, M., Beatty, J.: A parallel general implementation of Kohonen's Self-Organizing Map algorithm: performance and scalability. Neurocomputing **44–46**, 567–571 (2002)
8. Strupl, D., Neruda, R.: Parallelizing self-organizing maps. In: SOFSEM'97: Theory and Practice of Informatics, LNCS, vol. 1338, pp. 563–570. Springer (1997)
9. Ultsch, A.: Clustering with SOM: U*C. In: Proceedings of the 5th Workshop on Self-Organizing Maps (WSOM'05), pp. 75–82 (2005)

Sparse Online Self-Organizing Maps for Large Relational Data

Madalina Olteanu and Nathalie Villa-Vialaneix

Abstract During the last decades, self-organizing maps were proven to be useful tools for exploring data. While the original algorithm was designed for numerical vectors, the data became more and more complex, being frequently too rich to be described by a fixed set of numerical attributes. Several extensions of the original SOM were proposed in the literature for handling kernel or dissimilarity data. Most of them use the entire kernel/dissimilarity matrix, which requires at least quadratic complexity and becomes rapidly unfeasible for 100 000 inputs, for instance. In the present manuscript, we propose a sparse version of the online relational SOM, which sequentially increases the composition of the prototypes.

Keywords Relational data · Online relational SOM · Sparse approximations

1 Introduction

The self-organizing map (SOM) algorithm, [1], was proven, over the years, to be a powerful and convenient tool for clustering and visualizing data. While the original algorithm was designed for numerical vectors, the available data in the applications became more and more complex, being frequently too rich to be described by a fixed set of numerical attributes only. This is the case, for example, when data are described by relations between objects (individuals involved in a social network) or by measures of ressemblance/dissemblance (professional trajectories).

M. Olteanu (✉)
SAMM - Université Paris 1 Panthéon-Sorbonne, 90, Rue de Tolbiac, 75013 Paris, France
e-mail: madalina.olteanu@univ-paris1.fr

N. Villa-Vialaneix (✉)
INRA, UR 0875 MIAT, BP 52627, 31326 Castanet Tolosan, France
e-mail: nathalie.villa@toulouse.inra.fr

© Springer International Publishing Switzerland 2016
E. Merényi et al. (eds.), *Advances in Self-Organizing Maps and Learning Vector Quantization*, Advances in Intelligent Systems and Computing 428,
DOI 10.1007/978-3-319-28518-4_6

During the past twenty years, the SOM algorithm was extended for handling relational data, either described by kernels (see [2] for the online version and [3] for the batch version), or by dissimilarities (see [4] for the online version and [5] for the batch version). All these extensions are based on the same underlying principle: the dissimilarity or the kernel implicitly define an Euclidean (or pseudo-Euclidean) space in which the prototypes can be expressed as convex combinations of the embedded input data. However, when the goal is to explore large data sets, the relational approaches may become rapidly unfeasible. Indeed, complex relational data often have a large dimensionality. Moreover, kernel and relational SOM rely on the knowledge of the dissimilarity matrix for the entire data set, which generates at least quadratic complexity for the algorithms. As stressed in [5], algorithms will be slow for data sets with 10,000 observations and impossible to run on a normal computer for 100,000 input data. In addition to the complexity issue, expressing prototypes as convex combinations of the entire data set has a second drawback, as emphasized in [6]: the interpretability of the prototypes and of the model is lost.

In order to tackle these two issues, several approaches were introduced for relational data, all of them seeking for a sparse representation of the prototypes and a linear (in the number of observations) computational cost. [7] use the natural sparsity of the prototypes in batch relational k-means in order to reduce the complexity. The natural sparsity is enhanced by selecting the K (K fixed) closest inputs to each prototype. In [5], the complexity is reduced using iterative "patch clustering". First, the data are split into P patches of size n_P (P fixed). A prototype-based clustering algorithm in batch version (neural gas or SOM) is then run on a patch P_t and the resulting prototypes, which may be viewed as compressed representations of the data already seen, are added as new data points to the next patch, P_{t+1}. Moreover, the full vector of coefficients is replaced by the K closest input data (K fixed). With this method, linear time and constant space representation are obtained. Another technique consists in using the Nyström approximation [8] for the dissimilarity matrix. This technique also leads to a linear computational cost in the number of input data, but is strongly dependent on the intrinsic dimensionality of the given dissimilarity matrix, which has to be of low rank and entirely known in advance. All these cited approaches are batch algorithms.

In the online framework, [9] propose a bagging approach for kernel SOM. Data is split into B subsamples of size n_B (B fixed), the online kernel SOM is trained on each subsample and, after training, the most representative K observations are chosen for each prototype (K fixed). Eventually, a final map is trained on the resulting most representative observations. The algorithm has the advantage of being parallelizable, although it does not consider all the advantages of an online implementation.

In the present paper, we propose a sparse version of the online relational SOM algorithm, which takes further advantage of the online setting. Instead of expressing prototypes as convex combinations of the entire data set from the beginning, the size and the composition of the prototypes are sequentially increased with each new input fed to the algorithm. When the size of the prototypes becomes too large, prototypes are made sparse by deleting all the insignificant coefficients. Different approaches for selecting the most interesting observations are reported in [6]. In this manuscript, we use a slightly different technique, by interpreting the coefficients as a probability distribution and by selecting the most probable observation: a global probability mass v is fixed and the largest coefficients summing to v are kept. In this way, more flexibility is allowed to the prototypes which are no longer represented by a fixed number K of observations, but by the necessary number of observations allowing an "almost complete" knowledge of the composition of the prototypes (if v is chosen close to 1).

The rest of the paper is organized as follows: Sect. 2 recalls the online relational SOM, while Sect. 3 introduces the sparse version of the online relational SOM. The equivalent algorithm for kernels is briefly described in Sect. 4, while Sect. 5 contains some examples on real data-sets.

2 Online Relational SOM

In this section we shall briefly recall the principles of the online relational SOM (RSOM) algorithm, as introduced in [4]. Throughout the rest of the paper, let us suppose that the input data, $x_1, \ldots, x_N$, belong to some arbitrary space $\mathcal{G}$ and can be described through a dissimilarity measure δ, such that $\delta_{ij} = \delta(x_i, x_j)$. The dissimilarity measure is supposed to verify some basic assumptions: symmetry $(\delta_{ij} = \delta_{ji})$ and non-negativity $(\delta_{ij} \geq 0)$, for all $i, j = 1, \ldots, N$, and also $\delta_{ii} = 0$, for all $i = 1, \ldots, N$.

The online RSOM algorithm aims at mapping the input data onto a low dimensional grid (usually a two-dimensional rectangle), composed of U units, each of them described by a prototype p_u, $u = 1, \ldots, U$. The units are linked together by a neighborhood relationship H, expressed as a function of the distance between the units on the grid, $d(u, u')$. The distance on the grid, d, may be chosen, for example, as the length of the shortest path between the units. The U prototypes are initialized either as random convex combinations of the input data or randomly among the input data.

The extension of the original SOM algorithm is based on two key ideas:

- First, prototypes are written as (symbolic) convex combinations of the input data, $p_u = \sum_{i=1}^{N} \beta_{u,i} x_i$, with $\beta_{u,i} \geq 0$ and $\sum_{i=1}^{N} \beta_{u,i} = 1$, for all $u = 1, \ldots, U$. This definition is justified by the fact that, when a dissimilarity is given, it can be viewed as the dot product of the images by a mapping function ϕ into a pseudo-Euclidean space [10]: the prototypes are thus truly the convex combinations of $(\phi(x_i))_i$ in this space (see [4, 5] for further explanations).
- Second, the distance between an input data x_i and a prototype p_u can be written only in terms of the dissimilarity matrix of the input data and the coefficients $\beta_{u,i}$ as follows:

$$\|x_i - p_u\|^2 = \Delta_i \beta_u^T - \frac{1}{2}\beta_u \Delta \beta_u^T \ , \tag{1}$$

where $\Delta = (\delta_{ij})_{i,j=1,...,N}$, Δ_i represents the i-th row of the matrix Δ and $\beta_u = (\beta_{u,1}, \ldots, \beta_{u,N})$ is the vector of coefficients for the prototype p_u.

Expressing the prototypes as convex combinations of the input data and computing the distances between observations and prototypes as in Eq. (1) consists, in fact, in a generalization of the original SOM algorithm. Indeed, one can easily see that the two are equivalent if the dissimilarity δ is the squared Euclidean distance and if the prototypes of the original SOM are initialized within the convex hull of the input data.

  This general framework allowing an elegant writing of the algorithm for complex data described by dissimilarities was introduced initially for the online version of kernel SOM (KSOM) in [2]. Afterwards, extensions and rediscoveries were described for batch relational SOM in [5], batch kernel SOM in [3] and online relational SOM in [4]. A detailed and complete comparison of these methods and their equivalences may be found in [11].

  The distance computation in Eq. (1) may be theoretically justified in the very general setting of dissimilarities by extending the Hilbert embedding for kernels to a pseudo-Euclidean embedding, as shown, for example, in [5]

  The online relational SOM algorithm is summarized in Algorithm 1. The neighborhood function H is supposed to verify the following assumptions: $H : \mathbb{R} \to \mathbb{R}$, $H(0) = 1$ and $\lim_{x \to +\infty} H(x) = 0$. In the setting of Algorithm 1, H^t decreases piecewise linearly, while $\mu(t)$ vanishes at the rate $\frac{1}{t}$.

Algorithm 1 Online relational SOM

1: For all $u = 1, \ldots, U$ and $i = 1, \ldots, N$, initialize $\beta_{u,i}^0$ such that $\beta_{u,i}^0 \geq 0$ and $\sum_i^N \beta_{u,i}^0 = 1$.
2: **for** $t = 1, \ldots, T$ **do**
3:      Randomly choose an input x_i
4:      *Assignment step*: find the unit of the closest prototype

$$f^t(x_i) \leftarrow \arg\min_{u=1,...,U} \left[\Delta_i \left(\beta_u^{t-1}\right)^T - \frac{1}{2}\beta_u^{t-1}\Delta(\beta_u^{t-1})^T \right]$$

5:      *Representation step*: $\forall u = 1, \ldots, U$,

$$\beta_u^t \leftarrow \beta_u^{t-1} + \mu(t)H^t(d(f^t(x_i),u)) \left(\mathbf{1}_i - \beta_u^{t-1} \right)$$

  where $\mathbf{1}_i$ is a vector with a single non null coefficient at the ith position, equal to one.
6: **end for**

---

3 Sparse Online Relational SOM

Similarly to relational SOM, prototypes are written as convex combinations of the observations, but, in this case, they are restricted to the input data already fed to the algorithm and, more particularly, to the most significant of them. In order to guarantee the sparsity of the writing as well as similar properties with the original online relational SOM algorithm, several issues have to be verified.

1. Prototypes have to be initialized at random among the input data. Hence, the observations have to be randomly presented to the algorithm. The first U observations will be then used as initial values for the U prototypes.
2. The dissimilarity between a new input data and a prototype, written as a convex combination of the most significant past observations, has to be computed. This can be achieved using the following formula $\|x_k - p_u\|^2 = \sum_{j \in I(t)} \beta_{u,j} \delta(x_k, x_j) - \frac{1}{2} \sum_{i \in I(t)} \sum_{j \in I(t)} \beta_{u,i} \beta_{u,j} \delta(x_i, x_j)$, where $p_u = \sum_{j \in I(t)} \beta_{u,j} x_j$ and $I(t)$ contains the indices of the most significant inputs already fed to the algorithm before x_k is chosen.
3. Prototypes are sparse combinations of the input data. Hence, prototypes are periodically updated and the most coefficients only are selected. The updates may be performed throughout the iteration using either a deterministic design (the number of updates is fixed and updates are uniformly distributed during the learning of the map), or a random design (the updates are distributed according to some geometric distribution. The parameter of the geometric distribution may depend on the total number of iterations and on the size of the neighborhood). Sparsity could be achieved by selecting the first Q most important coefficients, where Q is a fixed integer. However, in order to allow for more flexibility in the expression and interpretability of the prototypes, the most significant coefficients are selected according to their value, by fixing a threshold: let $0 < \nu \leq 1$ be the selected threshold (if $\nu = 1$, the algorithm is no longer sparse, but the original one).

For $u = 1, \ldots, U$, the coefficients are ordered in descending order for each prototype $\beta_{u,(1)}, \ldots, \beta_{u,(\sharp I(t))}$, where $\beta_{u,(1)} = \max_{i \in I(t)} \beta_{u,i}$ and $\beta_{u,(\sharp I(t))} = \min_{i \in I(t)} \beta_{u,i}$. Consider N_u such that $N_u = \arg\min_{n=1,\ldots,\sharp I(t)} \{\sum_{i=1}^{n} \beta_{u,(i)} \geq \nu\}$. The most significant coefficients are updated as follows

$$\beta_{u,(i)} = \begin{cases} \frac{\beta_{u,(i)}}{\sum_{j=1}^{N_u} \beta_{u,(j)}}, & \text{if}(i) \leq N_u \\ 0, & \text{if } (i) > N_u \end{cases}$$

The sparse online relational SOM algorithm is summarized in Algorithm 2.

Algorithm 2 Sparse online RSOM

1: For all $u = 1, \ldots, U$, initialize p_u^0 among the first U input data: $\beta_u^0 = 1_u^U$, where 1_u^U is a vector of length U with a single non-null coefficient on the u-th position, equal to 1. Initialize $I(0) = \{1, \ldots, U\}$.

2: **for** $t = 1, \ldots, T$ **do**

3:     Randomly choose an input x_k, $k \in \{1, \ldots, N\}$.

4:     ***Assignment step***: find the unit of the closest prototype

$$
f^t(x_k) \leftarrow \arg\min_{u=1,\ldots,U} \left[\sum_{j \in I(t-1)} \beta_{u,j}^{t-1} \delta\left(x_k, x_j\right) - \frac{1}{2}\beta_u^{t-1} \Delta_{I(t-1)} \left(\beta_u^{t-1}\right)^T \right],
$$

where $\Delta_{I(t-1)} = \left(\delta\left(x_i, x_j\right)\right)_{i,j \in I(t-1)}$.

5:     ***Representation step***: $\forall u = 1, \ldots, U$

6:     **if** $k \in I(t-1)$, **then**

7:         $\beta_u^t \leftarrow \beta_u^{t-1} + \mu(t) H^t(d(f^t(x_k), u))\left(1_k - \beta_u^{t-1}\right)$

8:         $I(t) = I(t-1)$

9:     **else if** $k \notin I(t-1)$, **then**

10:         $\beta_u^t \leftarrow \left[1 - \mu(t)H^t(d(f^t(x_k), u))\right]\left(\beta_u^{t-1}, 0\right) + \mu(t)H^t(d(f^t(x_k), u))\underbrace{(0, \ldots, 0, 1)}_{\sharp I(t-1)}$

11:         $I(t) = I(t-1) \cup \{k\}$.

12:     **end if**

13:     ***Sparse representation***:

14:     **if** t is an update instant (deterministic or random design) **then**

15:         Sparsely update the prototypes: $\forall u = 1, \ldots, U$,

$$
\beta_{u,(1)}^t \geq \cdots \geq \beta_{u,\sharp I(t)}^t,
$$

$$
N_{t,u} = \arg\min_{n=1,\ldots,\sharp I(t)} \left\{ \sum_{i=1}^n \beta_{u,(i)}^t \geq \nu \right\}
$$

$$
\beta_{u,(i)}^t = \begin{cases} \dfrac{\beta_{u,(i)}^t}{\sum_{j=1}^{N_{t,u}} \beta_{u,(j)}^t} \,, & \text{if } (i) \leq N_{t,u} \\[2ex] 0 \,, & \text{if } (i) > N_{t,u} \end{cases}
$$

16:     **end if**

17: **end for**

---

4   The Kernel Version

In some cases, data may be described by a kernel, K, instead of a dissimilarity. We shall recall that a kernel is a symmetric similarity such that $K(x_i, x_i) = 0$ and which satisfies the following positive constraint: $\forall M > 0$, $\forall (x_i)_{i=1,\ldots,M} \in \mathcal{G}$, $\forall (\alpha_i)_{i=1,\ldots,M} \in \mathbb{R}$, $\sum_{i,j=1}^M \alpha_i \alpha_j K(x_i, x_j) \geq 0$. According to [12], there exists a Hilbert space $\mathcal{H}$, also called feature space, as well as a feature map $\psi : \mathcal{G} \to \mathcal{H}$, such that $K(x, x') = \langle \psi(x), \psi(x') \rangle_{\mathcal{H}}$. Similarly to the dissimilarity case, the prototypes are defined as convex combinations of (the images by ψ of) $(x_i)_i$. The distance between an input data x_k and some prototype p_u is then computed as the squared distance induced by the kernel $\|x_k - p_u\|^2 = K(x_k, x_k) - 2\sum_{i \in I(t)} \beta_{u,i} K(x_k, x_i) + \sum_{i,j \in I(t)} \beta_{u,i}\beta_{u,j} K(x_i, x_j)$. The sparse online relational SOM can thus be immediately adapted for kernels. Algorithm 2 has to be modified only in the assignment step which becomes

1: *Assignment step*: find the unit of the closest prototype

$$f^t(x_k) \leftarrow \arg \min_{u=1,\dots,U} \left[\beta_u^{t-1} \mathbf{K}_{I(t-1)} \left(\beta_u^{t-1} \right)^T - 2 \sum_{j \in I(t-1)} \beta_{u,j}^{t-1} K(x_k, x_j) \right],$$

where $\mathbf{K}_{I(t-1)} = \left(K(x_i, x_j) \right)_{i,j \in I(t-1)}$.

5 Examples

The sparse version introduced in the present manuscript was compared to the online relational SOM on two real data sets. For the sparse version, several values were considered for the threshold v. The sparse updates were performed either in a uniform deterministic design (fixed number of updates), or at random, according to a geometric distribution. The performances of the sparse RSOM and the online RSOM were then compared in terms of average computational time (in seconds), quantization and topographic errors and sparsity (number of non-zero coefficients). Scripts were all implemented under the free statistical software environement R.

Astraptes fulgerator. The first data set was introduced in [13]. In contains information on 465 Amazonian butterflies, each of them described by a sample of their DNA. Each input data is a DNA sequence of length 350. The Kimura distance for genetical sequences, as introduced in [14], was computed and the resulting distance matrix was used as input for relational and sparse relational SOM. For both algorithms, 100 different initializations with 2 500 iterations each were performed on a square grid of size 5×5. The results are summarized in Tables 1 and 2 for the deterministic and random designs respectively.

Professional trajectories. The second example comes from [15]. It contains information about 2 000 people having graduated high-school in 1998 and monitored during 94 months afterwards. For each individual, a categorical sequence of length 94, giving his monthly professional status is available. In all, there are nine possible situations, from permanent contracts to unemployment. The dissimilarity used for these data is the optimal matching (OM) distance, as introduced in [16]. Here, 100 different initializations with 10 000 iterations each were performed on a square grid of size 10×10. The sparse version was compared to the standard online relational SOM (itself run from 100 different initializations and 10 000 iterations). The results

Table 1 Average results for **Astraptes fulgerator** (100 random initializations)

nb. updates	ν	Comp. time (s)	Quantization err.	Topographic err.	nb. coefs
50	0.80	2.04	0.00087	0.0339	5.87
50	0.85	2.13	0.00076	0.0157	7.65
50	0.90	2.37	**0.00067**	0.0077	12.07
50	0.95	2.91	**0.00064**	0.0067	23.45
50	0.99	4.14	**0.00067**	0.0055	46.80
25	0.80	2.76	**0.00067**	0.0167	12.58
25	0.85	3.48	**0.00065**	0.0139	17.13
25	0.90	3.17	**0.00065**	0.0128	22.99
25	0.95	3.61	**0.00064**	0.0107	34.99
25	0.99	4.69	0.00070	**0.0041**	53.75
10	0.80	7.04	**0.00066**	0.0079	40.09
10	0.85	6.96	**0.00065**	0.0087	43.08
10	0.90	7.55	**0.00067**	0.0075	47.93
10	0.95	7.87	**0.00065**	0.0055	57.55
10	0.99	8.52	0.00068	0.0054	68.15
Online RSOM		12.18	**0.00067**	**0.0051**	

The first column contains the number of updates (deterministic design). The third column is the computational time (provided in seconds). The last column is the average number of non zero coefficients in the prototypes. The bolded values correspond to the results at least as good as the online RSOM

Table 2 Average results for **Astraptes fulgerator** (100 random initializations, updates were made with a random design)

nb. updates	ν	Comp. time (s)	Quantization err.	Topographic err.	nb. coefs
50	0.80	1.92	0.00093	0.0353	5.44
50	0.85	2.09	0.00078	0.0176	7.35
50	0.90	2.37	0.00069	0.0145	11.02
50	0.95	2.92	**0.00067**	0.0102	21.75
50	0.99	4.02	0.00068	0.0068	45.51
25	0.80	2.50	**0.00067**	0.0210	9.92
25	0.85	2.88	**0.00066**	0.0114	14.09
25	0.90	2.94	**0.00066**	0.0107	20.41
25	0.95	3.56	**0.00064**	0.0057	29.63
25	0.99	4.66	**0.00066**	0.0053	51.93
10	0.80	4.23	**0.00062**	0.0132	22.48
10	0.85	4.69	**0.00065**	0.0072	28.41
10	0.90	5.18	**0.00065**	0.0098	33.97
10	0.95	5.14	**0.00065**	**0.0051**	43.34
10	0.99	6.30	**0.00067**	**0.0033**	59.95
Online RSOM		12.18	**0.00067**	**0.0051**	

Table 3 Average results for "professional trajectories" (100 random initializations, updates were made with a deterministic design)

nb. updates	ν	Comp. time (s)	Quantization err.	Topographic err.	nb. coefs
100	0.80	111	29.5	0.384	**1.4**
100	0.85	130	27.8	0.348	1.8
100	0.90	147	25.5	0.277	2.9
100	0.95	215	21.8	0.112	11.3
100	0.99	480	**20.5**	0.084	40.4
50	0.80	157	25.6	0.247	2.6
50	0.85	174	23.8	0.177	4.4
50	0.90	223	22.1	0.109	9.8
50	0.95	307	21.0	0.086	23.3
50	0.99	672	**20.5**	0.080	52.9
25	0.80	247	22.6	0.124	7.3
25	0.85	278	21.6	0.102	12.2
25	0.90	339	21.0	0.089	20.1
25	0.95	470	**20.5**	0.090	34.0
25	0.99	800	**20.6**	0.078	60.9
Online RSOM		9126	**20.7**	**0.075**	

Simulations were all performed on a server with OS Debian 8 Jessie, 8 processors AMD Opteron 8384 with 4 cores each and 256 Go RAM

for the deterministic design are summarized in Table 3 (due to the lack of space, we do not report here the results with a random design, which are quite similar).

It is interesting to note that the sparsity has a strong influence on the computational time: increasing the number of updates tends to decrease the computational time since the prototypes are regularly cleared from unnecessary coefficients. The computational time compared to the standard version is at least 10 times smaller in the sparse version for this large dataset. On the contrary, the performances, measured in terms of quantization and topographic errors, can be affected by a too large sparsity but the best ones remain close to those of the standard version.

6 Conclusion and Future Work

A sparse version of the online relational SOM algorithm was proposed, by sequentially increasing the composition of the prototypes and sparsely updating them. The algorithm was compared with the online ROM on two real data sets and the sparse version appeared to achieve very similar performances as compared to the original algorithm, while improving computational time and prototype representation.

References

1. Kohonen, T.: Self-Organizing Maps, 3rd edn, vol. 30. Springer, Berlin, Heidelberg, New York (2001)
2. Mac Donald, D., Fyfe, C.: The kernel self organising map. In: Proceedings of 4th International Conference on knowledge-based Intelligence Engineering Systems and Applied Technologies, pp. 317–320 (2000)
3. Boulet, R., Jouve, B., Rossi, F., Villa, N.: Batch kernel SOM and related Laplacian methods for social network analysis. Neurocomputing **71**(7–9), 1257–1273 (2008)
4. Olteanu, M., Villa-Vialaneix, N.: On-line relational and multiple relational SOM. Neurocomputing **147**, 15–30 (2015)
5. Hammer, B., Hasenfuss, A.: Topographic mapping of large dissimilarity data sets. Neural Comput. **22**(9), 2229–2284 (2010)
6. Hofmann, D., Schleif, F., Paaßen, B., Hammer, B.: Learning interpretable kernelized prototype-based models. Neurocomputing **141**, 84–96 (2014)
7. Rossi, F., Hasenfuss, A., Hammer, B.: Accelerating relational clustering algorithms with sparse prototype representation. In: Proceedings of the 6th Workshop on Self-Organizing Maps (WSOM 07), Bielefield, Germany, Neuroinformatics Group, Bielefield University (2007)
8. Gisbrecht, A., Mokbel, B., Hammer, B.: The nystrom approximation for relational generative topographic mappings. NIPS Workshop on Challenges of Data Visualization (2010)
9. Mariette, J., Olteanu, M., Boelaert, J., Villa-Vialaneix, N.: Bagged kernel SOM. In: Villmann, T., Schleif, F., Kaden, M., Lange, M. (eds.) Advances in Self-Organizing Maps and Learning Vector Quantization (Proceedings of WSOM 2014). Volume 295 of Advances in Intelligent Systems and Computing, Mittweida, Germany. Springer, Berlin, Heidelberg, pp. 45–54 (2014)
10. Goldfarb, L.: A unified approach to pattern recognition. Pattern Recogn. **17**(5), 575–582 (1984)
11. Rossi, F.: How many dissimilarity/kernel self organizing map variants do we need? In: Villmann, T., Schleif, F., Kaden, M., Lange, M. (eds.) Advances in Self-Organizing Maps and Learning Vector Quantization (Proceedings of WSOM 2014). Volume 295 of Advances in Intelligent Systems and Computing, Mittweida, Germany. Springer, Berlin, Heidelberg, pp. 3–23 (2014)
12. Aronszajn, N.: Theory of reproducing kernels. Trans. Am. Math. Soc. **68**(3), 337–404 (1950)
13. Hebert, P.D.N., Penton, E.H., Burns, J.M., Janzen, D.H., Hallwachs, W.: Ten species in one: DNA barcoding reveals cryptic species in the neotropical skipper butterfly astraptes fulgerator. Genetic Analysis (2004)
14. Kimura, M.: A simple method for estimating evolutionary rates of base substitutions through comparative studies of nucleotide sequences. J. Mol. Evol. **16**(2), 111–120 (1980)
15. Rousset, P., Giret, J.F.: Classifying qualitative time series with SOM: the typology of career paths in France. In: Sandoval, F., Prieto, A., Cabestany, J., Graña, M. (eds.) Computational and Ambient Intelligence. Lecture Notes in Computer Science, vol. 4507, pp. 757–764. Springer, Berlin, Heidelberg (2007)
16. Abbott, A., Forest, J.: Optimal matching methods for historical sequences. J. Interdisc. Hist. **16**(3), 471–494 (1986)

Part II
Clustering and Time Series Analysis with Self-Organizing Maps and Neural Gas

A Neural Gas Based Approximate Spectral Clustering Ensemble

Yaser Moazzen and Kadim Taşdemir

Abstract The neural gas has been successfully used for prototype based clustering approaches. Its topology based quantization effectively aids in approximate spectral clustering (ASC) to define distinct similarity criteria which are optimally selected for the relevant application. To utilize the advantages of ASC by harnessing those criteria derived from different information types, we propose a neural gas based approximate spectral clustering ensemble (NGASCE). The NGASCE obtains a joint decision for accurate partitioning, by a 2-step ensemble approach derived from 1-step graph-based models. We show the outperformance of NGASCE on five datasets from UCI Machine Learning Repository.

1 Neural Gas Based Approximate Spectral Clustering

Spectral clustering, a graph-based approach depending on eigendecomposition of pairwise similarities of data points, has an ability to extract clusters of different characteristics without a parametric model [1–3]. However, it has high computational complexity (due to required eigendecomposition) and hence it is infeasible for large datasets. To overcome this challenge, approximate spectral clustering (ASC) approaches apply spectral clustering to data prototypes obtained by sampling or quantization [4–8]. Among them, neural gas [9] based ASC is shown more powerful than other sampling or quantization methods [7, 10]. In addition, ASC enables new similarity definitions which integrate different information types (such as distance, density, topology), producing an effective exploitation of available information for extraction of precise cluster structure [8]. On the one hand this results in

Y. Moazzen (✉)
Department of Electronics and Communication Engineering,
Istanbul Technical University, ITU Ayazaga Kampusu, Ayazaga, Istanbul, Turkey
e-mail: yaser.mg2000@gmail.com

K. Taşdemir (✉)
Department of Computer Engineering, Antalya International University,
Universite Cd. 2, 07190 Dosemealti, Antalya, Turkey
e-mail: kadim.tasdemir@antalya.edu.tr

© Springer International Publishing Switzerland 2016
E. Merényi et al. (eds.), *Advances in Self-Organizing Maps and Learning Vector Quantization*, Advances in Intelligent Systems and Computing 428,
DOI 10.1007/978-3-319-28518-4_7

diverse partitionings obtained by different information types tailored to the application requirements. On the other hand, it necessitates selection of the optimal similarity criterion, which is often hard to determine for real world applications (due to lack of class labels). A solution to overcome this necessity can be the use of cluster ensembles.

Cluster ensembles merge diverse partitionings obtained by different input or feature sets, distinct methods or the same method with several parameter settings, using various approaches such as majority voting, evidence accumulation, hyper graph operations, meta-clustering, or mixture models [11–14]. They eliminate the need to determine the optimal set, method or parameter values in addition to (usually) achieving a more accurate partitioning than those partitionings obtained by single methods individually. An approximate spectral clustering ensemble based on majority voting and meta clustering algorithm is proposed for segmentation of SAR images [15], using Nystrom approximation [4] and the traditional distance based Gaussian kernel similarity with different kernel parameter values to achieve diverse segmentation results.

In this paper, we propose a neural gas based approximate spectral clustering ensemble (NGASCE) which combines the advantages of approximate spectral clustering with different similarity criteria followed by an ensemble to exploit distinct information types to achieve an aggregated consensus decision for accurate partitioning. First, we obtain data prototypes by neural gas quantization which is shown outperforming over sampling approaches for ASC [7, 8]. Second, we produce ASC partitionings by the recent similarity definitions proposed in [8]. Third, we propose a 2-step ensemble approach based on 1-step graph-based models [11]. We then show the performance of NGASCE on five datasets from UCI Machine Learning Repository. The paper is outlined as follows: Sect. 2 briefly explains ASC, Sect. 3 describes the proposed NGASCE, Sect. 4 provides accuracy assessment and Sect. 5 concludes the paper.

2   Approximate Spectral Clustering

Approximate spectral clustering (ASC) applies spectral clustering on a reduced set of data representatives (prototypes) selected by sampling or quantization [4–8]. Namely, ASC has two steps: (i) selection of prototypes, (ii) their spectral clustering. For the first step, neural gas, a topology-based quantization, is shown powerful to achieve high accuracies with ASC [7, 8], because it enables manifold based similarity definitions (such as data topology and local density).

We now briefly explain the spectral clustering of the neural gas prototypes with different similarity criteria. Spectral clustering methods are associated with relaxed optimization of graph-cut problems based on a graph *Laplacian* matrix, L, with respect to an optimization criterion [1–3]. We employ the spectral method in [2] in our ensemble. We first obtain a weighted undirected graph $G = (V, S)$ where V represents the elements (prototypes) to be clustered and the edges S are their pairwise similarities to be determined with respect to some user-defined criterion. Ng et al. [2] defines the normalized Laplacian matrix, L_{norm}, as

$$L_{norm} = D^{-1/2} S D^{-1/2}, \tag{1}$$

based on S and its diagonal degree matrix D with $d_i = \sum_j s(i, j)$. The k eigenvectors $\{e_1, e_2, \ldots, e_k\}$ of L_{norm}, associated with the k highest eigenvalues are found. Then, the $n \times k$ matrix $E = [e_1 e_2 \ldots e_k]$ is constructed and $n \times k$ matrix U is obtained by normalizing the rows of E to have norm 1, i.e. $u_{ij} = \frac{e_{ij}}{\sqrt{\sum_k e_{ik}^2}}$. Finally, the n rows of U are clustered with the k-means algorithm into k clusters.

2.1 Similarity Measures

The criterion for the similarity matrix S plays a significant role to achieve an accurate cluster extraction. The pairwise similarities, $s(i, j)$s, are traditionally calculated by a Gaussian kernel based on the (Euclidean) distances, $d_{Euc}(p_i, p_j)$:

$$s_{Euc}(i, j) = \exp\left\{-d_{Euc}(x_i, x_j)/(2\sigma_i \sigma_j)\right\} \tag{2}$$

where σ_i is a decaying parameter (to be optimally found through experiments [2] or to bet set locally as the distance to the kth nearest neighbor of p_i [16]). For ASC, new information types such as topology, density can be embedded into S to define pairwise similarities of prototypes more effectively [7, 10, 17]. A recent approach [7] uses a similarity measure (CONN) [18], a weighted version of the induced Delaunay triangulation in [9], which exploits local density together with data topology on the prototype level. $CONN(i, j)$ shows the number of data points for which the prototypes p_i and p_j are the pair of the best-matching and the second-best-matching units. In other words, $CONN(i, j)$ represents the local density distribution inside the subregions $V_{ij} \cup V_{ji}$ of the Voronoi polygons V_i and V_j, (V_i is the set of data points v for which w_i is the closest prototype and V_{ij} is its subregion where w_i is the next closest prototype):

$$CONN(i, j) = |V_{ij} \cup V_{ji}| \tag{3}$$

The distance is integrated with CONN to produce a hybrid criterion S_{hyb} [17]

$$s_{hyb}(i, j) = s_{Euc}(i, j) \times \exp\left\{CONN(i, j)/\max_{i,j} CONN(i, j)\right\} \tag{4}$$

The hybrid S_{hyb} scales the distance based similarity with respect to CONN, producing a greater similarity upto a scale of e for the maximum $CONN(i, j)$.

Geodesic similarities are also proposed for ASC [8] based on various neighborhood graphs. A traditional way for this graph is the (mutual) $(k - nn)$ graph: if p_i and

p_j are among their k closest neighbors, they are neighbors. Their geodesic distance is the sum of the Euclidean distances (d_{Euc}) at their shortest path:

$$d_{geoknn}(p_i, p_j) = \sum_{lm \in SP_{knn}(p_i,p_j)} d_{Euc}(l,m) \qquad (5)$$

where $SP_{knn}(p_i, p_j)$ is the set of edges in the shortest path between p_i and p_j calculated with d_{Euc} and $k - nn$ graph. A data topology based alternative to reflect specific number of neighbors for each prototype is the CONN [8]. The geodesic distance d_{geoadj} based on CONN and the Euclidean distances d_{Euc} is

$$d_{geoadj}(p_i, p_j) = \sum_{lm \in SP_{adj}(p_i,p_j)} d_{Euc}(l,m) \qquad (6)$$

where $SP_{adj}(p_i, p_j)$ is the set of edges in the shortest path between p_i and p_j based on d_{Euc} and CONN. Alternatively, local density distribution is used for geodesic distance calculation. Namely, using density based distance

$$d_{CONN}(p_i, p_j) = e^{-\frac{CONN(i,j)}{\max_{y,z} CONN(y,z)}} \; if \; CONN(i,j) > 0 \qquad (7)$$

a geodesic distance using data topology and the data distribution is defined [10]:

$$d_{geoconn}(p_i, p_j) = \sum_{lm \in SP_{conn}(p_i,p_j)} d_{CONN}(l,m) \qquad (8)$$

with $SP_{conn}(p_i, p_j)$ is the set of edges in the shortest path between p_i and p_j with respect to d_{CONN} distance and CONN. A hybrid approach $d_{geohyb}(p_i, p_j)$ exploits all available information for ASC on the prototype level:

$$d_{geohyb}(p_i, p_j) = \sum_{lm \in SP_{hyb}(p_i,p_j)} d_{Euc}(l,m) d_{CONN}(l,m) \qquad (9)$$

Taşdemir et al. [8] analyses these similarity criteria for ASC of large datasets and shows improvement with geodesic approaches [10]. However, any single criterion may not be the optimal solution for each dataset. Therefore, instead of finding a best criterion for various applications, their ensemble will utilize advantages of different criterion to reach a consensus in clustering.

3   Neural Gas Based Clustering Ensemble (NGASCE)

Our neural gas based approximate spectral clustering ensemble (NGASCE) is based on a graph-based ensemble [11]. In NGASCE, we have diverse results with respect to various similarity criteria and k-means in the ASC algorithm. Alternative to the traditional approach of ensembling all results into one, we use a two-step ensemble process: we first ensemble different partitionings obtained by k-means runs for each similarity criterion; then, we ensemble the fused partitions of each similarity into final labels. Our two-step approach first addresses the randomness in k-means and then exploits distinct results obtained by different information types. Note that we ensemble the clustering labels at the prototype level and then determine the labels of the data points based on the ensembled labels of their prototypes. We can summarize NGASCE using n_s similarity criteria and n_{km} k-means runs as follows:

1. Obtain N_p neural gas prototypes and their $n_s n_{km}$ partitionings by ASC.
2. Obtain a similarity matrix S_{CE1} based on the number of identically labelled prototypes among n_{km} different partitionings of each similarity criterion: $S_{CE1} = \sum_{k=1}^{n_{km}} S_{n_s,k}$ where $S_{n_s,k}(p_i, p_j) = 1$ if p_i and p_j are in the same cluster, else $S_{n_s,k}(p_i, p_j) = 0$.
3. Apply spectral clustering using these S_{CE1} to obtain n_s first step ensemble partitionings for each similarity criterion
4. Obtain a similarity matrix S_{CE2} (similar to S_{CE1}) based on the resulting n_s ensemble partitionings
5. Apply spectral clustering using S_{CE2} to obtain the ensemble of prototypes.

4   Accuracy Assessment of NGASCE

We evaluate the proposed NGASCE with the datasets from UCI Machine Learning Repository, which have different characteristics and features. The Iris dataset has 150 samples with 4 features grouped into 3 classes. The Breast Cancer Wisconsin (BCWS) dataset has 9D 699 samples in two classes (benign or malignant). The Yeast dataset has 1484 samples with 8D features and 10 classes. The Statlog data is a remote sensing dataset with 4D features and has 6 classes. The Pen Digits dataset has 10992 samples and describe 10 digits with 16D features.

We first obtain neural gas prototypes using SOMtoolbox with default parameters. The number of prototypes is one tenth of the number of data points. Then these prototypes are clustered by spectral clustering using the similarity criteria described above and 20 different k-means step, resulting in 20 partitionings for each criterion. Table 1 shows the accuracies averaged over 20 runs obtained for each criterion, where accuracy is the percentage of correctly clustered data samples. For NGASCE, we obtain the first step ensemble by merging all k-means results to have a consensus at each criterion, and then we ensemble the merged labels of each criterion. The NGASCE accuracies are shown at Table 1.

It is important to note that to achieve the best single accuracy with the ensemble approach is already a success since this removes the necessity to select the best similarity criterion (which is often not possible due to unavailability of the class labels in real world problems). For all five datasets, the proposed NGASCE improves the clustering accuracy over the best accuracies obtained by individual methods

Table 1 Mean accuracies of neural gas based approximate spectral clustering ensemble (NGASCE)

	Iris	BCWS	Yeast	Statlog	Pen Digits
Similarity	150; 4D	699; 9D	1484; 8D	6435; 4D	10992; 16D
Criterion	3 classes	2 classes	10 classes	6 classes	10 classes
s_{Euc}	63.24 (7.1)	95.84 (0.6)	43.04 (1.9)	60.90 (3.3)	46.17 (14.9)
s_{CONN}	57.45 (9.3)	96.51 (0.8)	42.31 (4.4)	57.84 (14.9)	63.29 (10.6)
s_{hyb}	54.67 (2.3)	96.62 (0.6)	40.22 (3.6)	49.31 (10.6)	51.07 (12.9)
s_{geoknn}	89.47 (2.6)	93.87 (1.2)	34.36 (4.3)	65.77 (5.0)	68.47 (4.8)
s_{geoadj}	84.47 (9.6)	94.98 (0.5)	43.68 (3.2)	63.40 (5.8)	66.86 (5.4)
$s_{geoconn}$	86.76 (10.5)	95.04 (0.6)	43.54 (2.9)	54.61 (4.8)	53.00 (6.3)
s_{geohyb}	86.69 (10.5)	94.94 (0.4)	43.67 (2.9)	63.71 (6.2)	67.69 (5.5)
NGASCE1	85.60	96.25	**50.13**	71.55	81.05
NGASCE	**96.67**	**96.85**	48.85	**73.54**	**83.90**

The numbers of data points, features, and classes are provided for each dataset. NGASCE1 is the 1-step ensemble of all partitionings

Table 2 Adjusted Rand index (ARI) values of neural gas based approximate spectral clustering ensemble (NGASCE)

	Iris	BCWS	Yeast	Statlog	Pen Digits
Similarity	150; 4D	699; 9D	1484; 8D	6435; 4D	10992; 16D
Criterion	3 classes	2 classes	10 classes	6 classes	10 classes
s_{Euc}	0.492 (0.020)	0.793 (0.006)	0.154 (0.003)	0.521 (0.016)	0.299 (0.014)
s_{CONN}	0.465 (0.020)	0.868 (0.02)	0.132 (0.001)	0.388 (0.003)	0.474 (0.010)
s_{hyb}	0.502 (0.012)	0.869 (0.001)	0.123 (0.001)	0.288 (0.003)	0.366 (0.019)
s_{geoknn}	0.453 (0.005)	0.689 (0.048)	0.110 (0.005)	0.518 (0.011)	0.575 (0.004)
s_{geoadj}	0.694 (0.018)	0.807 (0.018)	0.156 (0.003)	0.342 (0.020)	0.419 (0.005)
$s_{geoconn}$	0.699 (0.017)	0.790 (0.003)	0.157 (0.002)	0.439 (0.004)	0.566 (0.009)
s_{geohyb}	0.747 (0.008)	0.788 (0.004)	0.158 (0.002)	0.447 (0.005)	0.575 (0.011)
NGASCE1	0.687	0.829	**0.173**	0.521	0.694
NGASCE	**0.765**	**0.881**	0.152	**0.558**	**0.710**

The numbers of data points, features, and classes are provided for each dataset. NGASCE1 is the 1-step ensemble of all partitionings

using different similarity criteria. In addition, this improvement is significant for four datasets: Iris (from 89.5%—with s_{geoknn}—to 96.7%), Yeast (from 43.7%—with s_{geoadj}—to 48.9%), Statlog (from 65.8%—with s_{geoknn}—to 73.5%) and Pen Digits (from 67.7%—with s_{geohyb}—to 83.9%). Moreover, the proposed two-step NGASCE has also significantly higher accuracies than the traditional 1-step ensemble approach for four of the five datasets. A similarly high performance is also obtained by adjusted Rand index [19] which is an evaluation measure considering the class sizes based on labeled samples (Table 2). Moreover, we evaluate the results using the

Table 3 Cluster validity index (CVI) values of neural gas based approximate spectral clustering ensemble (NGASCE)

Dataset	CVI	s_{Euc}	s_{CONN}	s_{hyb}	s_{geoknn}	s_{geoadj}	$s_{geoconn}$	s_{geohyb}	NGASCE1	NGASCE
Iris	SWC	0.06	0.09	0.09	0.25	0.28	0.27	0.28	0.24	0.31
	DBI	0.79	4.11	4.37	0.61	0.58	0.59	0.58	0.59	0.54
	GDI	1.28	0.39	0.17	2.61	2.48	2.48	2.48	2.49	2.69
	ConnI	0.47	0.14	0.11	0.68	0.72	0.76	0.76	0.77	0.83
BCWS	SWC	0.21	0.21	0.21	−0.01	0.21	0.21	0.21	0.21	0.22
	DBI	0.73	0.71	0.73	0.80	0.75	0.75	0.75	0.72	0.70
	GDI	2.02	1.92	1.94	1.71	2.04	2.03	2.04	2.05	2.11
	ConnI	0.92	0.97	0.95	0.88	0.94	0.94	0.94	0.95	0.97
Yeast	SWC	−0.20	−0.21	−0.22	−0.36	−0.25	−0.25	−0.24	0.12	0.01
	DBI	0.90	1.48	1.46	1.93	1.13	1.13	1.12	0.85	1.01
	GDI	0.04	0.11	0.11	0.32	0.45	0.46	0.46	0.65	0.47
	ConnI	0.30	0.15	0.11	0.24	0.33	0.33	0.33	0.35	0.33
Statlog	SWC	−0.27	−0.18	−0.17	−0.27	−0.30	−0.26	−0.31	0.00	0.00
	DBI	0.90	1.09	0.95	0.92	0.90	0.83	0.91	0.83	0.77
	GDI	1.30	0.41	0.49	1.27	0.97	0.99	0.97	1.32	1.35
	ConnI	0.46	0.41	0.78	0.46	0.47	0.47	0.48	0.78	0.79
Pen Digits	SWC	−0.57	−0.19	−0.31	−0.19	−0.26	−0.38	−0.25	0.00	0.09
	DBI	2.97	1.70	1.67	1.74	1.66	1.59	1.64	1.55	1.32
	GDI	0.52	0.46	0.48	0.63	0.72	0.60	0.73	0.77	0.91
	ConnI	0.60	0.35	0.53	0.46	0.48	0.46	0.48	0.68	0.72

intrinsic data characteristics calculated by various cluster validity indices (silhoutte width criterion-SWC, Davies-Bouldin index-DBI, generalized Dunn index-GDI and CONN index-ConnI) for the resulting partitionings and their ensembles. Leaving the detailed discussions on these validity indices to [20], we note that SWC, GDI and ConnI favors the clustering with the maximum value whereas DBI favors the one with the minimum value. The cluster validity indices provided at Table 3 often favor the proposed ensemble as well. The outperformance of the NGASCE with respect to these different evaluation criteria is promising for clustering of large datasets.

5 Conclusion

Neural gas based approximate spectral clustering is powerful for partitioning of large datasets, when a similarity criterion appropriate for the data characteristics has been selected [10]. However, it has been a long standing challenge to select the optimum criterion for a dataset. Ensemble methods can be of great help to merge distinct clustering decisions without the need of finding the optimum one. In this respect, we proposed a two-step prototype-level ensemble method for neural gas based approximate spectral clustering (NGASCE). Its success on the selected well-known datasets in this study indicate its potential to achieve high clustering accuracies. It is also shown successful for remote sensing image analysis [21]. Our future work is to reduce the computational complexity of the graph-based ensemble approach to make it feasible for ensemble at the data level.

Acknowledgments This work is funded by TUBITAK Career Integration Grant 112E195. Taşdemir is also funded by FP7 Marie Curie Career Integration Grant IAM4MARS.

References

1. Shi, J., Malik, J.: Normalized cuts and image segmentation. IEEE Trans. Pattern Anal. Mach. Intell. **22**(8), 888–905 (2000)
2. Ng, A., Jordan, M., Weiss, Y.: On spectral clustering: analysis and an algorithm. In: Dietterich, T., Becker, S., Ghahramani, Z. (eds.) Advances in Neural Information Processing Systems 14. MIT Press (2002)
3. Meila, M., Shi, J.: A random walks view of spectral segmentation. In: 8th International Workshop on Artificial Intelligence and Statistics (AISTATS) (2001)
4. Fowlkes, C., Belongie, S., Chung, F., Malik, J.: Spectral grouping using the Nyström method. IEEE Trans. Pattern Anal. Mach. Intell. **26**(2), 214–225 (2004)
5. Wang, L., Leckie, C., Ramamohanarao, K., Bezdek, J.C.: Approximate spectral clustering. In: Theeramunkong, T., Kijsirikul, B., Cercone, N., Ho, T.B. (eds.) PAKDD, Volume 5476 of Lecture Notes in Computer Science, pp. 134–146. Springer (2009)
6. Wang, L., Leckie, C., Kotagiri, R., Bezdek, J.: Approximate pairwise clustering for large data sets via sampling plus extension. Pattern Recogn. **44**(2), 222–235 (2011)
7. Taşdemir, K.: Vector quantization based approximate spectral clustering of large datasets. Pattern Recogn. **45**(8), 3034–3044 (2012)
8. Taşdemir, K., Yalcin, B., Yildirim, I.: Approximate spectral clustering with utilized similarity information using geodesic based hybrid distance measures. Pattern Recogn. Under revision(0):0 (2014)
9. Martinetz, T., Schulten, K.: Topology representing networks. Neural Netw. **7**(3), 507–522 (1994)
10. Taşdemir, K., Moazzen, Y., Yildirim, I. Geodesic based similarities for approximate spectral clustering. In: 22nd International Conference on Pattern Recognition, Stockholm, Sweden, 24–28 Aug 2014
11. Strehl, A., Ghosh, J.: Cluster ensembles—a knowledge reuse framework for combining multiple partitions. J. Mach. Learn. Res. **3**(3), 583–617 (2002)
12. Dudoit, S., Fridlyand, J.: Bagging to improve the accuracy of a clustering procedure. Bioinformatics **19**(9), 1090–1099 (2003)
13. Topchy, A., Jain, A.K., Punch, W.: A mixture model for clustering ensembles. In: Proceedings of SIAM International Conference Data Mining, pp. 379–390 (2004)
14. Fred, A.L.N., Jain, A.K.: Combining multiple clusterings using evidence accumulation. IEEE Trans. Pattern Anal. Mach. Intell. (PAMI) **27**(6), 835–850 (2005)
15. Zhang, X., Jiao, L., Liu, F., Bo, L., Gong, M.: Spectral clustering ensemble applied to SAR image segmentation. IEEE Trans. Geosci. Remote Sens. **46**(7), 2126–2136 (2008)
16. Zelnik-Manor, L., Perona, L.: Self-tuning spectral clustering. In: Advances in Neural Information Processing Systems (2004)
17. Taşdemir, K.: A hybrid similarity measure for approximate spectral clustering of remote sensing images. In: 2013 IEEE International Conference on Geoscience and Remote Sensing Symposium (IGARSS), pp. 3136–3139, July 2013

18. Taşdemir, K., Merényi, E.: Exploiting data topology in visualization and clustering of self-organizing maps. IEEE Trans. Neural Netw. **20**(4), 549–562 (2009)
19. Santos, J., Embrechts, M.: On the use of the adjusted rand index as a metric for evaluating supervised classification. In: International Conference on Artificial Neural Networks-ICANN (2009), Limassol, Cyprus, pp. 175–184, 14–17 Sept 2009
20. Taşdemir, K., Merényi, E.: A validity index for prototype-based clustering of data sets with complex cluster structures. IEEE Trans. Syst. Man Cybern. Part B: Cybern. **41**(4), 1039–1053 (2011)
21. Tasdemir, K., Moazzen, Y., Yildirim, I.: An approximate spectral clustering ensemble for high spatial resolution remote-sensing images. IEEE J. Sel. Top. Appl. Earth Obs. Remote Sens. **8**(5), 1996–2004 (2015)

Reliable Clustering Quality Estimation from Low to High Dimensional Data

Jean-Charles Lamirel

Abstract This paper presents new cluster quality indexes which can be efficiently applied for a low-to-high dimensional range of data and which are tolerant to noise. These indexes relies on feature maximization, which is an alternative measure to usual distributional measures relying on entropy or on Chi-square metric or vector-based measures such as Euclidean distance or correlation distance. Experiments compare the behavior of these new indexes with usual cluster quality indexes based on Euclidean distance on different kinds of test datasets for which ground truth is available. This comparison clearly highlights the superior accuracy and stability of the new method.

1 Introduction

Unsupervised classification or clustering is a data analysis technique which is increasingly widely-used in different areas of application. If the datasets to be analyzed have growing size, it is clearly unfeasible to get ground truth that permits to work on them in a supervised fashion. The main problem which then arises in clustering is to qualify the obtained results in terms of quality. A quality index is a criterion which makes it possible to decide which clustering method to use, to fix an optimal number of clusters and also to evaluate or develop a new method. Many approaches have been developed for that purpose as has been pointed out in [1, 20, 21, 24]. However, even if recent alternative approaches do exist [3, 10, 11], the usual quality indexes are mostly based on the concepts of dispersion of a cluster and dissimilarity between clusters. Computation of the latter criteria themselves relies on Euclidean distance. Most popular such indexes are the Dunn index [7], the Davis-Bouldin index [5], the Silhouette index [22], the Calinski-Harabasz index [4] and the Xie-Beni index [25]. They implement the afore mentioned concepts in slightly different ways.

J.-C. Lamirel (✉)
SYNALP Team, LORIA, Bâtiment B, 54506 Vandoeuvre Cedex, France
e-mail: lamirel@loria.fr

© Springer International Publishing Switzerland 2016
E. Merényi et al. (eds.), *Advances in Self-Organizing Maps and Learning Vector Quantization*, Advances in Intelligent Systems and Computing 428,
DOI 10.1007/978-3-319-28518-4_8

The Dunn index (DU) identifies clusters which are well separated and compact. It combines dissimilarity between clusters and their diameters to estimate the most reliable number of clusters. The Davies-Bouldin index (DB) is similar to the Dunn index and identifies clusters which are far from each other and compact. The Silhouette index (SI) computes a width depending on its membership in any cluster. A negative silhouette value for a given point means that the point is most suited to belong to a different cluster from the one it is allocated. The Calinski-Harabasz index (CH) computes a weighted ratio between the within-group scatter and the between group scatter. Well separated and compact clusters should maximize this ratio. The Xie-Beni index (XI) is a compromise between the approaches provided by the Dunn index and by the Calinski-Harabasz index.

As stated in [9, 14, 24] most of the presented indexes have the imensional space as well as they are unable to detect degenerated clusdefect to be sensitive to the noisy data and outliers. In [17], Lamirel et al. also observed that the proposed indexes are not suitable to analyze clustering results in high-dtering results. Also these indexes are not independent of the clustering method with which they are used. As an example, a clustering method which tends to optimize WGSS, like k-means [19], will also tend to naturally produce low value for that criteria which optimizes indexes output, but does not necessarily guarantee coherent results, as it was also demonstrated in [17]. Last but not least, as Hamerly et al. pointed out in [12], the experiments on these indexes in the literature are often performed on unrealistic test corpora made up of low dimensional data with a small number of "well-shaped" embedded virtual clusters. As an example, in their reference paper, Milligan and Cooper [20] compared 30 different methods for estimating the number of clusters. They classified CH and DB in the top 10, with CH the best but their experiments only used simulated data described in a low dimensional Euclidean space. The same remark can be made about the comparison performed in [24] or in [6]. However, Kassab et al. [13] used the Reuters test collection to shown that the aforementioned indexes are often unable to identify an optimal clustering model whenever the dataset is constituted by complex data which need to be represented in both high-dimensional and sparse description space, obviously with embedded non-Gaussian clusters, as is often the case with textual data. The silhouette index is considered one of the more reliable indexes among those mentioned above especially in the case of multidimensional data, mainly because it is not a diameter-based index optimized for Gaussian context. However, like the Dunn and Xie-Beni indexes, its main defect is that it is computationally expensive, which could represent a major drawback for use with large datasets of high-dimensional data.

There are also other altenatives to the usual indexes. For example, in 2009 Lago-Fernãndez et al. [15] proposed a method using negentropy which evaluates the gap between the cluster entropy and entropy of the normal distribution with the same covariance matrix, but again their experiments were only conducted on two-dimensional data. Also other recent indexes attempts were limited by the researchers'choice of complex parameters [24].

Our aim was to get rid of the method-index dependency problem and the issue of sensitivity to noise while also avoiding computation complexity, parameter settings and dealing with a high-dimensional context. To achieve goals, we exploited features of the data points attached to clusters instead of information carried by cluster centroids and replaced Euclidean distance with a more reliable quality estimator based on the feature maximization measure. This measure has been already successfully used by Lamirel et al. to solve complex high-dimensional classification problems with highly imbalanced and noisy data gathered in similar classes thanks to its very efficient feature selection and data resampling capabilities [18]. As a complement to this information, we shall show in the upcoming experimental section that cluster quality indexes relying on this measure do not possess any of the defects of usual approaches including computational complexity.

Section 2 presents a feature maximization measure and our proposed new indexes. Section 3 presents our experimental context. Section 4 our results before Sect. 5 draws our conclusion and ideas for future work.

2 Feature Maximization for Feature Selection

Feature maximization is an unbiased measure which can be used to estimate the quality of a classification whether it be supervised or unsupervised. In unsupervised classification (i.e. clustering), this measure exploits the properties (i.e. the features) of data points that can be attached to their nearest cluster after analysis without prior examination of the generated cluster profiles, like centroids. Its principal advantage is thus to be totally independent of the clustering method and of its operating mode.

Consider a partition C which results from a clustering method applied to a dataset D represented by a group of features F. The feature maximization measure favours clusters with a maximal feature F-measure. The feature F-measure $FF_c(f)$ of a feature f associated with a cluster c is defined as the harmonic mean of the feature recall $FR_c(f)$ and of the feature predominance $FP_c(f)$, which are themselves defined as follows:

$$FR_c(f) = \frac{\Sigma_{d \in c} W_d^f}{\Sigma_{c' \in C} \Sigma_{d \in c'} W_d^f} \quad FP_c(f) = \frac{\Sigma_{d \in c} W_d^f}{\Sigma_{f' \in F_c, d \in c} W_d^{f'}} \tag{1}$$

with

$$FF_c(f) = 2 \left(\frac{FR_c(f) \times FP_c(f)}{FR_c(f) + FP_c(f)} \right) \tag{2}$$

where W_d^f represents the weight of the feature f for the data d and F_c represents all the features present in the dataset associated with the cluster c.

There is some important similarities between Recall and Predominance used in the proposed approach and Recall and Precision used in information retrieval. We have already exploited this analogy more thoroughly in some of our former works,

like in [16], but the measures proposed here must be considered as generalizations of
such information retrieval measures which are no more based on agreement but on
influence of a feature materialized by a weight. Weight represents the importance of
a feature for a data and furthermore for a cluster. The choice of the weighting scheme
is not really constrained by the approach instead of producing positive values. Such
scheme is supposed to figure out the significance (i.e. semantic and importance) of
the feature for the data.

Feature recall is a scale independent measure but feature predominance is not.
We have however shown experimentally in [18] that the F-measure which is a combination of these two measures is only weakly influenced by feature scaling. Nevertheless, to guaranty full scale independent behavior for this measure, data must be
standardized.

Feature maximization measure can be exploited to generate a powerfull feature
selection process [18]. In the clustering context, this kind of selection process can be
defined as non-parametrized process based on clusters content in which a cluster feature is characterized using both its capacity to discriminate between clusters ($FP_c(f)$
index) and its ability to faithfully represent the cluster data ($FR_c(f)$ index). The set
S_c of features that are characteristic of a given cluster c belonging to a partition C is
translated by:

$$S_c = \{f \in F_c \mid FF_c(f) > \overline{FF}(f) \text{ and } FF_c(f) > \overline{FF}_D\} \text{ where} \tag{3}$$

$$\overline{FF}(f) = \Sigma_{c' \in C} \frac{FF_{c'(f)}}{|C_{/f}|} \text{ and } \overline{FF}_D = \Sigma_{f \in F} \frac{\overline{FF}(f)}{|F|} \tag{4}$$

where $C_{/f}$ represents the subset of C in which the feature f occurs.

Finally, the set of all selected features S_C is the subset of F defined by:

$$S_C = \cup_{c \in C} S_c. \tag{5}$$

In other words, the features judged relevant for a given cluster are those whose
representations are better than average in this cluster, and better than the average
representation of all the features in the partition, in terms of feature F-measure.
Features which never respect the second condition in any cluster were discarded.

A specific concept of contrast $G_c(f)$ can be defined to calculate the performance of
a retained feature f for a given cluster c. It is an indicator value which is proportional
to the ratio between the F-measure $FF_c(f)$ of a feature in the cluster c and the average
F-measure $\overline{FF}$ of this feature for the whole partition.[1] It can be expressed as:

$$G_c(f) = FF_c(f)/\overline{FF}(f) \tag{6}$$

---

[1] Using p-value highlighting the significance of a feature for a cluster by comparing its contrast
to unity contrast would be a potential alternative to the proposed approach. However, this method
would introduce unexpected Gaussian smoothing in the process.

The active features of a cluster are those for which the contrast is greater than 1. Moreover, the higher the contrast of a feature for one cluster, the better its performance in describing the cluster content.

As already mentioned before, the active features in a cluster are selected features for which the contrast is greater than 1 in that cluster. Conversely, the passive features in a cluster are selected features present in the cluster's data for which contrast is less than unity.[2] A simple way to exploit the features obtained is to use active selected features and their associated contrast for cluster labelling as we proposed in [18]. A more sophisticated method (as we shall propose hereafter) is to exploit information related to the activity and passivity of selected features in clusters to define clustering quality indexes identifying an optimal partition. This kind of partition is expected to maximize the contrast described by Eq. 6. This approach leads to the definition of two different indexes:

The PC index, whose principle corresponds by analogy to that of intra-cluster inertia in the usual models, is a macro-measure based on the maximization of the average weighted contrast of active features for optimal partition. For a partition comprising k clusters, it can be expressed as:

$$PC_k = \frac{1}{k} \sum_{i=1}^{k} \frac{1}{n_i} \sum_{f \in S_i} G_i(f) \tag{7}$$

The EC index, whose principle corresponds by analogy to that of the combination between intra-cluster inertia and inter-cluster inertia in the usual models, is based on the maximization of the average weighted compromise between the contrast of active features and the inverted contrast of passive features for optimal partition:

$$EC_k = \frac{1}{k} \sum_{i=1}^{k} \left(\frac{\frac{|s_i|}{n_i} \sum_{f \in S_i} G_i(f) + \frac{|\overline{s_i}|}{n_i} \sum_{h \in \overline{S_i}} \frac{1}{G_i(h)}}{|s_i| + |\overline{s_i}|} \right) \tag{8}$$

where n_i is the number of data associated with the cluster i, $|s_i|$ represents the number of active features in i, and $|\overline{s_i}|$, the number of passive features in the same cluster.

3   Experimental Data and Process

To objectively calculate the accuracy of our new indexes, we used several different datasets of varying dimensionality and size for which the optimal number of clusters (i.e. ground truth) is known in advance.

A part of the datasets came from the UCI machine learning repository [2] and is more usually exploited for classification tasks. The 4 selected UCI datasets represent

---

[2]As regards the principle of the method, this type of selected features inevitably have a contrast greater than 1 in some other cluster(s) (see Eq. 3 for details).

Table 1 Datasets overall characteristics (Binarization of IRIS dataset results in 12 binary features out of 4 real-valued features)

	IRIS	IRIS-b	WINE	PEN	ZOO	VRBF	R8	R52
Nbr. class	3	3	3	10	7	12–16	8	52
Nbr data	150	150	178	10992	101	2183	7674	9100
Nbr feat.	4	12	13	16	114	231	3497	7369

mostly low to middle dimensional datasets and small datasets (except for PEN dataset which is large). The ZOO dataset which includes variables with modalities was transformed into a binary file. IRIS is exploited both in standard and in binarized version to obtain clearer insight into the behavior of quality index on binary data.

The VERBF dataset is a dataset of French verbs which are described both by semantic features and by subcategorization frames. The ground truth of this dataset has been established both by linguists who studied different clustering results and by a gold standard based on the VerbNet classification, as in [23]. This binary dataset contains verbs described in a space of 231 Boolean features. It can be considered a typical middle size and middle dimensional dataset.

The R8 and R52 corpora were obtained by Cardoso Cachopo3 from the R10 and R90 datasets, which are derived from the Reuters 21578 collection.[3] The aim of these adjustments was to only retain data with a single label. R8 only considers monothematic documents and classes with at least one example of training and one of testing and is a reduction of the R10 corpus (the 10 most frequent classes) to 8 classes while R52 is a reduction of the R90 corpus (90 classes) to 52 classes. R8 and R52 are large and multidimensional datasets with respective sizes of 7674 and 9100 data and an associated bag of word description spaces of 1187 and 2618 words. These datasets can be considered large and high dimensional.

The R8 and R52 corpora were obtained by Cardoso Cachopo from the R10 and R90 datasets, which are derived from the Reuters 21578 collection.[4] The aim of these adjustments was to only retain data that had a single label. Considering only monothematic documents and classes that still had at least one example of training and one of test, R8 is a reduction of the R10 corpus (the 10 most frequent classes) to 8 classes and R52 is a reduction of the R90 corpus (90 classes) to 52 classes. The R8 and R52 are large and multidimensional datasets with respective size of 7674 and 9100 and associated bag of words description spaces of 1187 and 2618 words. This datasets can be considered as large and high dimensional datasets.

The summary of datasets overall characteristics is provided in Table 1.

We exploited 2 different usual clustering methods, namely k-means [19], a winner-take-all method, and GNG [8], a winner-take-most method with Hebbian learning. For text and/or binary datasets we also used the IGNGF neural clustering method [17] which has already been proven to outperform other clus-tering methods, including

---

[3]http://web.ist.utl.pt/~acardoso/datasets/.

[4]http://www.research.att.com/lewis/reuters21578.html.

spectral methods [23], on this kind of data. We have reported on the method that produced the best results in the following experiments.

As class labels were provided in all datasets and considering that the clustering method could only produce approximate results as compared to reference categorization, we also used purity measures to estimate the quality of the partition generated by the method as regards to category ground truth. Following [23], we use modified purity (mPUR) to evaluate the clusterings produced and this was computed as follows:

$$mPUR = \frac{|P|}{|D|} \tag{9}$$

where $P = \{d \in D \mid \text{prec}(c(d)) = g(d) \wedge |c(d)| > 1\}$ with D being the set of exploited data points, $c(d)$ a function that provides the cluster associated to data d and $g(d)$ a function that provides the gold class associated to data d. Clusters for which the prevalent class has only one element are considered as marginal and are thus ignored.

For the same reason, we also varied the number of clusters in a range up to 3 times that determined by the ground truth. An index which gave no indication of optimum in the expected range was considered to be out-of-range or diverging index (-out-). We finally obtained a process which consists of generating disturbance in the clustering results by randomly exchanging data between clusters to different fixed extents (10, 20, 30 %) whilst maintaining the original size of the clusters. This process simulated increasingly noisy clustering results and the aims was to estimate the robustness of the proposed estimators.

4   Results

The results are presented in Tables 2 and 3. Some complementary information is required regarding the validation process. In the tables, MaxP represents the number of clusters of the partition with highest mPur value (Eq. 9), or in some cases, the interval of partition sizes with highest stable mPur value. When a quality index identified an optimal model with MaxP clusters and MaxP differed from the number of categories established by ground truth, its estimation was still considered valid. This approach took into account the fact that clustering would quite systematically produce sub-optimal results as compared to ground truth. The partitions with the highest purity values were thus studied to deal with this kind of situation. For similar reason, all estimations in the interval range between the optimal k (ground truth) and MaxP values were also considered valid. When indexes were still increasing and decreasing (depending on whether they were maximizers or minimizers) when the number of clusters was more than 3 times the number of expected classes, there were considered out-of-range (-out- symbol in Tables 2 and 3).

Table 2 Overview of the indexes estimation results (Bold numbers represent valid estimations)

	IRIS	IRIS-b	WINE	PEN	ZOO	VRBF	R8	R52	Number of correct matches
DB	2	5	**5**	7	**8**	-out-	5	58	**2/8**
CH	2	**3**	6	8	4	7	**6**	-out-	**2/8**
DU	1	1	8	17	**8**	2	-out-	-out-	**1/8**
SI	**4**	2	7	14	4	-out-	-out-	**54**	**2/8**
XB	2	7	-out-	19	-out-	23	-out-	-out-	**0/8**
EC	**3**	**3**	**4**	9	7	18	-out-	-out-	**3/8**
PC	**3**	2	**4**	9	**7**	**15**	**6**	**52**	**6/8**
MaxP	3	3	5	11	10	12–16	6	50–55	
Method	K-means	K-means	GNG	GNG	IGNGF	IGNGF	IGNGF	IGNGF	

Table 3 Indexes estimation results in the presence of noise (UCI ZOO dataset)

	ZOO	ZOO Noise 10%	ZOO 20%	ZOO Noise 30%	Number of correct matches
DB	**8**	4	3	3	**1/4**
CH	4	5	3	3	**0/4**
DU	**8**	2	2	2	**1/4**
SI	14	-out-	-out-	-out-	**0/4**
XB	-out-	-out-	-out-	-out-	**0/4**
PC	6	4	11	**9**	**1/4**
EC	7	5	6	**9**	**2/4**
MaxP	10	7	10	10	
Method	IGNGF	IGNGF	IGNGF	IGNGF	

When considering the results presented in Table 2, it should first be noted that one of our tested indexes, the Xie-Beni (XB) index never provides any correct answers. These were either out of range (i.e. diverging) or answers (i.e. minimum value when this index was a minimizer) in the range of the variation of k, but too far from ground truth or even too far from optimal purity among the set of generated clustering models. Some indexes were in the low mid-range of correctness and provide unstable answers. This was the cases with the Davis-Bouldin (DB), Calinski-Harabasz (CH), Dunn (DU) and Silhouette (SI) indexes. When there was dimension growth, these indexes were found to become generally unable to provide any correct estimation. This phenomenon has already been observed in previous experiments with Davis-Bouldin (DB) and Calinski-Harabasz (CH) indexes [13]. Our PC index was found to perform slightly better than average but obviously remains a better low dimensional problem estimator than a high dimensional one. Help from passive features somehow seems mandatory to estimate an optimal model in the case of high dimensional problems. Hence, the EC index which exploited both active and passive features was found to have from far the best performance, whatever it faced with low or high dimensional estimation problem. Additionally, both the EC and PC indexes, were both found to be capable of dealing with binarized data in a transparent manner which is not the case of some of the usal indexes namely the Xie-Beni (XI) index, and to a lesser extend, Calinski-Harabasz (CH) and Silhouette (SI) indexes.

Interestingly, on the UCI ZOO dataset, the results of noise sensitivity analysis presented in Table 3 underline the fact that noise has a relatively limited effect on the operation of PC and EC indexes. The EC index was again found to have the most stable behavior in that context. As for the Silhouette index, this firstly delivered the wrong optimal k values on this dataset before getting out of range when the noise reached 20 % on clustering results. The Davis-Bouldin (DB) and Dunn indexes (DU) were found to shift from a correct to a wrong estimation as soon as noise began to appear.

In all our experiments, we observed that the quality estimation depends little on the clustering method. Morever, we noted that the computation time of the index was one of the lowest among the indexes studied. As an example, for the R52 dataset, the EC index computation time was 125 s as compared to 43,000 s for the Silhouette index using a standard laptop with 2.2 GHz quadricore processor and 8 GB of memory.

5 Conclusion

We have proposed a new set of indexes for clustering quality evaluation relying on feature maximization measurement. This method exploits the information derived from features which could be associated to clusters by means of their associated data. Our experiments showed that most of the usual quality estimators do not produce satisfactory results in a realistic data context and that they are additionally sensitive to noise and perform poorly with high dimensional data. Unlike the usual quality estimators, one of the main advantages of our proposed indexes is that they produce stable results in cases ranging from a low dimensional to high dimensional context and also require low computation time while easily dealing with binarized data. Their stable operating mode with clus-tering methods which could produce both different and imperfect results also constitutes an essential advantage. However, further experiments are required using both an extended set of clustering methods and a larger panel of high dimensional datasets to confirm this promising behavior.

Additionally, we plan to test the ability of our indexes to discriminate between correct and degenerated clustering results in the context of large and heterogeneous datasets.

References

1. Angel Latha Mary, S., Sivagami, A.N., Usha Rani, M.: Cluster validity measures dynamic clustering algorithms. ARPN J. Eng. Appl. Sci. **10**(9) (2015)
2. Bache, K., Lichman, M.: UCI Machine Learning Repository (http://archive.ics.uci.edu/ml). University of California, School of Information and Computer Science, Irvine (2013)
3. Bock, H.-H.: Probability model and hypothese testing in partitionning cluster analysis. In: Arabie, P., Hubert, L.J., De Soete, G. (eds.) Clustering and Classification, pp. 377–453. World Scientific, Singapore (1996)
4. Calinsky, T., Harabasz, J.: A dendrite method for cluster analysis. Commun. Stat. **3**(1), 1–27 (1974)
5. Davies, D.L., Bouldin, D.W.: A cluster separation measure. IEEE Trans. Pattern Anal. Mach. Intell., PAMI-1, 2:224–227 (1979)
6. Dimitriadou, E., Dolnicar, S., Weingessel, A.: An examination of indexes for determining the number of clusters in binary data sets. Psychometrika **67**(1), 137–159 (2002)
7. Dunn, J.: Well separated clusters and optimal fuzzy partitions. J. Cybern. **4**, 95–104 (1974)
8. Fritzke, B.: A growing neural gas network learns topologies. In: Tesauro, G., Touretzky, D.S., Leen, T.K. (ed.) Advances in Neural Information Processing Systems 7, pp. 625–632 (1995)
9. Guerra, L., Robles, V., Bielza, C., Larrañaga, P.: A comparison of clustering quality indices using outliers and noise. Intell. Data Anal. **16**, 703–715 (2012)
10. Gordon, A.D.: External validation in cluster analysis. Bull. Int. Stat. Inst. **51**(2), 353–356 (1997); Response to comments. Bull. Int. Stat. Inst. **51**(3), 414–415 (1998)
11. Halkidi, M., Batistakis, Y., Vazirgiannis, M.: On clustering validation techniques. J. Int. Inf. Syst. **17**(2/3), 147–155 (2001)
12. Hamerly, G., Elkan, C.: Learning the K in K-means. In: Neural Information Processing Systems (2003)
13. Kassab, R., Lamirel, J.-C.: Feature based cluster validation for high dimensional data. In: IASTED International Conference on Artificial Intelligence and Applications (AIA), pp. 97–103. Innsbruck, Austria (2008)
14. Kolesnikov, A., Trichina, E., Kauranne, T.: Estimating the number of clusters in a numerical data set via quantization error modeling. Pattern Recogn. **48**(3), 941–952 (2015)
15. Lago-Fernãndez, L.F., Corbacho, F.: Using the negentropy increment to determine the number of clusters. In: Cabestany, J., Sandoval, F., Prieto, A., Corchado, J.M., et al. (eds.) Bio-Inspired Systems: Computational and Ambient Intelligence, pp. 448–455. Springer, Berlin (2009)
16. Lamirel, J.-C., Francois, C., Al Shehabi, S., Hoffmann, M.: New classification quality estimators for analysis of documentary information: application to patent analysis and web mapping. Scientometrics **60**(3), 445–462 (2004)
17. Lamirel, J.-C., Mall, R., Cuxac, P., Safi, G.: Variations to incremental growing neural gas algorithm based on label maximization. In: Proceedings of IJCNN 2011, pp. 956–965, San Jose (2011)
18. Lamirel, J.-C., Cuxac, P., Chivukula, A.S., Hajlaoui, K.: Optimizing text classification through efficient feature selection based on quality metric. J. Intell. Inf. Syst., Spec. Issue PAKDD-QIMIE **2013**, 1–18 (2014)
19. MacQueen, J.B.: Some methods for classification and analysis of multivariate observations. In: Proceedings of 5th Berkeley Symposium on Mathematical Statistics and Probability (1), pp. 281–297. University of California Press (1967)
20. Milligan, G.W., Cooper, M.C.: An examination of procedures for determining the number of clusters in a dataset. Psychometrika **50**(2), 159–179 (1985)
21. Rendón, E., Abundez, I., Arizmendi, A., Quiroz, E.M.: Internal versus external cluster validation indexes. Int. J. Comput. Commun. **5**(1), 27–34 (2011)

22. Rousseeuw, P.J.: Silhouettes: a graphical aid to the interpretation and validation of cluster analysis. J. Comput. Appl. Math. **20**, 53–65 (1987)
23. Sun, L., Korhonen, A., Poibeau, T., Messiant, C.: Investigating the cross-linguistic potential of VerbNet-style classification Proceedings of ACL, pp. 1056–1064. Beijing (2010)
24. Yanchi, L., Zhongmou, L., Xiong, H., Gao, X., Wu, J.: Understanding of internal clustering validation measures. In: Proceedings of the 2010 IEEE International Conference on Data Mining, ICDM '10, pp. 911–916
25. Xie, X.L., Beni, G.: A validity measure for fuzzy clustering. IEEE Trans. Pattern Anal. Mach. Intell. **13**(8), 841–847 (1991)

Segment Growing Neural Gas
for Nonlinear Time Series Analysis

Jorge R. Vergara, Pablo A. Estévez and Álvaro Serrano

Abstract In this work we propose an extension to Growing Neural Gas (GNG) for dealing with the spatiotemporal quantization of time series. The two main changes to the original GNG algorithm are the following. First, the basic unit of the GNG network is changed from a node to a linear segment joining two nodes. Secondly, temporal connections between neighboring units in time are added. The proposed algorithm called Segment GNG (SGNG) is compared with the original GNG and Merge GNG algorithms using three benchmark time series: Rössler, Mackey-Glass and NH_3 Laser. The algorithms are applied to the quantization of trajectories in the state space representation of these time series. The results show that the SGNG outperforms both GNG and Merge GNG in terms of quantization error and temporal quantization error.

1 Introduction

Time series analysis has two main goals: (i) identify the dynamics of the data generating process, and (ii) predict future values based on the signal previous behavior [1]. Vector quantization is a tool that allows extracting prototypes, e.g. centroids of receptive fields. The Self-Organizing Map (SOM) [11] performs vector quantization through unsupervised learning and adds an output grid to achieve a topological

J.R. Vergara · P.A. Estévez (✉) · Á. Serrano
Department of Electrical Engineering, University of Chile, Santiago, Chile
e-mail: pestevez@ing.uchile.cl

J.R. Vergara
e-mail: jorgever@ing.uchile.cl

J.R. Vergara · P.A. Estévez
Millennium Institute of Astrophysics, Santiago, Chile

P.A. Estévez
Advanced Mining Technology Center, Santiago, Chile

© Springer International Publishing Switzerland 2016
E. Merényi et al. (eds.), *Advances in Self-Organizing Maps and Learning Vector Quantization*, Advances in Intelligent Systems and Computing 428,
DOI 10.1007/978-3-319-28518-4_9

107

ordered mapping, which allows us visualizing the topological relationships among prototypes. A variant is the Neural Gas (NG) algorithm [15], which gets rid of the output grid in order to achieve a good quantization for any kind of topology. In both SOM and NG, the number of prototypes is a user defined parameter. Growing NG (GNG) [8] and Growing SOM (GSOM) [2] start with two prototypes and grow adaptively during iterations.

The aforementioned algorithms were designed to represent the data spatial distribution but not the data temporal relationships. As a consequence the direct application of SOM and NG to time series is rather limited. Several works have been developed to include the temporal data relationships using feedback connections such as recursive SOM [21] and recurrence SOM [12]. Another family of models add compact temporal contexts such Merge SOM [19], Merge NG [18], Merge GNG [1], γ-SOM [5], γ-NG [4], and γ-GNG [6].

State et. al [17] and Coleca et. al [3] introduced an extension of the SOM for performing 3D hand and full body skeleton tracking. In this method the hand and body are represented by line and plane segments between nodes, that are adjusted adaptively. We found the concept of segments very useful for time series analysis, and herein we propose an extension to GNG where segments are the basic units instead of nodes. To obtain a spatiotemporal data representation we also introduce temporal connections between units. In addition our analysis is based on the spatiotemporal quantization of the state-space representation of time series, instead of quantizing the signal directly.

The remainder of this work is divided into 5 sections. Section 2 introduces the fundamental concepts used in our model. Section 3 presents the proposed extension to the GNG model. Section 4 shows the simulation results obtained with 3 benchmark time series. In Sect. 5 the conclusions are drawn.

2   Background

2.1   Delay Coordinate Embedding

The state space is the set of all states of a deterministic dynamical system. According to Takens' embedding theorem [20], it is possible to reproduce entirely the properties of such a system (topology and temporal structure) starting from one-dimensional time series. The time series correspond to a sequence of scalar measurements of the state space or a single state variable, x_t. To embed a time series, a delay coordinate vector is constructed as follows: $\boldsymbol{\phi}_n = [x_n, x_{(n-\zeta)}, x_{(n-2\zeta)}, \cdots, x_{(n-(m-1)\zeta)}]$, where the delay ζ and dimension m are the embedding parameters. Although the embedding theorems do not provide a way to estimate these parameters, there are some heuristic methods to do so. The parameter ζ is usually estimated by seeking for the delay that provides the first minimum of the average mutual information [7], while the dimension m is estimated by the false nearest neighbor algorithm [10].

2.2 Segments as Basic Units for the Self-Organizing Map

State et. al [17] and Coleca et. al [3], in their work on hand and full body skeleton tracking extended SOM by considering line and plane segments as basic units. A segment is the line joining two nodes $\mathbf{w}^i$ and $\mathbf{w}^j$ defined as $\mathbf{W}^{ij} = \overline{\mathbf{w}^i \mathbf{w}^j}$. The distance between a sample point $\mathbf{y}_n$ and the segment $\mathbf{W}^{ij}$ is obtained by projecting $\mathbf{y}_n$ over segment $\mathbf{W}^{ij}$ and then calculating the distance between $\mathbf{y}_n$ and its projection $\mathbf{p}$. Defining $\Delta\mathbf{w}^{ji} = \mathbf{w}^j - \mathbf{w}^i$, then $\mathbf{p}$ can be expressed as $\mathbf{p} = \mathbf{w}^i + \eta_{ji}\, \Delta\mathbf{w}^{ji}$, $0 \leq \eta_{ji} \leq 1$, with $\eta_{ji} + \eta_{ij} = 1$. Given the unit vector $\widehat{\Delta\mathbf{w}^{ji}}$, the coefficient η_{ji} is computed as:

$$\eta_{ji} = \frac{(\mathbf{y}_n - \mathbf{w}^i)}{\|\Delta\mathbf{w}^{ij}\|} \cdot \widehat{\Delta\mathbf{w}^{ji}}. \tag{1}$$

The square euclidean distance of $\mathbf{y}_n$ to the segment $\mathbf{W}^{ij}$ is:

$$\left\| d\left(\mathbf{y}_n, \mathbf{W}^{ij}\right) \right\|^2 = \left\| \mathbf{y}_n - \mathbf{w}^i \right\|^2 - \left\| \eta_{ji}\, \Delta\mathbf{w}^{ji} \right\|^2. \tag{2}$$

3 Spatiotemporal Extension of Growing Neural Gas

3.1 Proposed Method

Herein we propose an extension to the GNG algorithm for the spatiotemporal quantization of time series. The main changes to the original GNG algorithm are the following: (i) the basic unit is changed from a node (neuron) to a segment (connection between two nodes), and (ii) a register for keeping temporal connections is introduced. In what follows the details of the proposed algorithm are explained. The new algorithm is called Segment Growing Neural Gas (SGNG).

The SGNG algorithm seeks to approximate trajectories in the state space representation by linear segments. A segment $\mathbf{S}^i$ is defined as the line joining two nodes $\mathbf{s}^i_O, \mathbf{s}^i_F \in \mathbb{R}^m$ where $\mathbf{s}^i_O$ and $\mathbf{s}^i_F$ correspond to the initial and final points of segment $\mathbf{S}^i$ respectively. Segment $\mathbf{S}^i$ is used to identify and quantize portions of trajectories in the state space that could be locally approximated by a linear segment. A trajectory portion $\{\boldsymbol{\phi}\}_n^{n-\tau}$ is associated to sample $\boldsymbol{\phi}$ at time n and its τ past samples

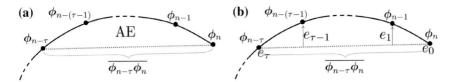

Fig. 1 **a** Area enclosed (AE) between a trajectory portion in the state space representation $\{\boldsymbol{\phi}\}_n^{n-\tau}$ and the line $\overline{\phi_{n-\tau}\phi_n}$. **b** AE estimation through the sum of distances (e_i, $i = 0, 1, \ldots, \tau$) between each sample from the trajectory portion $\{\boldsymbol{\phi}\}_n^{n-\tau}$ and the line $\overline{\phi_{n-\tau}\phi_n}$

$\{\boldsymbol{\phi}\}_n^{n-\tau} = \{\boldsymbol{\phi}_{n-\tau}, \boldsymbol{\phi}_{n-\tau+1}, \cdots, \boldsymbol{\phi}_n\}$. The size of a portion of trajectory that will be quantized by $\mathbf{S}^i$ is determined by the linearity of this portion. The linearity of a portion of a trajectory is evaluated as follows. The parameter τ starts with the value 1. Later on, this delay is increased iteratively until the area enclosed (AE) between the current trajectory portion $\{\boldsymbol{\phi}\}_n^{n-\tau}$ and the line joining the extreme points of this trajectory portion $\overline{\phi_{n-\tau}\phi_n}$ reaches a certain threshold E_{max} (See Fig. 1a).

To avoid the cost of computing the AE accurately, we approximate it through the sum of distances between each sample in the trajectory portion $\{\boldsymbol{\phi}\}_n^{n-\tau}$ and its projection onto the line $\overline{\phi_{n-\tau}\phi_n}$. Figure 1b) illustrates the above mentioned distances, $e_\tau, \cdots, e_0$.

DISTANCE MEASURE. To obtain the Best Matching Linear Segment (BMLS) for each trajectory portion $\{\boldsymbol{\phi}\}_n^{n-\tau}$, a distance measure that evaluates two features of each linear segment $\mathbf{S}^i$ is used: (i) the closeness between $\mathbf{S}^i$ and the trajectory portion $\{\boldsymbol{\phi}\}_n^{n-\tau}$ is measured through the spatial distance and (ii) the degree of parallelism between $\mathbf{S}^i$ and the line $\overline{\phi_{n-\tau}\phi_n}$ is measured through the cosine similarity. To measure the spatial distance, first the midpoint ($\boldsymbol{\rho}_n$) of the trajectory portion $\{\boldsymbol{\phi}\}_n^{n-\tau}$ is estimated. Secondly, the distance between this midpoint and the linear segment $\mathbf{S}^i$ is computed by using Eq. (2). The cosine similarity between $\mathbf{S}^i$ and the line $\overline{\phi_{n-\tau}\phi_n}$ is computed as:

$$sim(\mathbf{S^i}, \overline{\phi_{n-\tau}\phi_n}) = \frac{\Delta s^i \cdot \Delta \phi}{\|\Delta s^i\| \|\Delta \phi\|}, \tag{3}$$

where $\Delta s^i = \mathbf{s}_F^i - \mathbf{s}_O^i$ and $\Delta \phi = \phi_n - \phi_{n-\tau}$.

The combined distance measure used by SGNG is the following:

$$D\left(\theta, \{\boldsymbol{\phi}\}_n^{n-\tau}, \mathbf{S}^i\right) = \left(\theta\left(1 - sim(\mathbf{S}^i, \overline{\phi_{n-\tau}\phi_n})\right) + 1\right) d\left(\boldsymbol{\rho}_n, \mathbf{S}^i\right), \tag{4}$$

where θ is a parameter that controls the trade-off between the cosine value (parallelism) and the spatial distance (closeness). Once selected the BMLS at iteration n, a temporal link is created between this unit and the BMLS selected at iteration $n - 1$. The step by step SGNG algorithm is described in Algorithm 1.

Algorithm 1 Pseudo-code algorithm SGNG.

1: Create randomly two linear segments $\mathbf{S}^i = \{\mathbf{s}_O^i, \mathbf{s}_F^i\}$, $i = 1, 2$. Connect them spatially with a zero age edge. Set to zero their respective errors, $error^i$.

2: Create matrix of temporal connections for segments of the network. If there are already segments temporally connected, disconnect them.

3: Present sample ϕ_n to the network.

4: Find the maximum delay τ such that the AE of the sequence $\{\phi\}_n^{n-\tau}$ does not exceed a E_{max}.

5: Find the best matching linear segment (BMLS), I_n, and the second closest segment, J_n, using Eq. 4.

6: Update the BMLS's error: $error^{I_n} = error^{I_n} + D\left(\theta, \{\phi\}_n^{n-\tau}, \mathbf{S}^{I_n}\right)$, where $D\left(\theta, \{\phi\}_n^{n-\tau}, \mathbf{S}^{I_n}\right)$ is the distance obtained from Eq. 4.

7: Update BMLS's position using the following rule:

$$\mathbf{s}_O^{I_n} = \epsilon_w \left(\phi_{n-\tau} - \mathbf{s}_O^{I_n}\right) \quad \text{and} \quad \mathbf{s}_F^{I_n} = \epsilon_w \left(\phi_n - \mathbf{s}_F^{I_n}\right) \tag{5}$$

and update the position of its neighboring segments (i.e. all segments connected to the BMLS by an edge of topological connection) changing step-size ϵ_w to ϵ_n in Eq. 5.

8: Increment the age of all edges connecting the BMLS and its topological neighbors, $a_j = a_j + 1$.

9: If the BMLS and the second closest segment are connected by a topological edge, then set the age of that edge to zero. Otherwise create a topological edge between them.

10: If there are topological edges with an age larger than a_{max} then remove them. If after this operation, there are segments without topological edges remove them.

11: Create a temporal connection between the current BMLS (I_n) and the BMLS of the past iteration (I_{n-1}).

12: If the current iteration n is an integer multiple of λ, and the maximum number of segments not been reached, then insert a new segment. The parameter λ controls the number of iterations required before inserting a new segment. Insertion of a new segment, r, is done as follows:

    (a) Find segment u with the largest error.

    (b) Among the neighbors of u, find the segment v with the largest error.

    (c) Insert the new segment r between u and v as follows:

$$\mathbf{s}_O^r = 0.5\left(\mathbf{s}_O^u + \mathbf{s}_O^v\right) \quad \text{and} \quad \mathbf{s}_F^r = 0.5\left(\mathbf{s}_F^u + \mathbf{s}_F^v\right) \tag{6}$$

    (d) Create topological edges between u and r, and v and r, and then remove the topological edge between u and v.

    (e) Create a temporal connection between segments u and r.

    (f) Decrease the error of u and v as $error^u = (1 - \alpha)\, error^u$ and $error^v = (1 - \alpha)\, error^v$. Set the error of node r as $error^r = error^u$.

13: Decrease error for all segments j by a factor $(1 - \beta)$, $error^j = (1 - \beta)\, error^j$. Typically, $\alpha = 0.5$ and $\beta = 0.0005$.

14: Set $n \rightarrow n + 1$.

15: If $n < L$ go back to step 3. L is the cardinality of the time series in the state space.

16: If there are segments without temporal connections, remove them.

17: If the stopping criterion is not met, go back to step 2.

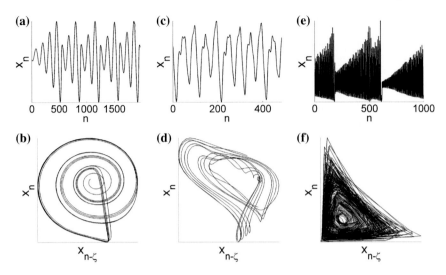

Fig. 2 Plots of one-dimensional time series (first row) and their respective 2D state space representation (second row): Rössler **a–b**, Mackey-Glass **c–d**, and Laser **e–f**

4   Simulation Results

4.1   Description of Datasets

Three datasets were used to evaluate the performance of the proposed algorithm. The first dataset is the Rössler system, which is a well-known example of a strange attractor [13]. It is defined by a system of 3 ordinary differential equations with nonlinear components [16]. A total of 1968 samples were drawn from the Rössler system. The delay embedding parameter was estimated as $\zeta = 33$ by mutual information. The Rössler's time series and its attractor are shown in Fig. 2a, b, respectively. The second dataset corresponds to the Mackey-Glass time series, which is defined by a differential equation but depending on the parameters chosen a wide variety of different behaviors are obtained, including chaotic solutions [14]. A total of 484 samples were drawn from this dataset. The delay embedding parameter was estimated as $\zeta = 18$ by mutual information. The Mackey-Glass time series and its attractor are shown in Fig. 2c, d, respectively. The laser time series corresponds to data set A[1] in the Santa Fe time series competition. This is a univariate time series, containing 1000 measurements from a FIR-Laser in a chaotic state. The delay embedding parameter was estimated as $\zeta = 3$ by mutual information. The Laser time series and its attractor are shown in Fig. 2e, f, respectively. Without losing generality in this work we use $m = 2$ as the dimension of the delay coordinate embedding vector i.e., the trajectories are in a two-dimensional space.

---

[1] Available at http://www-psych.stanford.edu/~andreas/Time-Series/SantaFe.html.

4.2 Parameter Setting and Performance Measurements

The proposed SGNG algorithm is compared with the GNG and Merge GNG (MGNG) algorithms on the three datasets described in the previous section. The common parameters for SGNG, GNG and Merge GNG were set using recommended values in the literature [6, 8, 19, 21]. These parameters are: $a_{max} = 60$, $\alpha = 0.5$, $\beta = 0.0005$, $\lambda = 100$, and $epoch = 1000$. Other parameters were varied using a grid with all possible combinations of the following values: $e_w = \{0.5; 0.05; 0.005\}$, $e_n = \{0.005; 0.0005; 0.0001\}$, maximal number of nodes or segments as a percentage of the length of the time series $maxN = \{5; 7; 11; 15, 20\%\}$. Each combination was repeated 5 times, and the combination having the best results was chosen for each time series. For the SGNG algorithm, the value $\theta = 5$ was used. The parameter E_{max} with the following values $E_{max} = \{0.1; 0.01; 0.001\}$ was included in the search grid. The performance measurements used are the quantization error (QE) and the temporal quantization error (TQE) [9, 21]. QE measures the average distance (mean error) of each sample of time series to its nearest quantization unit. In the case of GNG and Merge GNG their units are the neurons (nodes), while the unit of SGNG is a segment. TQE measures average dispersion of samples delayed ξ time steps associated to each quantization unit. TQE is formally defined as [9]:

$$TQE(\xi) = \frac{1}{N} \sum_{i=1}^{N} \sqrt{\sum_{j:I(j)=i} \frac{\left\| x_{j-\xi} - a_\xi^i \right\|^2}{win_i}}, \qquad (7)$$

where N is the number of units of quantification, win_i is the number of samples associated with the receptive field of the i-th unit, $I(j) = i$ is the index of the j-th sample belonging to the i-th unit ($j = 1, 2, \cdots, win_i$; $i = 1, 2, \cdots, N$), a_ξ^i it is the average of the samples belonging to the i-th unit. To compute the TQE for the SGNG algorithm, we must notice that the receptive field of a linear segment is not spherical. Therefore to compute the dispersion of samples in the receptive field of a linear segment, each sample is associated with its nearest node (the initial or final point). Then the TQE is computed using Eq. (7). Delays up to 50 samples ($\xi = 0, 1, \cdots, 50$) are used to compute TQE.

4.3 Rössler

Figure 3a shows a quantization of the Rössler attractor performed by GNG. It can be observed that GNG does not represent well the vertical trajectories. In addition at the center of the attractor, there are spurious connections between nodes belonging to different trajectories. Figure 3b shows a quantization of the Rössler attractor performed by Merge GNG. It can be observed that the vertical trajectories are now better represented compared to those of GNG. The center of the attractor is cleaner too with less spurious connections between nodes belonging to different trajectories.

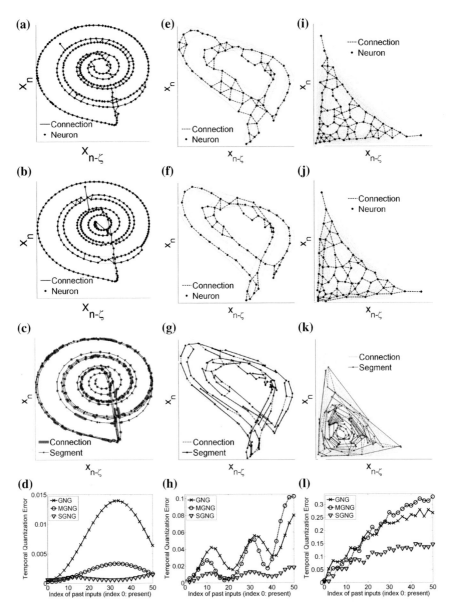

Fig. 3 Each column represents the results for a different dataset: Rössler, Mackey-Glass and Laser respectively. Rows 1-2-3 show the quantization results obtained by GNG, MGNG and SGNG respectively. The fourth row show the temporal quantization errors obtained for each dataset: **d** Rössler, **h** Mackey-Glass and **l** Laser

Figure 3c shows a quantization of the Rössler attractor performed by SGNG. It can be observed that the vertical trajectories are correctly quantized and it is even possible to distinguish different vertical trajectories. The quantization errors (QEs) for GNG, MGNG and SGNG are 0.0103, 0.0120 and 0.0015, respectively. For MGNG the largest contribution to QE comes from the intersections between vertical and horizontal trajectories. Figure 3d illustrates the TQEs for the three algorithms. GNG has a large TQE because it cannot distinguish samples with similar amplitudes belonging to different trajectories. Both MGNG and SGNG presents similar TQE values until 10 delays, but for larger number of delays, SGNG clearly outperforms MGNG.

4.4 Mackey-Glass

Figure 3e shows a quantization of the Mackey-Glass attractor performed by GNG. As GNG performs only a spatial quantization, it can be observed that there are many spurious connections between nodes belonging to different trajectories. As a consequence it is not possible to distinguish between the two main modes of behavior of the attractor. Figure 3f shows a quantization of the Mackey-Glass attractor performed by MGNG. It can be observed that the two modes of the attractor are clearly distinguished. But the resolution obtained is rather low, with several close trajectories represented as a single one. Figure 3g shows a quantization of the Mackey-Glass attractor performed by SGNG. It shows a better resolution than the other two algorithms, allowing us to distinguish between close trajectories. This can be clearly observed at the intersection of approximately perpendicular trajectories. The QEs for GNG, MGNG and SGNG are 0.0255, 0.0306 and 0.0101, respectively. The MGNG presents the highest QE among the 3 algorithms because its quantization has low resolution. In contrast, SGNG can differentiate between close trajectories. Figure 3h illustrates the TQEs for the three algorithms. SGNG obtains the best performance for most number of delays.

4.5 Laser

Figure 3i shows a quantization of the Laser attractor performed by GNG. It can be observed that there are many spurious connections between nodes belonging to different trajectories, so that they cannot be distinguished at all. Figure 3j shows a quantization of the Laser attractor performed by MGNG. The quantization in Fig. 3j possesses the same deficiencies as those in Fig. 3i. Figure 3k shows a quantization of the Laser attractor performed by SGNG. The proposed algorithm is the only one that captures well the dynamics of the laser attractor. The quantization errors (QEs) for GNG, MGNG and SGNG are 0.020, 0.0235, and 0.0116, respectively. Again MGNG presents the highest QE among the 3 algorithms, and SGNG the lowest one. Figure 3l illustrates the TQEs for the three algorithms. SGNG obtains the best performance for all number of delays.

5   Conclusions

We have proposed an extension to GNG that allows performing a spatiotemporal
quantization of time series. A key element in our proposal is changing the basic
unit from a node to a linear segment. The algorithm is able to identify the direction
of trajectories in the space state, and capture the dynamics of the time series, e.g.
an attractor. Segments are useful for the spatiotemporal quantization of time series
because they can be easily adapted to rapid changes in trajectories and even resolution. A second important element is the inclusion of temporal connections between
neighbors in time. This allows a higher accuracy in adjusting segments to trajectories
and smoother transitions between quantization levels. The results obtained show that
the proposed SGNG algorithm outperforms GNG and MGNG in terms of QE and
TQE in the three time series studied. As future work we plan to expand the algorithm
for its use with higher dimensional state space representations than 2D.

Acknowledgments This research was supported by Conicyt-Chile under grants Fondecyt 1140816,
Conicyt DPI20140090 and by the Ministry of Economy Development and Tourism of Chile under
grant IC12089 awarded to the Millennium Institute of Astrophysics.

References

1. Andreakis, A., Hoyningen-Huene, N., Beetz, M.: Incremental unsupervised time series analysis
   using merge growing neural gas. In: Advances in Self-Organizing Maps. Lecture Notes in
   Computer Science, vol. 5629, pp. 10–18. Springer, Berlin (2009)
2. Bauer, H.U., Villmann, T.: Growing a hypercubical output space in a self-organizing feature
   map. IEEE Trans. Neural Netw. **8**(2), 218–226 (1997)
3. Coleca, F., State, A., Klement, S., Barth, E., Martinetz, T.: Self-organizing maps for hand and
   full body tracking. Neurocomputing **147**, 174–184 (2015)
4. Estévez, P., Hernández, R., Pérez, C., Held, C.: Gamma-filter self-organising neural networks
   for unsupervised sequence processing. Electron. Lett. **47**(8), 494–496 (2011)
5. Estévez, P.A., Hernández, R.: Gamma som for temporal sequence processing. In: Advances in
   SOM's, pp. 63–71. Springer (2009)
6. Estévez, P., Vergara, J.: Nonlinear time series analysis by using gamma growing neural gas. In:
   Estévez, P.A., Príncipe, J.C., Zegers, P. (eds.) Advances in Self-Organizing Maps, Advances
   in Intelligent Systems and Computing, vol. 198, pp. 205–214. Springer, Berlin (2013)
7. Fraser, A.M., Swinney, H.L.: Independent coordinates for strange attractors from mutual information. Phys. Rev. A **33**(2), 1134 (1986)
8. Fritzke, B.: A growing neural gas network learns topologies. Adv. Neural Inf. Process. Syst. **7**,
   625–632 (1995)
9. Hammer, B., Micheli, A., Sperduti, A., Strickert, M.: Recursive self-organizing network models. Neural Netw. **17**(8–9), 1061–1085 (2004)
10. Kennel, M.B., Brown, R., Abarbanel, H.D.: Determining embedding dimension for phase-space
    reconstruction using a geometrical construction. Phys. Rev. A **45**(6), 3403 (1992)
11. Kohonen, T.: Self-organizing Maps. Springer, Heidelberg (1995)
12. Koskela, T., Varsta, M., Heikkonen, J., Kaski, K.: Temporal sequence processing using recurrent
    som. In: Proceedings KES '98. 1998 Second International Conference on, vol. 1, pp. 290–297
    (1998)
13. Lorenz, E.N.: Deterministic nonperiodic flow. J. Atmos. Sci. **20**(2), 130–141 (1963)

14. Mackey, M.C., Glass, L., et al.: Oscillation and chaos in physiological control systems. Science **197**(4300), 287–289 (1977)
15. Martinetz, T., Berkovich, S., Schulten, K.: 'Neural-gas' network for vector quantization and its application to time-series prediction. IEEE Trans. Neural Netw. **4**(4), 558–569 (1993)
16. Rössler, O.: An equation for continuous chaos. Phys. Lett. A **57**(5), 397–398 (1976)
17. State, A., Coleca, F., Barth, E., Martinetz, T.: Hand tracking with an extended self-organizing map. In: Advances in Self-Organizing Maps. Advances in Intelligent Systems and Computing, vol. 198, pp. 115–124. Springer, Berlin (2013)
18. Strickert, M., Hammer, B.: Neural gas for sequences. In: Proceedings of the Workshop on Self-Organizing Maps (WSOM03), pp. 53–57 (2003)
19. Strickert, M., Hammer, B.: Merge som for temporal data. Neurocomputing **64**, 39–71 (2005)
20. Takens, F.: Detecting strange attractors in turbulence. In: Lecture Notes in Math, vol. 898. Springer, New York (1981)
21. Voegtlin, T.: Recursive self-organizing maps. Neural Netw. **15**(8), 979–991 (2002)

Modeling Diversity in Ensembles for Time-Series Prediction Based on Self-Organizing Maps

Rigoberto Fonseca-Delgado and Pilar Gómez-Gil

Abstract A Self Organizing Map (SOM) projects high-dimensional feature vectors onto a low-dimensional space. If an appropriate feature vector is chosen, this ability may be used for measuring and adjusting different levels of diversity in the selection of models for building ensembles. In this paper, we present the results of using a SOM for selecting suitable models in ensembles used for long-term time series prediction. The temporal behavior of the predictors is represented by feature vectors built with a sequence of the errors achieved in each prediction step. Each neuron in the map represents a cluster of models with similar accuracy; the adjustment of diversity between models is achieved by measuring the distance between neurons on the map. Our experiments showed that this strategy generated ensembles with an appropriate level of diversity among their components, obtaining a better performance than just using a unique model.

1 Introduction

In the last years, it has been found that selecting and combining an appropriate set of models for univariate time series forecasting achieve better results than using only one model [1, 8]. However, to find the right model set to combine is not a trivial task [16]. In the ensemble research area exists a consensus about the strategy for the selection of models to combine: diversity and accuracy of the involved models are the main factors to consider [1, 15].

SOM neural network [13], following simple rules of competition and cooperation [12], has been used for building ensembles for time series forecasting. For example, Ni et al. developed SOMAR [19] which is a method that modifies SOM training to

R. Fonseca-Delgado (✉) · P. Gómez-Gil (✉)
Department of Computer Science, National Institute of Astrophisics,
Optics and Electronics, Tonantzintla Puebla, Mexico
e-mail: rfonseca@inaoep.mx

P. Gómez-Gil
e-mail: pgomez@acm.org

© Springer International Publishing Switzerland 2016
E. Merényi et al. (eds.), *Advances in Self-Organizing Maps and Learning Vector Quantization*, Advances in Intelligent Systems and Computing 428,
DOI 10.1007/978-3-319-28518-4_10

adjust auto-regressive models instead of neuron prototypes. A SOMAR extension, called NGMAR [21], uses a SOM variant for adjusting the weights of an ensemble of auto-regressive models. Koskela et al. proposed a recurrent SOM [14] and Chappell and Taylor a temporal Kohonen Map [6] which consider for training not to only the current input pattern, but also to the exponentially weighted past pattern. Merge SOM (MSOM), proposed by Strickert and Hammer [23], refers to a fusion of two properties characterizing the previous winner: the weight and the context of the last winner neuron are merged by a weighted linear combination [24]. Other method idealized as an probabilistic alternative to SOM is the Generative Topographic Mapping (GTM) [4]; the GTM Trough Time is one extension to GTM that performs simultaneous time series clustering and visualization [20].

We also have been working on using SOM as a guide for building ensembles for long-term forecasting of non-linear time series. In a previous work [11], we analyzed the impact of performing a model selection by the use of a SOM to find the maximum diversity among models. Using feature vectors built with errors generated in each prediction steps and meta-learning, we found that SOM was able to represent the individual accuracy and diversity among predictors. In [11] maximum diversity among models was represented by selecting models located in the farthest neuron from the best model. However, our results showed that, selecting models with maximum diversity was related with a poor and in some cases with the worst expected global accuracy [11].

Based on our past findings, this paper considers different levels of diversity among models using the neuron distances in the map. The experiments reported here showed that this strategy produced ensembles that achieved better results than selecting models based only on the expected accuracy. As in [11], a SOM divides the models into groups, using meta-features obtained from each involved model; each group corresponds with a neuron in a map of two dimensions. Let's call "A" the group containing the model with maximum expected accuracy. In [11], the farthest group of A, which represent the maximum diversity, was selected to build the ensemble. Here, we select neighboring neurons to A, and the level of diversity is adjusted managing the distance with respect to A. The k-best models to be combined are chosen from this neuron set. Once models are selected, their outputs are averaged to calculate the output of the prediction system.

For the experiments reported here, two base models where selected: (a) Non-linear Autoregressive with eXogenous inputs (NARX) model [17], implemented as NARX neural network [3] and (b) Autoregressive Integrated Moving Average (ARIMA) [5]. Several models were built changing the main parameters of base models. Four types of time series were used: a subset of the time series of the NN5 competition, an integration of the Mackey Glass equation [18], a time series generated with ARMA(2, 1) [5] and an integration of a sine function. These data was chosen in order to represent a variety of non-linear systems.

The paper is organized as follows: Sect. 2 describes the meta features used and the proposed method. Section 3 shows the experiments. Finally, Sect. 4 exposes the conclusions and future directions for this research.

2 Method Description

A time series Y is a sequence of observations y_t measured in constant time intervals. Multi-step ahead forecasting may be described as an estimation of a sequence of h future values based on current and past observations of Y, where the prediction horizon h is an integer greater than one. The present study follows the iterated strategy, which consists of estimating one value each time, using the previous predicted value for calculating the next prediction [8].

As we stated before, our goal is to analyze the effect of selecting and combining prediction models with different levels of diversity, where a greater diversity is related with a greater link distance in the neighborhood definition of a map representing such models. "Link distance" refers to the minimum number of steps separating one neuron from another. Diversity has been recognized as a very important characteristic in combination of models [9, 15], but an extreme diversity can be associated with poor results [11]. Therefore, a method able of adjusting the level of diversity is required. This research uses a SOM with one output layer organized in a two-dimensional array, following a hexagonal pattern, (Fig. 1). A neighborhood around a selected neuron is defined by a particular link distance among neurons; a neighborhood with a bigger link distance contains the neighborhoods with smaller link distances. For example, neighborhood with a link distance of 2 contains to a neighborhood with link distance 1; both are contained in the neighborhood with link distance of 3.

SOM is trained using as feature vector the concept of representative error r_e, introduced in [11]; it is a vector of size h, where h corresponds to a prediction horizon and each position is related with the error achieved in each forecasting step. In order to calculate r_e, the training time series of size n is split in two sets: training set (Y') which contains the first $n - h$ values and expected set (E), which contains the rest of h values. A model is generated using Y', which estimates the next h values in $\hat{Y}$, then vector r_e is defined as:

$$r_e = E - \hat{Y} \tag{1}$$

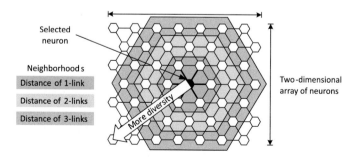

Fig. 1 Neighborhoods with different number of links from a selected neuron in a Self-Organizing Map (SOM) based on [3, 11]

Commonly, the training of neural networks starts with random weights. this implies that a network trained twice with the same Y' may achieve a different set of final weights, thus the trained model can have a different behavior each training. To avoid such instability, when the base model is a neural network, r_e is computed k iterations and the returned r_e is the average of these iterations. Metric Symmetric Mean Absolute Percentage Error (SMAPE) [2, 7] is used for evaluating the accuracy of the forecasting. It is defined as:

$$\text{SMAPE}\left(Y, \hat{Y}\right) = \frac{\sum_{t=1}^{h} \frac{|y_t - \hat{y}_t|}{\frac{1}{2}(|y_t| + |\hat{y}_t|)}}{h} \cdot 100 \qquad (2)$$

SMAPE $= 0$ means that the obtained prediction matches exactly with the expected output; the worst possible prediction implies a value of 200. SMAPE was chosen because it allows comparing different models with different time series regardless of their magnitudes.

Following is a toy example of the use of SOM for clustering representative errors. Suppose a time series whose next 4 expected values are $Y = \{1, 1, 1, 1\}$ that is, $h = 4$. Assuming that the only possible values for this time series are 0 or 1, there are 16 possible estimations. We can also assume that there are 16 prediction models, each one producing one of these estimations $\hat{Y}_i$, $i = \{1, 2, \ldots, 16\}$. For example the prediction of a particular model A is $\hat{Y}_A = \{1, 1, 1, 1\}$ having the best possible accuracy SMAPE $= 0$ with a representative error vector $r_e\left(\hat{Y}_A\right) = \{0, 0, 0, 0\}$. A SOM with 4 rows and 4 columns was trained 1000 epochs with all possible representative errors, resulting that each representative error in the training set was clustered in a different neuron. Figure 2 shows the trained map of this example; each neuron is tagged with an identification number and a representative error written in brackets. Neuron color corresponds to a SMAPE related with the representative error in the neuron. The neuron tagged with letter A has the best SMAPE and the neuron with letter W has the worst SMAPE of all predictions. The one-link neighborhood of the

Fig. 2 Example of a SOM organizing pre-defined representative errors. Each neuron has an identification number and a representative error in brackets

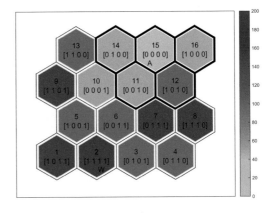

neuron with letter A is enclosed by a bold line. This example makes clear that models with similar SMAPE are neighbors and the models with the extreme SMAPE values 0 and 200 are distant on the map.

2.1 Proposed Method

The proposed method is an extension of [11], which contains two parts: extraction of meta-features and selection of models. Two types of meta-features are computed: performance estimation and representative error. The performance of each model is estimated using a Monte Carlo Cross-Validation (MCCV) [22] and the training time series. MCCV evaluates the same model training it with different sequences obtained from the same training time series; this process iterates k times. The estimation returned by MCCV is the average of these k iterations; for the experiments reported here, $k = 10$ [10, 11].

During the selection process, the representative errors of all models train a self-organizing map (SOM). The models are clustered into different groups, one group by neuron. This is done by assigning each model to the neuron with the minimum Euclidean distance between its weights and the representative error of such model. As it was showed by example in Fig. 2, models with similar representative errors are assigned to neighbor neurons and models with different representative errors are assigned to neurons far away each other.

Next step is to select the neuron that contains the model with the best expected accuracy, this neuron is tagged A. This assures that the ensemble contains the model with the best expected SMAPE. It is well known that the similarity between neurons in a trained SOM depends of the distance of the neurons on the map; this advantage is used for controlling the level of diversity among selected models. The distance of neurons is measured with a distance of l links, where l is the minimum number of steps required to travel from neuron A to neuron B. The neurons in the neighborhood of l links are the model source for selecting the k-best models to be combined. During the prediction process, the selected models are trained with the complete training sequence and their predictions are combined by average to obtain the final forecast.

With respect to the computational complexity of the proposed method, we noticed that the computational complexity for selecting models is much less than the complexity of obtaining meta-features. Indeed, the computational complexity of the calculation of meta-features is mainly dominated by the number of iterations k in the MCCV and computing of representative error. Currently, we are working with a formal estimation of this complexity, finding so far that a rough approximation of this value could be $O(Tkr)$, where T represents the total number of models involved in the selection, and r represents the operations required for training and using the involved prediction models.

3 Experimental Analysis

In this section, we present the results of analyzing different levels of diversity during the selection process; the number of links l varied from 1 to 5 and the number of selected models k varied from 1 to 25. Results are analyzed using the average of all SMAPE's achieved with the time series described next.

3.1 Time Series Set

This method was evaluated using four types of time series as in [11]. The first subset of time series was the reduced set provided by the NN5 prediction competition [7], which consisted of 11 time series, with 735 observations for training and 56 values for testing.

The second time series was an integration of the Mackey-Glass differential equation [18]:

$$dx(t)/dt = ax(t - \tau)/\left(1 + x(t - \tau)^{10}\right) - bx(t) \tag{3}$$

This function has a chaotic behavior when $a = 0.2$, $b = 0.1$, $\tau = 17$, $x_0 = 1.2$; the time step for integration was set to 0.1. The first 750 samples were used for training the model and the last 250 for testing. The third time series was generated using an ARMA(2, 1) model defined as:

$$y_t = 0.5y_{t-1} - 0.3y_{t-2} + \varepsilon_t + 0.2\varepsilon_{t-1} \tag{4}$$

where ε_t follows a Gaussian distribution with mean 0 and variance 0.1. The first 500 values formed the training set and the last 50 the test set. A fourth series was generated using a sine function with a time step size of $2\pi/64$; the first 750 observations were used for training and the next 250 values for testing.

3.2 Building the Prediction Models

Models to be selected for building the ensemble are generated by using different parameters in base models. In this work, two different base models were used: an Autoregressive Integrated Moving Average (ARIMA) [5], and a Non-linear Autoregressive with eXogenous inputs (NARX) [17]. Even though ARIMA models are lineal models, they are highly used as traditional forecasting methods and most of prediction works use ARIMA as a base case for ensembles [1]. We decided to include ARIMA because we consider that the ensemble should have the option of considering linear approximations. On the other hand, NARX was selected as base model because it is a non-linear model, which have proven to generate good approximations

Table 1 Parameters and settings for generating 81 different models

NARX as base model		ARIMA as base model	
Parameter	Settings	Parameter	Settings
Delay neurons	{3, 10, 25}	Auto-regressive terms p	{0, 1, 2}
Neurons in hidden layer	{10, 20, 30}	Non-seasonal difference d	{1, 2}
Training algorithm with matlab default values	trainbr, traincgf, trainlm	Lagged forecast errors q	{0, 1, 2}
		Seasonality	{0, 7, 12}

Here "trainbr" refers to Bayesian regulation back propagation (BP), "traincgf" refers to Conjugate gradient BP with Fletcher-Reeves updates, and "trainlm" refers to Leveberg-Marquardt BP

[8, 12]. Table 1 shows the parameters and settings used for generating 81 different models: 27 with NARX base and 54 with ARIMA base. The training was done using the Matlab Neural network toolbox with its default values [3].

3.3 Results

Next we present the prediction performances obtained by ensembles built using maps of different sizes {5 × 5, 6 × 6, 7 × 7, 8 × 8, 9 × 9, 10 × 10} and choosing some levels of diversities $l \in \{1, 2, \ldots, 5\}$. The initial neighborhood size was 3, and the training epochs were 12500. The forecast values of selected models were combined by average.

As an example, Fig. 3 presents two maps generated using the time series No. 3 of the NN5 reduced set. Figure 3a shows the map built with a SOM with 5 rows and 5 columns, while Fig. 3b shows a map of dimensions 7 × 7. The group tagged with a

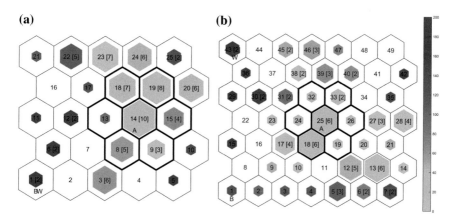

Fig. 3 **a** SOM (5 rows, 5 columns), **b** SOM (7 rows, 7 columns), both with the time series No. 3 of the NN5 reduced set. The color represents the average SMAPE of the models in each neuron

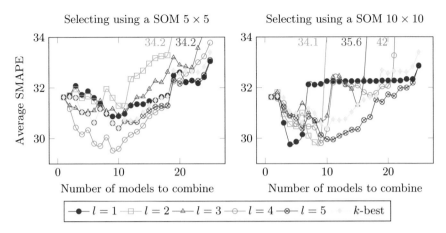

Fig. 4 Performance of selecting models with different number of links in the definition of *A* neighborhood and a different number of rows and columns in the SOM

letter "A" contains the model with the best expected SMAPE. The farthest group to *A* is referenced as neuron *B* and the neuron with a letter "W" contains the model with the worst expected SMAPE. Notice that when the size of SOM increases, the number of models in each group tends to decrease. It is also noticed that for the bigger SOM, models are distributed in groups far away from node *A*.

Figure 4 shows the average SMAPE obtained by the proposed method with two SOMs of sizes $\{5 \times 5, 10 \times 10\}$ and of links $l \in \{1, 2, \ldots, 5\}$. The results plotted are the average SMAPEs obtained using all the time series. This results are compared with the results of selecting and combining the k-best models based only on their expected accuracy without any clustering process. The number of k models to select for combining varied from 1 to 25, being $k = 10$ the case that obtained the best performance.

Table 2 summarizes the best results obtained by each SOM, ordered from best average SMAPE to the worst; Additionally, the last row shows the result of the baseline that is to use the model with the best expected accuracy after a MCCV. Notice that for all SOM sizes there is an improvement in the performance compared to using k-best models ($k = 1, 2, \ldots, 25$). Even though this improvement is small (1.17 for the best case), this value may be significant in several problems of forecasting. Notice in Fig. 4 that the average SMAPE was better with few models and an appropriate number of links. This is because the number of links influence the diversity among selected models. Selecting more than ten models degenerated in poor results, this is because the selection based on a SOM reduced the available models based on their diversity, then with a bigger k, the proposed method tends to select models with a poor expected SMAPE. The results suggest a relation between the size of the SOM and the number of links in the definition of *A* neighborhood required to achieve the best accuracy.

Table 2 Summary of the best results obtained by each SOM, ordered from best average SMAPE to the worst

SOM	Links	Number of models	Average of SMAPE	Improvement compared to k-best models
7×7	3	6	**29.24**	1.171
8×8	3	9	29.35	1.061
5×5	4	9	29.52	0.891
9×9	1	4	29.56	0.851
10×10	1	4	29.75	0.661
6×6	1	9	29.84	0.571
Benchmark				
k-best	–	$k = 10$	30.411	
Baseline				
Best model	–	1	31.632	

4 Conclusions

One key component on ensemble building is the diversity of involved models but an extreme diversity is related with poor results for the long-term prediction of time series [11]. This work presented a method for selecting models based on a self-organizing map, with the faculty of adjusting the level of diversity, which is related with the distance between neurons on the map. The proposed method was evaluated with different levels of diversity and compared with selecting models considering only the expected accuracy. In general, the proposed method achieved better results using the appropriate number of links for adjusting the diversity than selecting models considering only the expected accuracy. The results also suggest a relation between the size of the SOM an the number of links required to obtain the best results. When the size of the SOM increases the models tends to be more distributed in the trained map, however groups with similar average SMAPE tend to be neighbors. Currently, we are working with a formal definition of the computational complexity of the proposed algorithm. As future work, we will analyze how to define an appropriate SOM size and the correct number of links.

Acknowledgments R. Fonseca thanks the National Council of Science and Technology (CONA-CYT), México, for the scholarship granted to him, No. 234540. This research has been partially supported by CONACYT, project grant No. CB-2010-155250.

References

1. Andrawis, R.R., Atiya, A.F., El-Shishiny, H.: Forecast combinations of computational intelligence and linear models for the nn5 time series forecasting competition. Int. J. Forecast. **27**(3), 672–688 (2011)
2. Armstrong, J.S.: Long-range Forecasting from Crystall Ball to Computer, 2nd edn. Wiley (1985)
3. Beale, M.H., Hagan, M.T., Demuth, Howard, B.: Neural Network Toolbox User's Guide R2012b. MathWorks (2012)
4. Bishop, C.M., Svensen, M., Williams, C.K.I.: GTM: the generative topographic mapping. Neural Comput. **10**(1), 215–234 (1998)
5. Box, G.E.P., Jenkins, G.M., Reinsel, G.C.: Time Series Analysis Forecasting and Control, 3rd edn. Prentice-Hall International (1994)
6. Chappell, G.J., Taylor, J.G.: The temporal kohonen map. Neural Netw. **6**(3), 441–445 (1993)
7. Crone, S.F.: Competition instructions. Web (2010). http://www.neural-forecasting-competition.com/instructions.htm
8. Crone, S.F., Hibon, M., Nikolopoulos, K.: Advances in forecasting with neural networks? Empirical evidence from the NN3 competition on time series prediction. Int. J. Forecast. **27**(3), 635–660 (2011)
9. Cunningham, P., Carney, J.: Diversity versus quality in classification ensembles based on feature selection. In: Lopez de Mantaras, R., Plaza, E. (eds.) Machine Learning: ECML 2000, Lecture Notes in Computer Science, vol. 1810, pp. 109–116. Springer, Berlin Heidelberg (2000)
10. Fonseca-Delgado, R., Gómez-Gil, P.: An assessment of ten-fold and monte carlo cross validations for time series forecasting. In: 2013 10th International Conference on Electrical Engineering, Computing Science and Automatic Control (CCE), pp. 215–220 (2013)
11. Fonseca-Delgado, R., Gómez-Gil, P.: Selecting and combining models with self-organizing maps for long-term forecasting of chaotic time series. In: 2014 International Joint Conference on Neural Networks (IJCNN), pp. 2616–2623 (2014)
12. Haykin, S.: Neural Networks A Comprehensive Foundation, 2 edn. Pearson Prentice Hall (1999)
13. Kohonen, T.: The self-organizing map. Proc. IEEE **78**(9), 1464–1480 (1990)
14. Koskela, T., Varsta, M., Heikkonen, J., Kaski, K.: Temporal sequence processing using recurrent som. In: 1998 Second International Conference on Knowledge-Based Intelligent Electronic Systems, 1998. Proceedings KES '98, Apr, vol. 1, pp. 290–297 (1998)
15. Kuncheva, L., Whitaker, C.: Measures of diversity in classifier ensembles and their relationship with the ensemble accuracy. Mach. Learn. **51**(2), 181–207 (2003)
16. Lemke, C., Gabrys, B.: Meta-learning for time series forecasting and forecast combination. Neurocomputing **73**(10–12), 2006–2016 (2010)
17. Leontaritis, I.J., Billings, S.A.: Input-output parametric models for non-linear systems part ii: stochastic non-linear systems. Int. J. Control **41**(2), 329–344 (1985)
18. Mackey, M.C., Glass, L.: Oscillation and chaos in physiological control systems. Science **197**(4300), 287–289 (1977)
19. Ni, H., Yin, H.: A self-organising mixture autoregressive network for fx time series modelling and prediction. Neurocomputing **72**(16–18), 3529–3537 (2009), financial Engineering Computational and Ambient Intelligence (IWANN 2007)
20. Olier, I., Vellido, A.: Advances in clustering and visualization of time series using gtm through time. Neural Netw. **21**(7), 904–913 (2008)
21. Ouyang, Y., Yin, H.: A neural gas mixture autoregressive network for modelling and forecasting fx time series. Neurocomputing **135**, 171–179 (2014)
22. Picard, R.R., Cook, R.D.: Cross-validation of regression models. J. Am. Stat. Assoc. **79**(387), 575–583 (1984)
23. Strickert, M., Hammer, B.: Neural gas for sequences. In: Proceedings of the Workshop on Self-Organizing Maps (WSOM03), pp. 53–57 (2003)
24. Strickert, M., Hammer, B.: Merge som for temporal data. Neurocomputing **64**, 39–71 (2005), trends in Neurocomputing: 12th European Symposium on Artificial Neural Networks 2004

Part III
Applications in Control, Planning, and Dimensionality Reduction, and Hardware for Self-Organizing Maps

Modular Self-Organizing Control
for Linear and Nonlinear Systems

Paulo Henrique Muniz Ferreira and Aluízio Fausto Ribeiro Araújo

Abstract Nowadays, a good control system must meet some complex requirements. Two important ones are: quick and accurate responses to sudden changes in systems. This paper presents a control strategy for Self-Organizing Maps (SOM) that can do so. The proposed SOM-based control has a multiple-module architecture and learns from feedback on errors which enables it to generate appropriate controllers. Simulations of the mass-spring-damper system and the inverted pendulum validated the model. In the experiments, the systems had time-varying parameters. The results from the method proposed were compared with conventional methods and previous self-organizing control and suggest that the proposed control is suitable for controlling linear and nonlinear systems which undergo sudden changes.

Keywords Self-organization · Adaptive control · Nonlinear time-varying system

1 Introduction

The current trend in control systems is characterized by an increase in complex requirements. For example, there may be a need to deal with an unknown nonlinear system with multiple-input and multiple-output (MIMO), and a time-varying environment, process, or plant. Very often, classical controllers are not able to meet such complex requirements. The capacity to learn and adapt themselves to new situations are typical properties of intelligent controllers, which are suitable alternatives that can deal with nonlinearity, unknown plants, and parameter variation. Moreover, a learning controller could learn from a reduced set of training patterns. Considering this context, we propose a simple SOM-based multiple-model scheme to control.

P.H.M. Ferreira · A.F.R. Araújo (✉)
Center of Informatics, Federal University of Pernambuco (UFPE), Av. Jornalista Anibal
Fernandes, Recife CEP 50740-560, Brazil
e-mail: aluizioa@cin.ufpe.br

P.H.M. Ferreira
e-mail: phmf@cin.ufpe.br

© Springer International Publishing Switzerland 2016
E. Merényi et al. (eds.), *Advances in Self-Organizing Maps and Learning
Vector Quantization*, Advances in Intelligent Systems and Computing 428,
DOI 10.1007/978-3-319-28518-4_11

In such a context, control system parameters and architecture can change smoothly considering SOM topology preserving mapping. A self-organizing control can effectively handle a significant volume of data and redundant information, both of which are common in control systems. Furthermore, it can learn on-line when it allows a SOM-based controller to add new knowledge when necessary. Finally, a SOM-based controller might generalize from a limited and manageable number of patterns necessary to learn.

Using a single controller may not be an efficient strategy for controlling nonlinear and time varying systems [1]. Making use of multiple model controllers is a viable approach in these cases [1] because each model can respond easily and precisely for a region of the control space.

In this work, we used Self-Organizing Maps to divide the control system space into subspaces in which each region has a particular controller. Additionally, SOM can identify a current operational subspace and, thus, can determine a suitable controller. Furthermore, multiple-model SOM-based controllers can generalize some local controllers. In this way, this intelligent control system can present low sensitivity to variations in system parameters and it can respond appropriately to parameters not considered during the training phase. This paper also presents a comparative study of the proposed model with other controls in two experiments (Sect. 4).

2   Problem Formulation

Let there be an equation defining a dynamic system [2].

$$y(t+1) = f(w(t), x(t), u(t)) \tag{1}$$

where $y(t)$ is the output, $x(t)$ is the state and $u(t)$ is the input of a given system, and $w(t)$ is the parameter of function f at time t. f can be a system that is SISO or MIMO, linear or nonlinear, and it can have time-varying parameters.

Nonlinear time-varying systems can be handled using a two-phase control process [2]: (i) a phase for identifying the dynamics of the system and (ii) a control phase in which an appropriate control action is generated to achieve given goals.

In order to design a controller for this problem, the function f and its parameters $w(t)$, which are normally unknown, can be approximated by neural network algorithms from system input and output data. Hence, both f and $w(t)$ remain unknown and an approximation function $\tilde{f}$ is constructed to design the controller. We can enumerate two alternatives to do so. The first option entails a single nonlinear model $\tilde{f}$ to approximate f, which is commonly called a global model. The second approach divides the space into local regions, each of which locally modeling the function f. Mathematically, f is redefined as a series of functions $\tilde{f}_r$ where $r = 1, \ldots, N$. The complete model consists of the union of all functions $\tilde{f}_r$ [3].

$$\tilde{f}(x(t), u(t)) = \bigcup_{r=1,...,N} \tilde{f}_r(x(t), u(t)) \tag{2}$$

Potentially, local controllers can be simpler than a single overall controller. For example, linear controllers can be used locally.

SOM can be used to divide the system behavior's space into regions, to identify them, and to parametrize the local controllers. Hence, self-organizing controllers may present a loss of accuracy due to a discretization of their parameter space. Such an inaccuracy can be overcome by increasing the number of training samples or considering a larger number of prototypes. Moreover, one can design the controller from a variant of SOM which has an interpolation capacity [4].

In this article, we compare our proposal with another self-organizing control called *Self-Organizing Adaptive Controller* (SOAC) [5]. Both algorithms share two important features: rapid response to sudden changes in the controlled system and learning general behavior from a small set of training patterns. In an introductory paper [5], SOAC presented promising results for the same problem. SOAC is a multiple-module control using a variant of the SOM structure (*modular network SOM*) in which each module consists of a predictor/controller pair. The performance of the SOAC was better than that of multiple paired forward-inverse models (MPFIM), a well-known multiple-pair control strategy [6]. The main limitations of SOAC are its fixed topological structure and that the behavior and influence of some equations in the learning predictor procedure are not easy to understand.

3 SOM-Based Control

Our proposed solution can be seen as a *feedback error learning* approach to generate controllers, similar to what SOAC does. In spite of there being features in common with SOAC, our approach has a different procedure for identifying the dynamics and determining what the most suitable control action is. The control system is designed to meet two important requirements for a system in time-varying environments: (1) A rapid response to unexpected changes in the controlled system and (2) the capacity to establish general behavior learned from a small set of training patterns. The first requirement can be met by modeling the multiple local models and the second requirement by the capacity of the SOM to generalize.

For our proposed control, each weight vector w is divided into two parts w^{in} and w^{out}. The w^{in} is a vector with a number of previous control inputs and state sequence estimations, and it is used to identify the current system configuration while w^{out} represents an estimate of the system parameters. Such an approach is based on the *Vector-Quantized Temporal Associative Memory* (VQTAM) [7].

To identify the plant, SOM has to learn different behaviors of the system expressed by state and command control time sequences. In the SOM-based control, identifying the system considers the viewpoint of local models in which the input space is

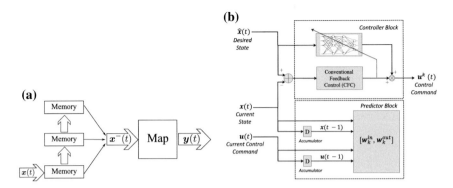

Fig. 1 SOM modified for the control problem with feedback error learning. **a** Concatenated input patterns. **b** SOM node modified

partitioned into smaller operating regions. That is, for a particular plant configuration learned by the SOM-based control, the weight vectors encode parameters of different operating regions. This can result in local control models being simpler than a single control model for a plant configuration. In SOAC, each module has a unique function to model the behavior of the dynamics of the system for a given configuration.

The concatenation of the input vectors, in accordance with their temporal order, forms the actual input of the network (Fig. 1a). Such a concatenation of the input patterns considers a time window containing a pattern sequence. In the control problem, the time window is defined as follows:

$$x_{in}^-(t) = [x(t), u(t), x(t-1), u(t-1), \ldots, x(t-p+1), u(t-p+1)] \quad (3)$$

where the pair of vectors $[x(t), u(t)]$ represents the system state and the system control input at time t within a predefined window size p.

In the training phase, each $x_{in}^-(t)$ has a complementary associated vector $x_{out}(t)$. This vector $x_{out}(t)$ encodes to the system parameters p crucial to establishes its behavior $x_{in}^-(t)$. The training pattern $x_{in}^-(t)$ is used to adjust w^{in} and $x_{out}(t)$ is used to adapt w^{out}. The winner node, k^*, can be determined using x_{in}^- (Eq. 4) or x_{out} (Eq. 5). In the execution phase, the competition uses x_{in}^- because, very often, there is no information available about the configuration of the system.

$$k^*(t) = \arg\min_k \left\| x_{in}^-(t) - w_k^{in}(t) \right\| \quad (4)$$

$$k^*(t) = \arg\min_k \left\| x_{out}(t) - w_k^{out}(t) \right\| \quad (5)$$

Thus, the weight vector updating of node k follows the equations:

$$w_k^{in}(t+1) = w_k^{in}(t) + \alpha(t)h_{k^*,k}(t)\left[x_{in}(t) - w_k^{in}(t)\right] \qquad (6)$$

$$w_k^{out}(t+1) = w_k^{out}(t) + \alpha(t)h_{k^*,k}(t)\left[x_{out}(t) - w_k^{out}(t)\right] \qquad (7)$$

where the neighborhood is $h_{bmm,k}(t) = exp(-\frac{\left\|\varepsilon^k(t) - \varepsilon^{k^*}(t)\right\|^2}{2\sigma^2(t)})$, for which ε^k and ε^{k^*} are the coordinates of the module k and of the winner in the map and $\sigma(t) = \sigma_\infty + (\sigma_0 - \sigma_\infty)exp(-\frac{t}{\tau_\sigma})$. $\alpha(t) = \alpha_\infty + (\alpha_0 - \alpha_\infty)exp(-\frac{t}{\tau_\alpha})$. In the execution phase, the SOM-based control uses feedback error learning to generate the controllers. The learning of SOM-based control occurs online, i.e., for each node k, there is a weight vector $^c w^k$, which defines a function $^c f^k$:

$$u^k(t) = {}^c f^k({}^c w^k, \hat{x}(t)) \qquad (8)$$

where $\hat{x}(t)$ is the desired system state.

The final control command is calculated by adding the conventional feedback control output (cfc) and the SOM-based control signal generated by the winner node:

$$u(t) = u^{k^*}(t) + u^{cfc}(t) \qquad (9)$$

The adjustments for each $^c w^k$ are determined by Eq. 10. If the controllers are designed by an analytical procedure, it uses an estimate of the system parameters w^{out} in module k.

$$\Delta^c w^k = \eta \cdot \phi^k \frac{\partial\,^c f^k}{\partial\,^c w^k} \cdot {}^{cfc} u(t) \qquad (10)$$

where ϕ^k is the responsibility signal (Eq. 11).

$$\phi^k = \frac{exp[\frac{-\left\|\varepsilon^k - \varepsilon^{k^*}\right\|^2}{2\sigma_\infty^2}]}{\sum_{k'} exp[\frac{-\left\|\varepsilon^{k'} - \varepsilon^{k^*}\right\|^2}{2\sigma_\infty^2}]} \qquad (11)$$

where ε^k and ε^{k^*} are the coordinates of the module k and of the winner in the map. σ_∞ is the final radius value of the neighborhood function.

A SOM network node is present in Fig. 1b. In the experiments below, the chosen time window size is $p = 2$.

4 Simulations

We validated the proposed control system for two different problems: the mass-spring-damper (MSD) system and the inverted pendulum. For the MSD case, we evaluated and analyzed the trained control modules ruling upon a linear time-varying system [8]. On the other hand, the inverted pendulum is evaluated and analyzed as a nonlinear and time-varying problem [9]. Both experiments assess the capacity of the models to generalize.

4.1 Spring-Mass-Damper System

The MSD is described by a second order linear differential equation:

$$m\ddot{x}(t) + b_i\dot{x}(t) + k_ix(t) = F(t) \tag{12}$$

where m, b_i and k_i are the mass [kg], the damping coefficient [$\frac{kg}{s}$], and the spring constant [$\frac{kg}{s^2}$]. The subscript i is the label for each particular system configuration i. We used nine alternatives for training and six others for testing. A single configuration was used for both training and testing. The mass is constant ($m = 1$ kg) in all experiments while parameters b_i and k_i may be 2, 6, or 10, thereby determining the training system configurations (p_i, ($i = 1, 2, \ldots, 9$)). The testing values for b_i and k_i are $p_A = [6, 6]$, $p_B = [6.1, 6.8]$, $p_C = [4.3, 6.5]$, $p_D = [5, 4]$, $p_E = [8.7, 4]$ and $p_F = [8.8, 9]$.

The sampling rate of the training test is constant 2001 pairs of system states and control commands, after, we apply each pair in each training system configuration. Hence, tuples {*current state, control command, next state, the training system configuration*} form the training set. The total of tuples is $9 \times 2001 = 18009$. For our experiments, the sampling interval of each variable is $-0.6 \leq x$ (position) ≤ 0.6, $-4 \leq y$ (velocity) ≤ 4 and $-2000 \leq u$ (control) ≤ 2000. These intervals were chosen after observing the behavior of these variables when the system was controlled by a PID controller. To train SOM, all variables were normalized between 0 and 1, however this was not the case for SOAC as in [5].

For the system test, we used a randomly chosen 30-s state sequence in which the system configuration changes every 5 s. The test configuration order was always $[p_A, p_B, p_C, p_D, p_E, p_F]$. The stochastic *Ornstein-Uhlenbeck* process generated reference for the state trajectory. This process is suitable for evaluating the speed and accuracy of the response. An exact solution for this process is:

$$S_{t+i} = S_i e^{-\lambda\delta} + \mu(1 - e^{-\lambda\delta}) + \sigma\sqrt{\frac{1 - e^{-2\lambda\delta}}{2\lambda}} N_{0,1} \tag{13}$$

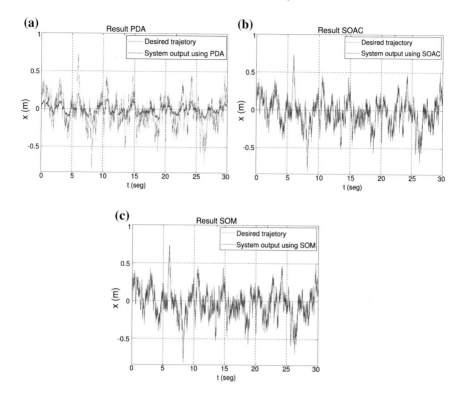

Fig. 2 Best result for MSD system. **a** PDA (RMSE = 0.14569). **b** SOAC (RMSE = 0.021073). **c** SOM (RMSE = 0.018233)

where $\lambda = 3$ is the mean reversion rate, $\mu = 0$ is the average value, $\sigma = 0.5$ is the volatility, $\delta = 0.001$ is the time step, and $N_{0,1}$ is a random value from a normal distribution with $\mu_N = 0$ and $\sigma_N = 1$. All simulations were conducted using a fourth-order Runge-Kutta method with a step size $h = 0.001$, as in [5]. Finally, the control accuracy was evaluated by the *Root Mean Square Error* (RMSE) between the system output and the desired trajectory.

PDA details: The Proportional Derivative Accelerative (PDA) was chosen as the classical feedback control. The chose parameters were $^{cfc}W = [k_x, k_{\dot{x}}, k_{\ddot{x}}] = [5, 10, 0.5]$ as in [5]. The result of the PDA control is shown in Fig. 2a.

SOAC details: For SOAC, the predictor inputs are position x, velocity v and control command u. The predictor output is an acceleration estimate. The controller inputs are the desired position x_d, the desired velocity v_d and the desired acceleration a_d. Both the predictor and the controller consist of a linear network [5]. The SOAC configuration with the best results was 49 (7×7) modules [5]. The randomly chosen parameters were (N (stop criterion) $= 20$, $\epsilon = 0.1$, $\sigma_0 = 10$, $\sigma_{\inf} = 1$, $\tau = 80$, η

(predictor) $= 0.0002$, and η (controller) $= 0.000001$). The best result for SOAC is presented in Fig. 2b.

SOM-based details: The window size considered was $p = 2$. x_{in}^- consists of $[x(t), u(t), x(t-1), u(t-1)]$, where $x(t)$ is the vector with position $x(t)$, velocity $v(t)$ and acceleration $a(t)$, and control command $u(t)$. $^c w^k$ are weight-vectors of a linear network, similar to that of SOAC. The map size is 9×9. The competition of the training phase used Eq. 4. All other parameters were chosen empirically (N (stopping criterion) $= 50$, $\sigma_0 = 70$, $\sigma_{inf} = 1$, $\tau_\sigma = 5$, $\alpha_0 = 0.9$, $\alpha_\infty = 0.0001$, $\tau_\alpha = 30$ and η (controller) $= 0.000001$). The best SOM result is presented in Fig. 2c.

Comparing the methods: The SOAC and the SOM-based results are compared. The PDA result is deterministic and poorer than SOAC and SOM-based algorithms. For SOM-based model and SOAC, 10 trials of learning were performed. The RMSE for each controller is: PDA $= 0, 14569$, SOAC $= 0, 021801 \pm 6, 4058 \cdot 10^{-4}$ and SOM-based $= 0, 018322 \pm 4, 8662 \cdot 10^{-5}$.

In this experiment, the two neural control methods were able to respond adequately after changing system parameters. Both controlled the plant for parameters not presented during the training phase. SOM-based method did better as shown by the RMSE average and variance. We argue that the SOM-based control can identify behavior more accurately and it can adjust the PDA control command more effectively.

4.2 Inverted Pendulum

The second experiment aims to control a pendulum fixed on a motorized cart that can move on a rail (Fig. 3). The control objective is to keep the pendulum in the inverted state and the cart has to reach a pre-determined position.

The inverted pendulum system is described as:

$$(M+m)\ddot{x} + ml\ddot{\theta}cos(\theta) - ml\dot{\theta}^2 sen(\theta) + f\dot{x} = a * u \qquad (14)$$

$$ml\ddot{x}cos(\theta) + (I + ml^2)\ddot{\theta} - mglsen(\theta) + C\dot{\theta} = 0 \qquad (15)$$

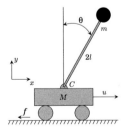

Fig. 3 The inverted pendulum system

where x is the position of the cart [m], θ is the angle of the pendulum [rad], $M = 5, 0$ is the mass of the cart [kg], m is the mass of the pendulum [kg], l is the length of the center of mass on the pendulum [m], $f = 10, 0$ and $C = 0,0004$ is the co-efficient of viscous friction of the cart [kg/s] and of the pendulum [kgm^2/s], respectively, $g = 9.8$ is the gravitational acceleration [m/s^2], $a = 25$ is a gain [N/V], and I is the moment of inertia $I = ml^2/3$ [kgm^2]. A state variable is represented as $x = [x, \theta, \dot{x}, \dot{\theta}]^T$ and the control variable as u. The simulations were performed using the fourth-order Runge-Kutta method with a step size $h = 0.01$ [s].

Nine system configurations were used for training and another nine, for testing. The training system configurations are $[m_i, l_i] = [0.2, 1.0, 1.8] \times [0.6, 1.2, 1.8]$. The testing values are: $p_A = [1, 1.2]$, $p_B = [0.9, 1.245]$, $p_C = [0.8, 1.134]$, $p_D = [0.92, 0.915]$, $p_E = [1.82, 0.915]$, $p_F = [1.48, 1.257]$, $p_G = [1.244, 1.71]$, $p_H = [0.648, 1.749]$, and $p_I = [0.2, 1.2]$.

The desired position of the cart changed every 10 s during the tests. The possible desired values are $[1, -1, 0]$. The initial state is $x_0 = [0, 0, 0, 0]^T$ and the first desired value is $x = 1$. The configuration of the testing system also changes every 10 s. The order of the test configurations is $[p_A, p_B, p_C, p_D, p_E, p_F, p_G, p_H, p_I]$.

For the training set, we uniformly sampled 2001 system state and control command pairs, and then, we applied them in each training configuration so as to generate the next state. Hence, the training set comprises tuples {*current state, control command, next state, the training system configuration*}. The total of tuples is $9 \times 2001 = 18009$. The sampling range of each variable is: $-0.6 \le x \le 0.6$, $-0.6 \le \theta \le 0.6$, $-4 \le \dot{x} \le 4$, $-4 \le \dot{\theta} \le 4$ and $-2000 \le u \le 2000$. These ranges were chosen from observations of the behavior of these variables when a Linear-Quadratic Regulator (LQR) controlled the system. For SOM-based training, all variables were normalized between 0 and 1.

SOAC details: The predictors consist of a linear network with current state $x(t)$ and control command $u(t)$ as inputs and next state estimation $x(t + \Delta t)$ as output where $\Delta t = 0, 01$. In this experiment, SOAC was designed using analytical local controllers, which are optimal strategies previously calculated. An LQR is generated for each module from the estimate of system parameters $\tilde{p}^k$. In the execution phase, only the control command generated by the winner module was applied in the system. The SOAC was configured with a two-dimensional map of 81 (9×9) modules.

SOAC experiments were performed 15 times. SOAC presented a typical unstable behavior for all trials. The typical dynamics is shown in Fig. 4. Additional experiments suggest that the instability is caused by the failure to identify the current behavior of the system.

SOM-based details: The window size was $p = 2$. x_{in}^- consists of $[x(t), u(t), x(t-1), u(t-1)]^T$, where $x(t)$ is the system state vector and $u(t)$, the control command. After the learning phase, w^{out} is used to generate the LQR control for each SOM node. The map size was 10×10. The competition during the training phase used w^{out} (Eq. 5). A simulation result of this control is shown in Fig. 5. This experiment was performed 15 times. Three simulations had results similar to those shown in Fig. 5. In all other simulations, the system presented unstable states.

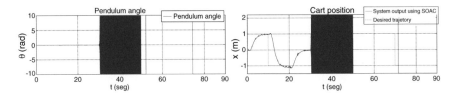

Fig. 4 SOAC result for inverted pendulum system. From the configuration P_D onwards, the system became unstable and this influenced the control of the following test configurations

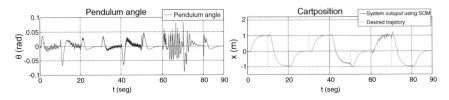

Fig. 5 SOM result for inverted pendulum system

5   Conclusion

The main idea introduced in this paper is to place a SOM in a multiple model control scheme. Multiple controllers are more likely to respond more appropriately than a single controller for time-varying systems [5]. A SOM-based control can autonomously set up the control modules while considering only a small sample set of training patterns. The SOM-based control strategy was used for linear and nonlinear plants: the mass-spring-damper (MSD) and the inverted pendulum. The MSD system is linear, second order and stable in an open loop configuration. The inverted pendulum system is nonlinear. In both problems, the parameters of the plant are time varying. The results compared the SOM-based control with the SOAC method. SOM-based control did better in both experiments in terms of accuracy and stability.

We suggest investigating improvements for SOM-based controls that can be applied to tracking target references in the inverted pendulum system. A study of the SOAC and SOM-based control to control other plants may also be interesting. For instance, these control strategies applied in a real-time complex system such as a robotic system.

Acknowledgments The authors would like to thank CNPq for supporting this research.

References

1. Narendra, K.S., Balakrishnan, J.: Adaptive control using multiple models. IEEE Trans. Autom. Control **42**, 171–187 (1997)
2. Thampi, G., Principe, J.C., Cho, J., Motter, M.: Adaptive inverse control using som based multiple models. In: Proceedings of the Portuguese Conference on Automatic Control, pp. 278–282 (2002)
3. Principe, J.C., Wang, L., Motter, M.A.: Local dynamic modeling with self-organizing maps and applications to nonlinear system identification and control. Proc. IEEE **86**, 2240–2258 (1998)
4. Göppert, J., Rosentiel, W.: The continuous interpolating self-organizing map. Neural Process. Lett. **5**, 185–192 (1997)
5. Minatohara, T., Furukawa, T.: The self-organizing adaptive controller. Int. J. Innovative Comput. Inf. Control **7**, 1933–1947 (2011)
6. Wolpert, D.M., Kawato, M.: Multiple paired forward and inverse models for motor control. Neural Netw. **11**(7), 1317–1329 (1998)
7. Barreto, G., Araújo, A.F.R.: Identification and control of dynamical systems using the self-organizing map. IEEE Trans. Neural Netw. **15**, 1244–1259 (2004)
8. Gu, D.W., Petkov, P.H., Konstantinov, M.M.: Robust control of a mass-damper-spring system. In: Robust Control Design with MATLAB, pp. 101–162. Springer (2005)
9. Boubaker, O.: The inverted pendulum benchmark in nonlinear control theory: a survey. Int. J. Adv. Robotic Syst. **10**(233) (2013)

On Self-Organizing Map and Rapidly-Exploring Random Graph in Multi-Goal Planning

Jan Faigl

Abstract This paper reports on ongoing work towards an extension of the self-organizing maps for the traveling salesman problem to more challenging problems of multi-goal trajectory planning for complex robots with a high-dimensional configuration space. The main challenge of this problem is that the distance function needed to find a sequence of the visits to the goals is not known a priori and it is not easy to compute. To address this challenge, we propose to utilize the unsupervised learning in a trade-off between the exploration of the distance function and exploitation of its current model. The proposed approach is based on steering the sampling process in a randomized sampling-based motion planning technique to create a suitable motion planning roadmap, which represents the required distance function. The presented results shows the proposed approach quickly provides an admissible solution, which may be further improved by additional samples of the configuration space.

1 Introduction

Self-Organizing Map (SOM) is a type of neural network that can provide a non-linear mapping of a high dimensional input space into a lower dimensional output space. In addition to data processing, visualization, and classification, it has also been successfully applied in optimization routing problems, in particular, the Traveling Salesman Problem (TSP). The TSP is a well-defined optimization problem arising from many practical scenarios and several SOM-based approaches have been proposed, e.g., see [2, 14]. In our case, the TSP is a problem formulation for robotic tasks like inspection, surveillance, and data collection where a mobile robot is requested to visit a set of locations, e.g., to perform an operation or take a sensor measurement [3, 4, 8, 11].

J. Faigl (✉)
Czech Technical University in Prague, Department of Computer Science,
Technická 2, 166 27 Prague, Czech Republic
e-mail: faiglj@fel.cvut.cz

© Springer International Publishing Switzerland 2016
E. Merényi et al. (eds.), *Advances in Self-Organizing Maps and Learning
Vector Quantization*, Advances in Intelligent Systems and Computing 428,
DOI 10.1007/978-3-319-28518-4_12

The most straightforward application of SOM to the TSP is in Euclidean instances, where the problem stands to find a closed shortest tour connecting a given set of goal locations (cities). In robotics, the problem is to find a shortest path connecting the locations such that the path is collision free. This make an application of SOM to the TSP a bit more challenging because a pure Euclidean distance cannot be simply used in the computation of distances between neuron weights and the presented goal location (signal) to the network; otherwise a poor solution would be found [5]. The distance corresponds to the length of the shortest path between two locations, which can be PSPACE-hard in 3D environment. Hence, the problem is called the *Multi-Goal Motion Planning* (MGMP) problem rather than the TSP to emphasize difficulty of distance queries.

Randomized sampling-based approaches are motion planning techniques for planning in high-dimensional configuration space $\mathcal{C}$ that provide the so-called motion planning roadmap, which is a graph representing collision free configurations in $\mathcal{C}$ [9]. A combination of the roadmap with SOM for a graph input [13] has been proposed in [6] to solve the MGMP by SOM. In this decoupled approach, the roadmap (graph) is constructed independently on the planning problem, and therefore, a complete graph is unnecessarily dense.

In this paper, we report our recent results on application of SOM in the roadmap generation and solution of the MGMP problem. The main idea of the proposed approach is based on combining principles of the optimal motion planning algorithm called *Rapidly-exploring Random Graph* (RRG) [7] with the SOM adaptation principles to simultaneously determine the sequence of the goal visits together with trajectories connecting the goals in the tour. The core of the proposed approach is a utilization of the SOM adaptation to steer a randomized sampling of $\mathcal{C}$ to increase the number of samples in the most promising areas to quickly find a solution and eventually improve quality of the final trajectory.

A feasibility of this idea has been reported in [12], where it has been employed in finding multi-goal trajectories for a hexapod walking robot. The proposed SOM-based algorithm needs a lower number of the roadmap expansions to find a first feasible solution of the MGMP problem in comparison to a straightforward MGMP solver based on a given sequence of visits to the goal locations.

Here, we focus on two main aspects of the proposed approach: (1) a detailed evaluation of the idea of SOM-based expansion of the roadmap to find an initial solution of the MGMP; and (2) improving the quality of the final solution with increasing number of the roadmap expansions. Based on the evaluation, we propose a hybrid approach that consists of the initial construction of the roadmap by SOM to find the first feasible solution followed by a consecutive roadmap improvement to find a shorter trajectory.

The paper is organized in the following way. The problem statement, notion of the configuration space $\mathcal{C}$, and related background is presented in the next section. The key idea of the SOM-based steering of the roadmap expansions using the RRG is briefly described in Sect. 3. Considered MGMP solvers are presented in Sect. 4 and results of their evaluation are in Sect. 5. Concluding remarks and future work are summarized in Sect. 6.

2 Problem Statement

The problem addressed by the proposed approach is motivated by autonomous data collection with a hexapod walking robot operating in a rough environment to collect samples, e.g., images, of the requested areas of interest, see Fig. 1. The robot has six legs, each with three joints that gives 18 control degrees of freedom, which together with the robot position and orientation in the 3D environment gives 24 dimensional vector fully describing the position of the robot body in the environment. Therefore, it is controlled by designed gait patterns and a set of motion primitives to simplify the motion control and planning [12]. In addition and without loss of generality, the robot pose (x, y, θ) is considered as the robot position on a surface x, y with orientation θ.

The working environment $\mathcal{W} \subset \mathbb{R}^3$ is represented as a set of obstacles $\mathcal{O} \subset \mathcal{W}$. The configuration space $\mathcal{C}$ describes all possible configurations of the robot in $\mathcal{W}$ and can be defined as follows. Let the robot body at q be $\mathcal{A}(q)$, then the configuration q is a collision free if $\mathcal{A}(q) \cap \mathcal{O} = \emptyset$. All configurations for which the robot is in a collision with the obstacles $\mathcal{O}$ are denoted as $\mathcal{C}_{obst}$, $\mathcal{C}_{obst} \subseteq \mathcal{C}$. The point of our interest to find a solution of the MGMP is a collision free part of $\mathcal{C}$, which can be denoted as $\mathcal{C}_{free} = \mathrm{cl}(\mathcal{C} \setminus \mathcal{C}_{obst})$, where $\mathrm{cl}(.)$ is the set closure.

A collision free path from some starting configuration q_{start} to a goal configuration q_{goal} is a continuous curve κ in $\mathcal{C}_{free}$, such that $\kappa : [0, 1] \to \mathcal{C}_{free}$ with $\kappa(0) = q_{start}$ and $d(\kappa(1), q_{end}) < \epsilon$. The end point $\kappa(1)$ of the path found by a motion planner will unlikely be exactly the requested goal location, and therefore, we rather admit an admissible distance ϵ of the path to the requested goal [7], e.g., 5 cm. Then, such a collision free path is called an admissible path.

Similarly to a simple trajectory, a multi-goal trajectory visiting a set of n goal locations $\mathcal{G} = (g_1, \ldots, g_n)$ can be defined as follows. Let the sequence of the visits to the locations be $(v_1, v_2, \ldots, v_n)$ for which $v_i \in \mathcal{G}$ and $\bigcup_{1 < i \leq n} v_i = \mathcal{G}$. Then, an admissible multi-goal trajectory is a closed trajectory $\tau : [0, 1] \to \mathcal{C}_{free}$ such that $\tau(0) = \tau(1) = q_{start}$ and for which there exists n points on τ such that $0 \leq t_1 \leq t_2 \leq \cdots \leq t_n$ and $d(\tau(t_i), v_i) < \epsilon$.

Having the aforementioned preliminaries, the MGMP problem can be formulated as follows: *For the given goal locations $\mathcal{G}$, configuration space $\mathcal{C}$, an admissible distance ϵ, and a monotonic, bounded, and strictly positive cost*

Fig. 1 Robot, its geometrical model, and visualized 3D environment

function c: find an admissible (according to ϵ) trajectory τ^* such that $c(\tau^*) = \min\{c(\tau) \mid \tau$ is admissible multi-goal trajectory$\}$.

2.1 Randomized Sampling-Based Motion Planners

Sampling based motion planning techniques have been proposed to address diffi-culty of explicit representation of $\mathcal{C}_{free}$ for a complex shape of the robot body and its high-dimensional $\mathcal{C}$ [9]. These techniques sample $\mathcal{C}_{free}$ into a finite number of configurations that are connected into a graph, where an edge represents a collision free trajectory between two configurations. Hence, $\mathcal{C}_{free}$ is represented by a graph and the key problem is how to efficiently create the graph (roadmap) in which the requested trajectory can be found, e.g., by a graph search technique.

In this work, we consider RRG [7] to create a graph $\mathbf{G}_{RRG} = (\mathbf{V}_{RRG}, \mathbf{E}_{RRG})$, which represents the motion planning roadmap. The set of vertices $\mathbf{V}_{RRG}$ are particular con-figurations of the robot $q \in \mathcal{C}_{free}$ and an edge $e \in \mathbf{E}_{RRG}$ describes a feasible collision free motion between two configurations $v_i, v_j \in \mathbf{V}_{RRG}, i \neq j$. The graph is incremen-tally constructed by the RRG algorithm as a result of the graph expansion from the nearest vertex of the graph towards a random sample by applying a particular control command. The main steps of the RRG expansion are depicted in Fig. 2, further details can be found in [7].

2.2 Basic Background of Self-Organizing Map for the TSP

The proposed MGMP solvers are based on SOM for the TSP, in particular, a variant for a graph input [13]. The neural network is structured in two layers. The first layer servers for presenting goal locations to be visited and towards which the network is

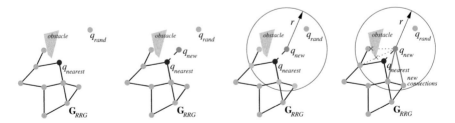

Fig. 2 An expansion of the RRG roadmap (from *left* to *right*): First, a random (collision free) configuration q_{rand} is sampled and the nearest vertex $q_{nearest} \in \mathbf{V}_{RRG}$ is determined; Then, the most suited control command is applied to expand the roadmap towards q_{rand} by a collision free trajectory and a new configuration q_{new} is added to the roadmap; To further improve the roadmap, all vertices within a ball with a particular radius r (see [7]) centered on q_{new} are connected with q_{new} by a collision free trajectory

adapted using the self-organizing principles. The output layer consists of m units, $\mathcal{N} = \{v_1, \ldots, v_m\}$, which represent neurons weights, where m is set according to the number of goal locations n, e.g., $m = 2.5n$. The units are organized into one-dimensional array that represents a sequence of configurations in $\mathcal{C}_{free}$. The learning procedure can be summarized as follows:

1. *Initialization*—Create a ring of connected neurons $\mathcal{N} = \{v_1, \ldots, v_m\}$.
2. *Randomization*—Create a random permutation of goals $\Pi(\mathcal{G}) \leftarrow$ permute$(\mathcal{G})$.
3. *Winner selection*—Select the best matching neuron v^* to the currently presented goal $g \in \Pi(\mathcal{G})$; $v^* \leftarrow \text{argmin}_{v \in \mathcal{N}} d(v, g)$.
4. *Adaptation*—Adapt the winner v^* and its neighbouring nodes v_j within the distance k (in the number of nodes) using the neighbouring function $f(\sigma, k) = \mu e^{(-k^2/\sigma^2)}$ for $k < 0.2m$ and $f(\sigma, 0) = 0$ otherwise. Remove g from the permutation, $\Pi(\mathcal{G}) \leftarrow \Pi(\mathcal{G}) \setminus \{g\}$, and If $|\Pi(\mathcal{G})| > 0$ go to Step 3.
5. *Update* the number of the learning epochs and neighbouring function variance.
6. *Termination condition*—If termination condition is met, stop the adaptation. Otherwise go to Step 2.
7. *Final tour construction:*—Traverse the output layer and use the associated goals to the last winners to construct the final goal tour.

The adaptation of neurons can be imagined as a movement of the neurons towards the presented goal location. For a graph input, the neurons weights are restricted to be at the graph edges or vertices and the adaptation can be imagined as neurons movements along the graph edges [13]. Thus, for an adaptation in the roadmap $\mathbf{G}_{RRG}$ with spatially close vertices (such that provided by the RRG), we can consider the neuron weights as a particular configuration represented by the closest vertex from V_{RRG}.

Notice, even though we can use SOM to find a solution of the MGMP on $\mathbf{G}_{RRG}$ like in [6]; here, we are rather interested in employing the adaptation procedure to grow and improve the roadmap $\mathbf{G}_{RRG}$ by the RRG expansions.

3 SOM-based Steering of Randomized Sampling in RRG

The fundamental issue of applying SOM to the given problem is that the selection of the winner node to a presented location g is based on computing a distance $d(v, g)$ between nodes $v \in \mathcal{N}$ and g. Such a distance corresponds to the length of the trajectory from v to g, which is obviously not known due to a sparse coverage of $\mathcal{C}$ by $\mathbf{G}_{RRG}$, especially at the beginning of the learning. In [12], we propose to address this issue by the approximation that combines Euclidean distance and the current knowledge about $\mathcal{C}_{free}$ stored in the incrementally built $\mathbf{G}_{RRG}$.

Regarding a collision free and feasible trajectory in $\mathcal{W}$, the current roadmap $\mathbf{G}_{RRG}$ provides a much more realistic estimation of the expected distance $d(v, g)$ than a pure Euclidean distance. Therefore, a part of $d(v, g)$ is based on a trajectory in $\mathbf{G}_{RRG}$ from v towards the vertex $w_{v,g}$ that is found as

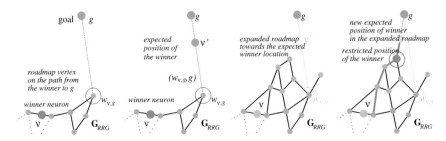

Fig. 3 SOM adaptation with $\mathbf{G}_{RRG}$ expansions, from *left* to *right*: First, vertex $w_{v,g}$ is found in $\mathbf{R}_{RRG}$ using (1) for the current winner v (green disc); The expected position v' of the neuron after the adaptation is determined; which is then utilized together with g in the RRG expansion of $\mathbf{G}_{RRG}$; Finally, new $w_{v,g}$ is found and v is updated to the nearest vertex to the expected position of v

$$w_{v,g} = \operatorname{argmin}_{v \in \mathbf{V}_{RRG}} (c(\kappa_{v,v}) + |(v, g)|^2), \qquad (1)$$

where $c(\kappa_{v,v})$ is the trajectory cost from the intermediate vertex w determined in $\mathbf{G}_{RRG}$ and $|(v, g)|$ is the Euclidean distance from v to g. Thus, the path from v to g consists of the trajectory $\kappa_{v,w_{v,g}}$ in $\mathbf{G}_{RRG}$ and a straight line segment from $w_{v,g}$ to g. Notice, the cost found in the roadmap should be preferred and the influence of the Euclidean distance should be suppressed, that is why it is in power of two in (1). The found path is utilized in adaptation of neurons to g.

However, the path over the vertex $w_{v,g}$ cannot be directly used for a new position of the adapted neuron because the expected position of the neuron may be out of the current roadmap $\mathbf{G}_{RRG}$. Therefore, the expected position of the neuron after the adaptation is determined and the roadmap is expanded towards it and the location g using the RRG expansion accompanied by the goal bias and goal zooming techniques [10] (in which a random sample is substituted by the given location and sampled around the location, respectively). Then, the vertex $w_{v,g}$ is determined again in the updated roadmap and a new expected position of the neuron being adapted is restricted to the nearest vertex of $\mathbf{G}_{RRG}$. Hence, the approximation together with the proposed adaptation of neurons turns out to a steering strategy to randomized sampling in the RRG. The process is schematically visualized in Fig. 3.

4 Solvers for the Multi-Goal Motion Planning Problem

The proposed approach to solve the MGMP problem consists of two steps. First, a roadmap $\mathbf{G}_{RRG}$ is created. An admissible solution of the MGMP problem is found if all locations $g \in \mathcal{G}$ have its corresponding (nearest) configuration $v_g \in \mathbf{G}_{RRG}$ in less than ϵ distance from the particular g and there exists a trajectory in $\mathbf{G}_{RRG}$ that connects all the locations $\mathcal{G}$. The final shortest multi-goal trajectory is found in $\mathbf{G}_{RRG}$ as a solution of the TSP using Chained Lin-Kernighan heuristic [1].

An admissible trajectory can be found in $\mathbf{G}_{RRG}$ if all vertices representing the goal locations are connected. The quality of the final trajectory depends on the roadmap and basically a denser roadmap may provide shorter trajectories at the cost of more demanding computations. The key to efficiently find a good trajectory is in the construction of the roadmap. Various methods how to steer the expansion of $\mathbf{G}_{RRG}$ can be proposed. The SOM-based steering of the RRG has been firstly introduced in [12]. The idea has been further investigated and the improved method is presented here. Moreover, we considered the proposed idea utilized in SOM steering also in a direct construction of the roadmap to verify the added value of the unsupervised learning. The proposed roadmap construction methods are briefly summarized in the following paragraphs.

Naive construction of the roadmap is based on iterative roadmap expansions towards the locations $\mathcal{G}$ that are alternating in a sequence found as a solution of the Euclidean TSP. Each location is iteratively used in the goal zooming technique for 5 expansions and the process is repeated until the maximum number of expansions M is not reached. The ball expansions of the RRG are activated after 100 alternations of the whole sequence, to reduce the computational burden and improve convergence of the roadmap to an admissible solution.

SOM expansion is based on the steering strategy described in Sect. 3 that is accompanied by additional expansions towards the presented location $g \in \mathcal{G}$ to the network, which support a fast convergence of the roadmap to $\mathcal{G}$. If g is not yet connected with the roadmap, 20 expansions towards g are performed using g in goal zooming prior adaptation of the winner neuron towards g. After that, the proposed SOM steering is employed. Similarly to *Naive* method, the ball expansions of the RRG are suppressed for the first 10 learning epochs.

Rand variant of the roadmap construction is based on additional expansions to $\mathcal{G}$ used in the SOM method. It is similar to the *Naive* method, but the sequence of locations $\mathcal{G}$ is a random permutation as in SOM. Each location $g \in \mathcal{G}$ is used in goal zooming for 20 expansions. Then, the algorithm continues with the next location in the sequence. Once all locations are used, a new permutation of $\mathcal{G}$ is created and the process is repeated up to M roadmap expansions are performed.

MST method represents an existing approach for the MGMP [11] based on an iterative determination of the Minimum Spanning Tree (MST) as approximation of the TSP. The MST is initially determined using Euclidean distances that is iteratively refined using an "optimal" motion planner to find corresponding trajectories for all MST edges until all the edges represent admissible trajectories. An optimal motion planning is too computationaly demanding for the hexapod robot, and therefore, the MST is used to steer roadmap expansion. For each MST edge without a corresponding trajectory in the roadmap, 20 expansions towards the edge's endpoints are performed for every iteration of the MST refinement. This is repeated until an admissible multi-goal trajectory is found.

Because the SOM method provides a first admissible solution very quickly, two hybrid approaches are proposed: **Naive-SOM** and **Rand-SOM**. The SOM method is utilized to find the first admissible solution. Then, the *Naive* and *Rand* approaches are used up to M expansions of the roadmap, respectively. In a similar way, the MST is utilized in the **Rand-MST** to further improve initial solution provided by the MST-based method.

5 Evaluation Results

The roadmap expansion strategies have been evaluated for a hexapod walking robot and several scenarios of the MGMP problem in the environment called *potholes*, see Fig. 1. A particular difficulty of the problem depends on spatial distribution of the goal locations in the environment. Therefore, 20 random problem instances are created in the given environment. Each instance is solved 20 times by each particular algorithm because all algorithms are stochastic, and the results are presented as average values accompanied by standard deviations. We considered problems with 10 goal locations ($n = 10$) as sufficient to demonstrate difficulty of constructing roadmap for the multi-goal trajectory planning. Particular algorithms have been evaluated for different parameters; however, only selected results are presented because of the space limit. The total number of the evaluated scenarios was more than twenty thousands. Examples of constructed roadmaps, the first admissible solution found by SOM, and the final found solution found by the Rand variant are shown in Fig. 4.

The most time consuming step in the solution of the MGMP problem is a single roadmap expansion, which, in the case of the RRG, is a more computationally demanding with increasing number of roadmap vertices. Moreover, it is even more demanding in the improving phase, where expansions are performed for vertices in the ball around the last added vertex to the roadmap. Therefore, the number of performed roadmap expansions is the main performance indicator.

The first evaluation is focused on the performance of the roadmap expansion strategies in finding the first admissible solution with the maximal number of expansions restricted to 100000. The results for 400 trials on 20 problems solved by each approach are depicted in Table 1 (values are computed from admissible solutions).

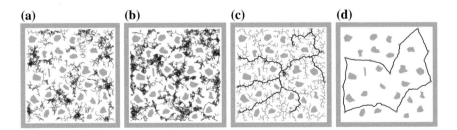

Fig. 4 Build roadmaps by *Naive* and *SOM-based* approaches after performing M expansions. A path found by the *Rand* approach after 204 357 expansions. Obstacles are in brown, goals are represented as green discs, roadmap edges are purple segments, and a multi-goal trajectory is in black. **a** Naive, M = 10000. **b** Naive, M = 20000. **c** SOM, M = 597. **d** Rand, M = 204357

Table 1 Roadmap construction for determining a first admissible solution

Method	Naive	SOM	Rand	MST
Success rate	54 %	**93** %	61 %	45 %
Average number of the RRG expansions	85 468	**14 258**	66 375	70 241
Average number of the roadmap vertices	24 698	**5 662**	25 781	37 815
Average number of the roadmap edges	142 472	**16 218**	109 096	84 372
Average required CPU time [s]*	39	**13**	56	209

*Indicative values because several machines of different configurations have been used

Here, we can observe that the same randomized schema utilized in SOM and Rand strategies provide different performance. Moreover, the MST-based approach proposed in [11] does not provide significant advantage over Rand and its more demanding because of determination of the MST. The results support the evidence that the proposed SOM-based steering significantly improves the performance in finding the first admissible solution. The main results is that SOM provides fastest admissible solutions with a high success rate.

Notice, the number of the roadmap vertices is always lower than the number of expansions. A higher number of vertices indicates a successful expansion of the roadmap and similarly a higher number of edges indicates a denser roadmap as a result of the improving step of the RRG.

The next evaluation has been focused on the quality improvement of the found multi-goal trajectory according to increasing maximal number of the performed RRG expansions. We found out that the proposed SOM improves solutions only slowly with more expansions, and therefore, we consider it only in finding the first admissible solution in the hybrid approaches *Naive-SOM* and *Rand-SOM*. The quality of the trajectory is considered as a ratio of the trajectory length to the best found solution for the particular problem determined from all the performed trials. This allows to aggregate results for various problem instances, for which trajectories may be significantly different. Thus, values of the ratio close to 1 indicate the particular approach provides relatively high quality solutions among the evaluated algorithms. The results for increasing number of roadmap expansions are depicted in Fig. 5.

Discussion—Based on the performed evaluation of the steering strategies of the randomized sampling in the RRG, the results support that the proposed SOM-based strategy provides the first admissible solution with a significantly less number of expansions than other strategies. However, the solution quality does not improve with more expansions and thus the current form of the strategy is suitable only for finding an admissible solution. On the other hand, the proposed combination of the SOM and randomized expansions in the hybrid solvers provide benefits of the both approaches and it seems to be a suitable technique to provide the first solution quickly and further quality improvements.

An important lesson learned from the presented evaluation is that the way how the roadmap is initially created significantly affects the ability to find an admissible solution quickly. Here, the SOM adaptation provides an efficient trade-off between

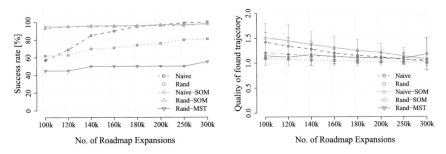

Fig. 5 Success rate and quality of the found trajectories

exploration of $\mathcal{C}$ and exploitation of the current $\mathbf{G}_{RRG}$ towards connecting the required goal locations. However, once the locations are connected in the roadmap, the adaptation process only moves neurons along the roadmap and does not explore possible shortcuts to improve the solution.

6 Conclusion

An evaluation of four multi-goal trajectory planners is presented in this paper. The results indicate the proposed SOM-based roadmap expansion improves finding the first admissible solution. However, a planner solely based on the SOM strategy does not improve the found solution, but the solution can be improved by additional expansions of the roadmap. Although the current achieved results does not meet the expectation of a motion planner solely based on SOM, it support feasibility of the SOM-based simultaneous building of the distance function approximation together with its utilization in the multi-goal trajectory planning.

Regarding the applied SOM based principles, the whole graph $\mathbf{G}_{RRG}$ can be considered as a growing neural network, where the adaptation rules can be used to remove not promising configurations and thus reduce the number of vertices of the graph. Besides, they can also be utilized to further exploration of the configuration space to improve quality of the found solution. Consideration of these extensions is a subject of our further work.

Acknowledgments The presented work is supported by the Czech Science Foundation (GAČR) under research project No. 13-18316P. Computational resources were provided by the MetaCentrum under the program LM2010005 and the CERIT-SC under the program Centre CERIT Scientific Cloud, part of the Operational Program Research and Development for Innovations, Reg. No. CZ.1.05/3.2.00/08.0144.

References

1. Applegate, D., Cook, W., Rohe, A.: Chained Lin-Kernighan for large traveling salesman problems. Informs J. Comput **15**(1), 82–92 (2003)
2. Créput, J.C., Koukam, A.: A memetic neural network for the Euclidean traveling salesman problem. Neurocomputing **72**(4–6), 1250–1264 (2009)
3. Elinas, P.: Multi-goal planning for an autonomous blasthole drill. In: Gerevini, A., Howe, A.E., Cesta, A., Refanidis, I. (eds.) ICAPS. AAAI (2009)
4. Faigl, J., Hollinger, G.: Unifying multi-goal path planning for autonomous data collection. In: IROS, pp. 2937–2942 (2014)
5. Faigl, J., Kulich, M., Vonásek, V., Přeučil, L.: An application of self-organizing map in the non-euclidean traveling salesman problem. Neurocomputing **74**(5), 671–679 (2011)
6. Janoušek, P., Faigl, J.: Speeding up coverage queries in 3d multi-goal path planning. In: ICRA, pp. 5067–5072 (2013)
7. Karaman, S., Frazzoli, E.: Sampling-based algorithms for optimal motion planning. Int. J. Rob. Res. **30**(7), 846–894 (2011)
8. Lattanzi, L., Cristalli, C.: An efficient motion planning algorithm for robot multi-goal tasks. In: ISIE, pp. 1–6. IEEE (2013)
9. Lavalle, S.M.: Planning Algorithms. Cambridge University Press (2006)
10. Lavalle, S.M., Kuffner, J.J.: Rapidly-exploring random trees: progress and prospects. Algorithmic and Computational Robotics: New Directions, pp. 293–308 (2001)
11. Saha, M., Roughgarden, T., Latombe, J.C., Sánchez-Ante, G.: Planning tours of robotic arms among partitioned goals. Int. J. Rob. Res. **25**(3), 207–223 (2006)
12. Vanek, P., Faigl, J., Masri, D.: Multi-goal trajectory planning with motion primitives for hexapod walking robot. In: ICINCO, pp. 599–604 (2014)
13. Yamakawa, T., Horio, K., Hoshino, M.: Self-organizing map with input data represented as graph. In: Neural Information Processing, pp. 907–914 (2006)
14. Zhang, J., Feng, X., Zhou, B., Ren, D.: An overall-regional competitive self-organizing map neural network for the euclidean traveling salesman problem. Neurocomputing **89**, 1–11 (2012)

Dimensionality Reduction Hybridizations with Multi-dimensional Scaling

Oliver Kramer

Abstract Dimensionality reduction is the task of mapping high-dimensional patterns to low-dimensional spaces while maintaining important information. In this paper, we introduce a hybrid dimensionality reduction method that is based on the weighted average of the normalized distance matrices of two or more embeddings. Multi-dimensional scaling embeds the weighted average distance matrix in a low-dimensional space. The approach allows the hybridization of arbitrary point-wise embeddings. Instances of the hybrid algorithm template use principal component analysis, multi-dimensional scaling, and locally linear embedding. The variants are experimentally compared using three dimensionality reduction measures, i.e., the Shepard-Kruskal scaling, a co-ranking matrix measure, and the nearest neighbor regression error in presence of label information. The results show that the hybrid approaches outperform their native pendants in the majority of the experiments.

1 Introduction

Dimensionality reduction (DR) has an important part to play in many machine learning applications. It finds applications in preprocessing for machine learning techniques that perform best in low-dimensional data spaces and for visualization of high-dimensional data. Objective of most DR algorithms is to reduce the dimensionalities of patterns while maintaining distance information. In this paper, we employ the normalized pattern distance matrix oriented to Shepard-Kruskal scaling for the algorithmic method and the evaluation of the DR result. We propose an algorithm that allows the hybridization of various DR results based on weighted distances matrices and multi-dimensional scaling (MDS) [6]. The weighted hybridization offers the freedom to take advantage of the capabilities of the native methods. We compare the new approach to the native approaches w.r.t. the measures Shepard-Kruskal scaling,

O. Kramer (✉)
Computational Intelligence Group Department of Computing Science,
University of Oldenburg, Oldenburg, Germany
e-mail: oliver.kramer@uni-oldenburg.de

© Springer International Publishing Switzerland 2016
E. Merényi et al. (eds.), *Advances in Self-Organizing Maps and Learning
Vector Quantization*, Advances in Intelligent Systems and Computing 428,
DOI 10.1007/978-3-319-28518-4_13

155

the co-ranking matrix measure, and the nearest neighbor regression error for labeled
benchmark data.

This paper is structured as follows. In Sect. 2, we give a short introduction to DR
with an emphasis on MDS. Section 3 presents three DR quality measures, in particular
the distance-based DR measure, which is similar to the Shepard-Kruskal scaling [2].
The normalized distance matrix is basis of the hybrid approach introduced in Sect. 4.
An experimental analysis is presented in Sect. 5. Conclusions are drawn in Sect. 6.

2   Methodological Basis

2.1   DR Reduction

In DR, the task is to embed high-dimensional patterns $\mathbf{X} = \{\mathbf{x}_1, \ldots, \mathbf{x}_N\}$ with $\mathbf{x}_i \in \mathbb{R}^d$ into low-dimensional spaces by learning an explicit mapping $\mathbf{F} : \mathbb{R}^d \to \mathbb{R}^q$, by
finding low-dimensional counterparts $\hat{\mathbf{X}} = \{\hat{\mathbf{x}}_1, \ldots, \hat{\mathbf{x}}_N\}$ with $\hat{\mathbf{x}}_i \in \mathbb{R}^q$ with $q < d$
that conserve useful information of their high-dimensional pendant, or by finding
a set of codebook vectors (usually fewer than N) that represent the data like in
self-organizing maps (SOMs), e.g., see [12]. The DR problem has intensively been
studied in the past, see [2, 7], but is still a promising research area due to the growing
importance of high-dimensional data.

2.2   MDS

Principal component analysis (PCA) [3, 4], locally linear embedding (LLE) [13], and
isometric mapping (ISOMAP) [15] are famous methods for the point-wise embedding of patterns. ISOMAP and LLE are based on MDS, which estimates the coor-
dinates of a set of points while only the distances are known. Let $\mathbf{D}_\mathbf{X} = (d_{ij})$ be
the distance matrix of the set of patterns with d_{ij} being the distance between two
patterns $\mathbf{x}_i$ and $\mathbf{x}_j$. Given all pairwise distances d_{ij} with $i, j = 1, \ldots, N$ and $i \neq j$,
MDS computes the corresponding low-dimensional representations. For this sake, a
matrix $\mathbf{B} = (b_{ij})$ is computed with

$$b_{ij} = -\frac{1}{2}[d_{ij}^2 - \frac{1}{N}\sum_{k=1}^{N} d_{kj}^2 - \frac{1}{N}\sum_{k=1}^{N} d_{ik}^2 + \frac{1}{N^2}\sum_{k=1}^{N}\sum_{l=1}^{N} d_{kl}^2]. \tag{1}$$

The points are computed via an eigendecomposition of $\mathbf{B}$ with Cholesky or singular
value decomposition resulting in eigenvalues λ_i and corresponding eigenvectors
$\gamma_i = (\gamma_{ij})$. It holds $\sum_{j=1}^{N} \gamma_{ij}^2 = \lambda_i$. The embeddings in a q-dimensional space are
the eigenvectors of the q-largest eigenvalues $\hat{\mathbf{x}}_i = \gamma_i \sqrt{\lambda_i}$.

2.3 LLE

For non-linear manifolds, LLE by Roweis and Saul [13] is a powerful DR approach. LLE assumes the local linearity of manifolds and is appropriate for the hybridization as it computes point-wise embeddings. First, LLE computes weights that allow a linear reconstruction of point $\mathbf{x}_i$ from its k-nearest neighbors minimizing the cost function

$$E(\mathbf{w}) = \sum_{i=1}^{N} \|\mathbf{x}_i - \sum_{j=1}^{k} w_{ij}\mathbf{x}_j\|^2 \tag{2}$$

with weights $w_{ij} \in \mathbb{R}$.

3 Dimensionality Reduction Measures

In the experimental part, we will analyze the introduced hybrid methods w.r.t. three DR quality measures that are introduced in the following.

3.1 Shepard-Kruskal Scaling

A reasonable DR quality measure that reflects the objective to maintain distances is the Shepard-Kruskal scaling [2]. We formulate a normalized variant of the Shepard-Kruskal scaling in the following. Let $\mathbf{D_X}$ be the distance matrix in data space and $\mathbf{D_{\hat{X}}}$ be the distance matrix in the low-dimensional space. Both are normalized, i.e., each component is divided by the maximal component of the whole matrix. The Frobenius norm of the differences of the normalized distance matrixes

$$E_{ks} = \|\mathbf{D_X} - \mathbf{D_{\hat{X}}}\|_F^2 \tag{3}$$

is the Shepard-Kruskal measure variant we employ. If the deviation of normalized pattern-wise distances is zero, the high-dimensional distances are optimally maintained in the low-dimensions space. Normalization is required because the absolute pattern coordinates computed by different methods may vary significantly.

3.2 Co-ranking Matrix

A traditional DR quality measure we will use for comparison in the experimental section is the co-ranking matrix [8, 9] concentrating on measuring the maintenance of neighborhoods. The co-ranking matrix is based on the comparison of ranks w.r.t. distance-based sorting of patterns in data space and in the low-dimensional space. The co-ranking matrix is employed to define a measure $E_{nx} \in [0, 1]$ corresponding to the ratio of neighbors of patterns occurring in a k-neighborhood in data space and in the low-dimensional space. High values for E_{nx} show that the neighborhood relations are preserved.

3.3 Nearest Neighbor Error

As our last measure, we compare the embeddings w.r.t. a measure that is based on the k-nearest neighbor (kNN) regression error, which can only be computed for labeled data. If $(\mathbf{x}_1, y_1), \ldots, (\mathbf{x}_N, y_N)$ are pattern-label pairs with $y_i \in \mathbb{R}$, we can define the kNN regression error as mean squared error

$$E_{\mathbf{x}} = \frac{1}{N} \sum_{i=1}^{N} (f(\mathbf{x}_i) - y_i)^2 \tag{4}$$

with kNN model f. Based on the computed embeddings, the q-dimensional pattern-label pairs $(\hat{\mathbf{x}}_1, y_1), \ldots, (\hat{\mathbf{x}}_N, y_N)$ can be used to evaluate an equivalent kNN error $E_{\hat{\mathbf{x}}}$. The ratio

$$E_{kNN} = E_{\hat{\mathbf{x}}}/E_{\mathbf{x}} \tag{5}$$

of both errors is an indicator how well the DR methods perform as preprocessing method for regression tasks. It is smaller than 1.0, if the kNN error $E_{\hat{\mathbf{x}}}$ from the low-dimensional space is lower than the original kNN error $E_{\mathbf{x}}$ from data space.

4 Hybrid Embedding

4.1 Approach

In this section, we introduce an approach that hybridizes embeddings of two (and potentially more) DR methods. The concept is based on three main steps: 1. The computation of two embeddings $\hat{\mathbf{X}}^1$ and $\hat{\mathbf{X}}^2$, 2. The computation of a weighted sum of the low-dimensional distance matrices of Step 1, and last, 3. The MDS embedding using the novel distance matrix. Figure 1 shows the pseudo-code of this approach.

Let $\mathbf{D}_{\hat{\mathbf{X}}}^1$ be the normalized distance matrix of the embedding of the first DR method, and $\mathbf{D}_{\hat{\mathbf{X}}}^2$ the corresponding normalized distance matrix of the second method. The

---

Algorithm 1: HYBRID MDS EMBEDDING

Require: data set $\mathbf{X}$

1: embed $\mathbf{X}$ with DR method 1 $\rightarrow \hat{\mathbf{X}}^1$
2: embed $\mathbf{X}$ with DR method 2 $\rightarrow \hat{\mathbf{X}}^2$
3: compute normalized distance matrices $\mathbf{D}_{\hat{\mathbf{X}}}^1$ and $\mathbf{D}_{\hat{\mathbf{X}}}^2$
4: hybrid distance matrix $\mathbf{D}_{\hat{\mathbf{X}}}' = \alpha \cdot \mathbf{D}_{\hat{\mathbf{X}}}^1 + (1-\alpha) \cdot \mathbf{D}_{\hat{\mathbf{X}}}^2$
5: $\hat{\mathbf{X}} \rightarrow$ MDS embedding with $\mathbf{D}_{\hat{\mathbf{X}}}'$
6: **return** $\hat{\mathbf{X}}$

Fig. 1 Pseudo-code of hybrid MDS embedding approach

idea of the hybridization approach is to compute a weighted average distance matrix and employ this for embedding with MDS. The new distance matrix is

$$\mathbf{D}_{\hat{\mathbf{X}}}' = \alpha \cdot \mathbf{D}_{\hat{\mathbf{X}}}^1 + (1-\alpha) \cdot \mathbf{D}_{\hat{\mathbf{X}}}^2 \qquad (6)$$

with weight $\alpha \in [0, 1]$. The novel weighted distance matrix $\mathbf{D}_{\hat{\mathbf{X}}}'$ is the input to MDS resulting in a novel embedding $\hat{\mathbf{X}}'$. The optimal weight α can be found with grid search or optimization methods like evolutionary algorithms subject to the objective to minimize measure E_{ks} introduced in Eq. 3.

4.2 Related Work

Ensembles, i.e., the combination of more than one technique, are a common approach in machine learning. Interestingly, not many ensemble approaches have been introduced for DR problems. One of the few examples has been introduced by Moon and Qi [10], who present a hybrid of support vector machines and independent component analysis. Submanifold approaches exist that concentrate on the independent embedding of clusters, e.g., the hybrid manifold clustering approach by Kramer [5]. To the best of our knowledge, there are no methods that concentrate on the hybridization of multiple DR results based on MDS.

5 Experimental Evaluation

In this section, we experimentally analyze the hybrid embedding approach. For this sake, we concentrate on a deeper analysis on three benchmark problems, see Appendix A w.r.t. the three measures E_{ks}, E_{nx}, and E_{kNN}. We compare MDS, PCA, LLE, and the three hybrids MDS-PCA, MDS-LLE, and LLE-PCA. We employ the data set size $N = 500$, which we assume to be a sufficient size for a conceptual

Table 1 Comparison of native (MDS, PCA, LLE) and hybrid approaches (MDS-PCA, MDS-LLE, and LLE-PCA) w.r.t. measures E_{ks}, E_{nx}, and E_{kNN}

Problem	Measure	MDS	PCA	LLE	MDS-PCA	MDS-LLE	LLE-PCA
Swiss Roll	E_{ks}	51.259	56.120	201.689	**50.959**	51.256	56.120
	E_{nx}	0.590	0.596	0.259	0.598	0.657	**0.658**
	E_{kNN}	24.566	65.493	127.192	**14.906**	24.569	37.089
Housing	E_{ks}	1.328	**0.015**	118.298	1.414	2.184	1.703
	E_{nx}	0.871	**0.989**	0.655	0.900	0.859	0.889
	E_{kNN}	1.067	1.024	1.032	1.036	**1.011**	**1.011**
Friedman	E_{ks}	228.0221	309.802	366.304	228.039	**227.308**	299.872
	E_{nx}	0.120	0.129	0.117	0.130	0.125	**0.131**
	E_{kNN}	1.087	1.039	1.093	1.0243	1.075	**1.008**

evaluation of the introduced concepts. For LLE, E_{nx}, and E_{kNN} we use the same neighborhood size $k = 10$.

5.1 Quantitative Analysis

Table 1 shows a comparison between the native methods and the optimal values achieved by the three hybrids. The experiments reveal that the hybrid methods achieve better results than the native ones in the majority of the experimental settings. On the Swiss Roll, the MDS-PCA achieves the lowest distance matrix value E_{ks} and the lowest kNN error E_{kNN}, while the MDS-LLE variant achieves the highest co-ranking matrix value E_{nx}. On the Housing data set, linear conditions let the PCA achieve the best E_{ks} and a nearly optimal co-ranking matrix value of 0.989, but MDS-LLE and LLE-PCA achieve a lower nearest neighbor error. Among the hybrid variants, MDS-PCA achieves the lowest E_{ks} and the highest co-ranking matrix measure. MDS-LLE achieves the best distance matrix measure on Friedman, while LLE-PCA outperforms the other methods considering E_{nx} and E_{kNN}. The results show that the employed measures are consistent to each other, i.e., optimal values w.r.t. one measure are often optimal w.r.t. another measure.

Figure 2 compares the DR quality measures E_{ks}, E_{nx}, and the kNN error E_{kNN} on the data sets Swiss Roll, Housing, and Friedman w.r.t. various settings for α, i.e., $\alpha = 0.0, 0.05, \ldots, 1.0$. The results show that better settings for α other than 0 or 1 (i.e., the pure methods) do not exist in every case. But the MDS-PCA variant achieves local optima on the Swiss Roll for E_{kNN} and for E_{nx} and E_{kNN} on Housing and Friedman. The MDS-LLE hybrid is mostly increasing E_{ks} for increasing α with few local optimal values. For E_{kNN} on Housing and E_{nx} on all problems, there are optimal values achieved by the hybrids. The LLE-PCA hybrids outperform the pure methods w.r.t. E_{ks} on Friedman and w.r.t. E_{nx} and E_{kNN} on all three problems. The

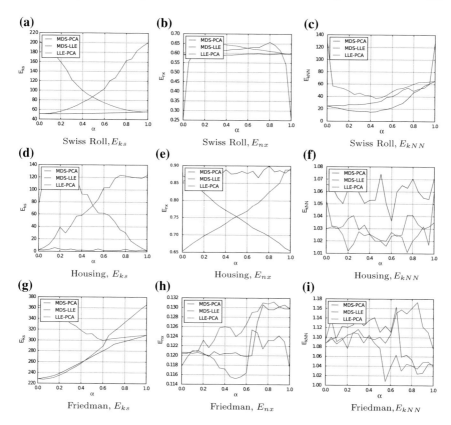

Fig. 2 Analysis of hybrid algorithm w.r.t. parameter α on Swiss Roll, Housing, and Friedman. Lower values are better for E_{ks} and E_{kNN}, higher values are better for E_{nx}. **a** Swiss Roll, E_{ks}. **b** Swiss Roll, E_{nx}. **c** Swiss Roll, E_{kNN}. **d** Housing, E_{ks}. **e** Housing, E_{nx}. **f** Housing, E_{kNN}. **g** Friedman, E_{ks}. **h** Friedman, E_{nx}. **i** Friedman, E_{kNN}

main observation is that often the pure DR methods are outperformed by the hybrid variants, albeit in some cases the original methods perform best.

5.2 Visualization of Swiss Roll Embedding

In the following, we visualize the result of the new hybrid approach on the Swiss Roll. Figure 3a shows the Swiss Roll data, while Fig. 3b, c compare the Swiss Roll MDS embedding to an embedding generated with a MDS-PCA hybrid. The MDS-PCA embedding shares similarities with the native MDS embedding, but the

(a)   **(b)**   **(c)**

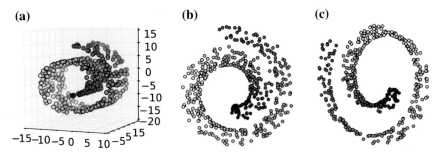

Fig. 3 Swiss Roll, its MDS embedding, and an MDS-PCA embedding. **a** Original data. **b** MDS.
c MDS-PCA

neighborhoods are preserved even more accurately—in particular, the manifold becomes narrower. The linear characteristics of the PCA embedding influence the result of the hybrid MDS-PCA embedding.

6   Conclusions

We introduced a hybrid method that allows merging the DR outputs of multiple methods. The hybridization is based on the employment of a hybrid normalized Kruskal-Shephard scaling matrix that is embedded into the target space with MDS. An experimental comparison on a small problem benchmark set reveals that in many cases the hybrids outperform their native pendants. The Kruskal-Shephard scaling is also used as measure and turns out to be consistent with the co-ranking matrix measure and the nearest neighbor error when mapping from the low-dimensional space, if label information is available.

A prospective research direction is the extension of the experimental analysis to further hybrid variants, e.g., based on ISOMAP, random projection [1], and further benchmark problems. Further, it will be interesting to put a focus on the analysis of hybrids that take into account the results of more than two embeddings. The approach can be applied to SOMs, if the second method (which might also be a SOM) employs the same number of codebook vectors.

Appendix A: Benchmark Problems

The experiments in this paper are based on the following data sets:

- The Swiss Roll is a simple artificial data set with $d = 3$ that allows a visualization of neighborhoods with colored patterns and label information based on pattern colors from SCIKIT-LEARN [11].

- The Housing data set, also known as California Housing from the STATLIB repository [14] comprises 20,640 8-dimensional patterns and one label.
- Friedman is the high-dimensional regression problem Friedman #1 generated with SCIKIT-LEARN [11] with $d = 500$.

References

1. Dasgupta, S.: Experiments with random projection. In: Uncertainty in Artificial Intelligence (UAI), pp. 143–151. Morgan Kaufmann Publishers Inc. (2000)
2. Hastie, T., Tibshirani, R., Friedman, J.: The Elements of Statistical Learning. Springer, Berlin (2009)
3. Hotelling, H.: Analysis of a complex of statistical variables into principal components. J. Educ. Psychol. **24**(6), 417–441 (1933)
4. Jolliffe, I.: Principal component analysis. In: Springer Series in Statistics. Springer, New York (1986)
5. Kramer, O.: Hybrid manifold clustering with evolutionary tuning. In: Applications of Evolutionary Computation, EvoApplications, pp. 481–490 (2015)
6. Kruskal, J.B.: Nonmetric multidimensional scaling: a numerical method. Psychometrika **29**, 1–27 (1964)
7. Lee, J.A., Verleysen, M.: Nonlinear Dimensionality Reduction. Springer (2007)
8. Lee, J.A., Verleysen, M.: Quality assessment of dimensionality reduction: rank-based criteria. Neurocomputing **72**(7–9), 1431–1443 (2009)
9. Lueks, W., Mokbel, B., Biehl, M., Hammer, B.: How to evaluate dimensionality reduction?—Improving the co-ranking matrix. CoRR (2011)
10. Moon, S., Qi, H.: Hybrid dimensionality reduction method based on support vector machine and independent component analysis. IEEE Trans. Neural Netw. Learn. Syst. **23**(5), 749–761 (2012)
11. Pedregosa, F., Varoquaux, G., Gramfort, A., Michel, V., Thirion, B., Grisel, O., Blondel, M., Prettenhofer, P., Weiss, R., Dubourg, V., Vanderplas, J., Passos, A., Cournapeau, D., Brucher, M., Perrot, M., Duchesnay, E.: Scikit-learn: machine learning in python. J. Mach. Learn. Res. **12**, 2825–2830 (2011)
12. Rossi, F.: How many dissimilarity/kernel self organizing map variants do we need? In: Advances in Self-Organizing Maps and Learning Vector Quantization, Workshop on Self-Organizing Maps (WSOM) (2014)
13. Roweis, S.T., Saul, L.K.: Nonlinear dimensionality reduction by locally linear embedding. Science **290**, 2323–2326 (2000)
14. StatLib-Datasets Archive. http://lib.stat.cmu.edu/datasets/ (2014)
15. Tenenbaum, J.B., Silva, V.D., Langford, J.C.: A global geometric framework for nonlinear dimensionality reduction. Science **290**, 2319–2323 (2000)

A Scalable Flexible SOM NoC-Based Hardware Architecture

Mehdi Abadi, Slavisa Jovanovic, Khaled Ben Khalifa, Serge Weber
and Mohamed Hédi Bedoui

Abstract In this paper, a parallel hardware implementation of a self-organizing map (SOM) is presented. Practical scalability and flexibility are the main architecture features which are obtained by using a Network-on-chip (NoC) approach for communication between neurons. The presented hardware architecture allows online learning and can be easily adapted for a large variety of applications without a considerable design effort. A hardware 5×5 SOM was validated through the FPGA implementation and its performances at a working frequency of 200 MHz for a 32-element input vector reach 724 MCUPS in the learning and 1168 MCPS in the recall phase.

1 Introduction

Since their introduction, Self-Organizing Maps (SOMs) have been largely used in many applications [1]. A SOM is an unsupervised learning neural network which is mainly used to reduce and classify high-dimensional input data sets to ease their interpretation and processing. A SOM can be implemented either in software (SW), hardware (HW) or mixed hardware-software platforms (HW/SW). Even though the

M. Abadi (✉) · S. Jovanovic (✉) · S. Weber
UMR 7198, Institut Jean Lamour, Université de Lorraine, Nancy, France
e-mail: mehdi.abadi@univ-lorraine.fr

S. Jovanovic
e-mail: slavisa.jovanovic@univ-lorraine.fr

S. Weber
e-mail: serge.weber@univ-lorraine.fr

M. Abadi · K. Ben Khalifa (✉) · M.H. Bedoui
LR12ES06, Laboratoire de Technologie Et Imagerie Médicale,
Université de Monastir, Monastir, Tunisia
e-mail: khaled.benkhalifa@issatso.rnu.tn

M. Abadi
Ecole Nationale d'Ingénieurs de Sousse, Université de Sousse,
Sousse, Tunisia

© Springer International Publishing Switzerland 2016
E. Merényi et al. (eds.), *Advances in Self-Organizing Maps and Learning Vector Quantization*, Advances in Intelligent Systems and Computing 428,
DOI 10.1007/978-3-319-28518-4_14

software solutions provide more flexibility, the hardware implementations exploit inherent parallelism of SOM networks and may be preferred to the software ones especially in real-time applications characterized with tight temporal constraints.

The hardware SOM implementations are typically application specific. Each application has its own specificities needing different SOM parameters: input layer size, number of neurons in the output layer, timing constrains, memory requirements, etc. In most of cases, the adaptation of a hardware SOM architecture for an other application is time consuming and needs considerable design efforts. Another important issue which make difficult the reuse of an existing hardware SOM implementation are the communication links between neurons. Generally, these communication links are established at the point-to-point basis and are hard wired and if we want to add some additional neurons (and thus new connections to them) we have to completely modify the way neurones are connected. To have a scalable and application-independent hardware SOM implementation, it is necessary to add more flexibility to the existing HW design approaches. A way of doing this is to completely decouple computation from communication. In this paper, we propose a Network-on-a-chip (NoC) based solution, where a NoC is used for communication purposes between neurons.

This paper is organized as follows: Sect. 2 presents the state of the art in the domain of hardware SOM implementations. Section 3 presents the proposed method and describes the modifications that should be made to an existing hardware SOM implementation to make it scalable. Section 4 presents some obtained results whereas some conclusions and perspectives are drawn in Sect. 5.

2   Related Work

The first reported SOM implementations were in software using processor-based architectures [2]. The performances of initial single-core microprocessor architectures have been recently boosted by the increasing parallelism of many-cores multiprocessor chips (MPSoC), but are still suffering from sequential processing and high power consumption with respect to the application-specific solutions. However, the SOM software implementations are flexible and easy to implement and are usually used beforehand a hardware implementation especially in the design exploration phase to give rapidly insights about the HW design choices to take. However, for hard real-time embedded applications, hardware solutions based on the use of Field Programmable Gate Arrays (FPGAs) or Application Specific Integrated Circuits (ASICs) may be preferred.

The FPGA solutions are a good trade-off between cost, design effort, performances and reduced time-to-market. However, the FPGAs can only be used to implement digital counterparts of SOMs, no analog design is supported. If the high performances, low power consumption or low area occupancy are targeted, an ASIC is preferable to an FPGA implementation. There are some ASIC implementations of SOMs that we found in literature [3–5]. An ASIC implementation gives the best performances but is costly, demands high design efforts and has little or no flexibility.

In a hardware design, each choice has a cost and must be carefully considered. The floating point operators are resources greedy and are often avoided in hardware implementations, unless there are no solutions to obtain needed precision. Besides the arithmetic precision, in HW SOM implementations the choice of norm for distance calculation can influence the overall complexity of the hardware. Therefore, we found some architectures using Manhattan or Euclidean distance [6–8]. The first one is preferred to the latter due to the lower computation requirements and thus lower power consumption. Some studies showed that for high-dimensional vectors the effect of choosing the L1 operator is negligible on the SOM performances [9].

Another important choice to take in HW SOM implementations is the type of neighbourhood function (NF) to use, whose function determines which neurons' coefficients in the vicinity of the winning one should be updated. In the original SOM algorithm, a Gaussian neighborhood function is used, but its hardware implementation demands complex arithmetic operations and is usually realized as an analog integrated circuit [4]. It is often approximated with other functions such as: rectangular, triangular, shift-register based [5, 7]. The shift-register solution of the NF greatly simplifies its implementation by replacing the resources consuming multipliers with simple shift registers and is widely used in digital SOM implementations [6–8, 10].

All presented HW architectures have a two-level structure: a massively parallel distance processing elements (PEs) layer usually connected with hard links to a global circuit used for winner neuron search and weight update operations. This type of connection may be advantageous in small SOM networks but in large ones, the increasing linking complexity considerably limits their clock frequency and thus the overall performances usually expressed in MCUPS/MCPS (million of connections and updates per second respectively). Manalakos et al. proposed in [10] a parallel HW SOM systolic architecture design in which an input vector traverses all neurons in a pipelined manner, forming that way shorter links and thus a faster HW.

The lack of flexibility of hardware SOMs, which is mainly due to the point-to-point communication between neurons and especially in large SOM networks, can be overcome with the use of a Network-on-chip (NoC). NoCs are presented as an alternative to traditional shared bus allowing the connection of several PEs on a single chip [11, 12]. They enjoy an explicit parallelism, high bandwidth and a high degree of modularity, which makes them very suitable for distributed architectures such as SOM networks.

3 Proposed Architecture

3.1 Self-Organizing Map (SOM)

The architecture of a SOM can be described with a two-dimensional distribution of $L \times K$ neurons. Each neuron has a weight vector $\overrightarrow{m}$ of dimension D, which is continuously compared to the input vector $\overrightarrow{X}$:

$$\vec{X} = \{\xi_1, \xi_2, \ldots, \xi_D\} \in \mathfrak{R}^D \tag{1}$$

Each neuron calculates the distance between its weights $\vec{m}_{l,k}$ ($0 \leq l \leq L - 1, 0 \leq k \leq K - 1$) and the input vector $\vec{X}$. In general, the calculated distance is the Euclidean distance:

$$\left\| \vec{X} - \vec{m}_{l,k} \right\| = \sqrt{(\xi_1 - \mu_{l,k1})^2 + (\xi_2 - \mu_{l,k2})^2 + \cdots + (\xi_D - \mu_{l,kD})^2} \tag{2}$$

Therefore, the winner neuron, which has the vector $\vec{m}_c$ closest to the input vector $\vec{X}$, is identified.

$$c = \operatorname*{argmin}_{l,k} \left\| \vec{X} - \vec{m}_{l,k} \right\| \tag{3}$$

During the learning phase, the winner's weights and the weights of the neurons in its vicinity are updated as described by the following equation:

$$\vec{m}_{l,k}(t + 1) = \vec{m}_{l,k}(t) + h_{c,l,k}(t) \left[\vec{X}(t) - \vec{m}_{l,k}(t) \right] \tag{4}$$

where $h_{c,l,k}(t)$ is the neighborhood function defined as follows:

$$h_{c,l,k}(t) = \alpha(t) \times \exp(-\frac{\left\| \vec{r}_c - \vec{r}_{l,k} \right\|}{2\sigma^2(t)}) \tag{5}$$

With $\alpha(t)$ learning rate; $\sigma(t)$ Neighbourhood rate; $\vec{r}_c$ position of the winning neuron; $\vec{r}_{l,k}$ position of the neuron with index (l, k).

3.2 Network on Chip (NoC)

The structure of a 2D mesh NoC is shown in Fig. 1a. The packets are transported from a source to a target through a network of routers and interconnection channels (Link). The network is composed of processing elements (PEs) and routers. Each router is associated to a PE via a network interface whose primary function is to pack (before sending) and unpack (after receiving) data exchanged between PEs. Figure 1d illustrates the structure of packets circulating in the network using the wormhole switching technique. Each packet is composed of flits: header (opening the communication and "showing" the route to other flits), body (containing the data) and tail flit(closing the communication). The router (see Fig. 1b) is composed of a crossbar which establishes multiple links between inputs and outputs of the router according to the predefined routing algorithm and scheduling policy. The crossbar and the arbitration of packets in the router are handled with a Control

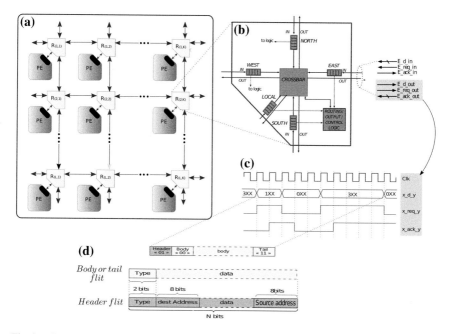

Fig. 1 **a** Structure of a 2D mesh NoC. **b** NoC router architecture. **c** Alternate bit protocol. **d** Message structure

Logic Block (CLB). Each router has input and/or output buffers, whose role is to temporarily accept flits before their transmission to either the local PE or to one of the neighboring routers. Figure 1b shows the interconnection signals of an isolated router. The connections between neighboring routers and the local PE are carried out with six signals (3 for each transmission and reception): a bidirectional data bus, a transfer request signal and an acknowledgment signal denoted with the infixes $\_d\_$, $\_req\_$ and $\_ack\_$ respectively. Message exchange between a router and its neighbors follows the alternate bit protocol illustrated in Fig. 1c. Sending a flit through a port x_d_y ($x \in \{W, S, E, N\}$, $y \in \{in, out\}$) is accompanied with the request signal on the same port x_req_y. Upon the reception of the acknowledge signal on the same port x_ack_y, the sending of the next flit can be proceeded. A change in a control signal ($\_req\_$ and $\_ack\_$ signals) is indicated by inverting its preceding value. If the request signal has the same binary level as the acknowledge signal or vice versa, the sending or receiving of a flit is successful and the router can proceed to the next one. Otherwise, it is blocked. It should be noted that the sending or receiving of a flit consumes 2 clock cycles.

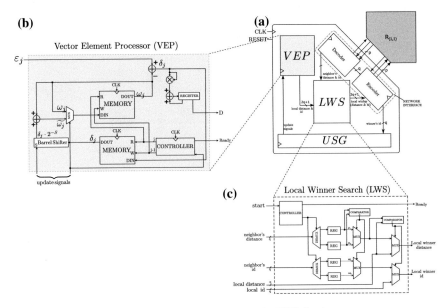

Fig. 2 **a** Architectures of the SOM-NoC's PE, **b** VEP and **c** LWS circuit

3.3 SOM-NoC

The PEs of a $L \times K$ NoC were adapted for SOM computation. The architecture of a SOM-NoC's PE is presented in Fig. 2a. It consists of 3 circuits: a Vector Element Processing (VEP) whose role is to calculate the distance and to update the weights of the corresponding neuron during the adaptation phase (see Fig. 2b); a Local Winner Search (LWS) circuit presented in Fig. 2c which carries out the comparison of the local distance and the received neighbor's distances; an Update Signal Generator (USG) is the circuit preparing update signals during the adaptation phase; and a Network Interface (NI) ensuring the sending to and receiving of data from neighboring PEs (neurons). Each PE has an identity (its address in the network) which determines the instructions its NI needs to execute during the winner search operation. The top left and the bottom right PEs have some additional functions: they initiate the winner search operation and winner id diffusion respectively.

Each PE behaves as a neuron: in the competition phase, it calculates the Euclidean distance between the input vector and its weights and send it through the NI to the nearest neighbors. The distance is propagated through the network in a systolic manner as presented in Fig. 3. Each computed distance crosses two neighboring routers before arriving at the PE's node. Upon reception of the neighboring PEs' distances, each PE compares them (with the LWS circuit) to the local one to locally determine the identity of the winner neuron. The NI is in charge to send the locally determined minimum distance and the corresponding neuron's id to the neighboring PEs. At the bottom right PE node of the network, the identity of the wining neuron is

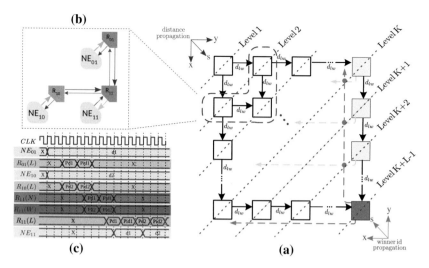

Fig. 3 Systolic architecture of SOM

known (see Fig. 3a). This identity is then broadcast to all nodes of the network while the winner neuron, as well as its neighbors, start the update of their weights. In this HW architecture, the used neighborhood function is a simple shift function carried out with a barrel shifter.

The execution time of the learning operation T_i is calculated using the competition and adaptation times T_c and T_a respectively:

$$T_i = T_c + T_a \qquad (6)$$

The distance calculation is carried out simultaneously in all PEs. Then, the propagation and comparison at the PE's level are also conducted in parallel as presented in Fig. 3c. From Fig. 3, it can be seen that the time needed to send (or propagate) a calculated distance from one PE to its neighbor PE is equal to the time to cross 2 routers in the network (2 hops). Therefore, the time needed for the competition phase is given by the expression:

$$T_c = T_{cd} + \{(T_p + T_{cmp}) \times (N_{stg} - 1)\} \qquad (7)$$

where T_{cd}, T_p, T_{cmp} and N_{stg} are distance calculation, propagation and comparison time and the level number respectively.

The calculation of the Euclidean distance between the input and the weight vector is done sequentially element-by-element using the VEP shown in Fig. 2a. Each elementary operation is performed in a single clock cycle. The intermediary results are stored in the accumulator, while the final results of the adaptation phase are stored in the local memory for further reuse. As it is presented in Fig. 4, each element of the input vector is processed in a single clock cycle and the final distance is ready after an additional clock cycle. The distance calculation time is given by:

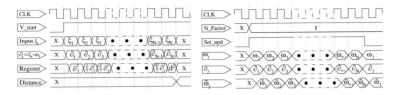

Fig. 4 Timings representing distance calculation (*left*) and update of weights (*right*)

$$T_{cd} = (N + 1) \times T_{CLK} \tag{8}$$

where N is the input vector size and T_{CLK} is the clock period.

The propagation time of a flit between two PEs in the presented NoC without collision (2 flits demanding the same output channel) is calculated by:

$$T_p = T_{pk} + \{T_r \times N_r\} + T_{dpk} \tag{9}$$

where T_{pk}, T_{dpk} are packing and unpacking time, T_r is the router latency and N_r is the number of routers to the final destination.

The adaptation phase involves two steps. Broadcasting of the identity of the winner neuron is ensured by neurons located at the end of each row of the network. The necessary broadcasting time is determined by the equation:

$$T_{bc} = (T_{pk} + T_{dpk} + 1) \times (K + L - 2) + 4T_r \tag{10}$$

Once the identity of the winner neuron is available, the USG circuit takes into account its position and generates the update signals. If a PE is concerned with the update signals, its VEP circuit starts the update phase in $T_{upd} = N \cdot T_{CLK}$ clock cycles. Figure 4 presents the timings of the update phase. The update of a weight vector element takes one clock cycle.

4   Results and Discussion

A 5×5 SOM-NoC using 32-element input vectors was used for performance evaluation. The parameters of the implemented architecture are presented in Table 1. The circuit was synthesized on a Xilinx Virtex-6 FPGA board. The obtained maximum working frequency is 200 MHz.

The performances of each step as well as the overall performances of the architecture are shown in Table 1.

Table 1 Parameters and performances of the proposed SOM-NoC architecture

Parameter	Value	Description
$L \times K$	5×5	SOM size
N	32	Vector size
N_{stg}	9	Level number
p	12	Data size
q	34	Flit size
R	5	Neighbour radius
T_r	3	Router latency
T_{pk}	2	Packing latency
T_{dpk}	2	Unpacking latency
Description	Step equation (clock cycles)	Time (ns)
1 Distance calculation	$T_{cd} = N + 1$	165
2 Global winner search	$T_{sw} = T_p \times (N_{stg} - 1) =$ $13 \times (N_{stg} - 1)$	520
3 Broadcasting of winner's id	$T_{bc} = (T_{dpk} + T_{pk} + 1) \times$ $(L + K - 2) + 4T_r$	260
4 Update weight	$T_{upd} = N$	160
5 Recall	$T_c = T_{cd} + T_{sw}$	685
6 Adaptation	$T_a = T_{bc} + T_{upd}$	420
7 Learning	$T_i = T_c + T_a$	1105
MCPS	$\frac{L \times K \times N}{T_c} \times 10^{-6}$	1168
MCUPS	$\frac{L \times K \times N}{T_i} \times 10^{-6}$	724

The distance calculation (step 1) time depends on the input vector dimension. For a 32-element input vector, it takes 33 clock cycles, which is equivalent to 165 ns for a 200 MHz working frequency. Similarly, in the weight update phase (step 4) all PEs operate in parallel, that means the number of clock cycles needed for this phase is proportional to the input vector dimension (32). On the other hand, we note that the winner neuron search (step 2) is the most time consuming step (almost 50 % of the overall learning time). This search is done sequentially in a systolic manner as explained in Sect. 3.3 and greatly depends on the network size ($L \times K$) and thus on the number of levels of propagation which is equal to $L + K - 1$. Moreover, the winner id broadcasting phase (step 3) depends on the number of lines of the network.

The performance of the presented architecture depends on the network size and the input vector dimension. Figure 5a shows the estimated performances in terms of MCPS and MCUPS as a function of the network size for a 32-element input vector. It can be seen that the performances increase non-linearly with the network size. As the network size increases and accordingly the number of available neurons, the communication time becomes preponderant to the computation one thus limiting the

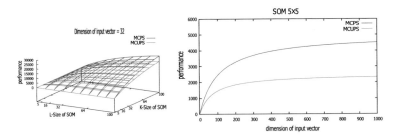

Fig. 5 **a** Performances of a SOM-NoC as a function of the network size (*left*) and **b** of a 5 × 5 SOM-NoC as a function of the input vector dimension (*right*)

increase in performances with regard to the number of neurons. On the other hand, for the same network size while increasing the input vector dimension, the computation becomes dominant over the communication and a rapid growth of performances as presented in Fig. 5b can be observed. Moreover, we also note that for the input vector dimension above 400 elements, the performances have a much slower rate of growth which can be explained with a very time consuming distance calculation and weight update phases which are both done sequentially on the input vector. If the calculating distance and update time exceed the time needed for the competition and broadcasting steps, there is no significant performance gain with the input vector dimension increase. Table 2 shows the comparison of the proposed HW architecture with the results reported in [7, 8, 10]. In order to make these results comparable, we presented the estimated performances (based on data from Table 1) of our architecture for a 16 × 16 SOM using 2048-element input vectors.

Table 2 Performance comparison

Work	Size	Input vector	Communication	Architecture	Frequency	MCUPS	Scalability
Lachmair et al. [8]	6050	194	Bus	Software core i7	NA	1628	Yes
Lachmair et al. [8]	6050	194	Bus	SIMD gNBXe processor	NA	20604	No
Hikawa and Maeda [7]	16 × 16	3	P to P	Parallel FPGA	33 MHz	25344	No
Manolakos and Logaras [10]	100	2048	P to P	Systolic array FPGA	148 MHz	3467	No
This work	16 × 16	2048	NoC	Sequential systolic FPGA	200 MHz	22555	Yes

5 Conclusion and Perspectives

We presented in this paper an FPGA implementation of a Network-on-Chip-based hardware Self-Organizing Map. Each neuron is associated to one NoC router allowing it to exchange data with other neurons of the network during both the learning and recall phases. The presented architecture is highly scalable and flexible and can be easily adapted to a large variety of applications demanding different working parameters. The implemented neurons support different input vector dimensions. The most time consuming phase is the winner search phase which is also done sequentially. The proposed architecture (SOM and NoC) is described in VHDL and its performances are evaluated for different network and input vector sizes. It has been showed that the presented architecture in its current state is most suitable for large input vector dimensions where communication time is neglected with regard to the computation one. The performance improvement of an order of magnitude in the recall phase can easily be obtained by exploring the architecture pipelining or by using faster NoC routers.

References

1. Kohonen, T.: Self-organizing map (2001)
2. Kolinummi, P., et al.: Parallel implementation of SOM on the partial tree shape neurocomputer. Neural Process. Lett. **12**(2), 171–182 (2000)
3. Talaska, T., et al.: Analog programmable distance calculation circuit for winner takes all neural network realized in the cmos technology (2015)
4. Abuelma'Ati, M., et al.: A reconfigurable gaussian/triangular basis functions computation circuit. Analog Integr. Circ. Sig. Process **47**(1), 53–64 (2006)
5. Kolasa, M., Długosz, R., Pedrycz, W., Szulc, M.: A programmable triangular neighborhood function for a kohonen self-organizing map implemented on chip. Neural Netw. **25**, 146–160 (2012)
6. Ramirez-Agundis, A., et al.: A HW design of a massive-parallel, modular NN-based VQ for RT video coding. Microprocess. Microsyst. **32**(1), 33–44 (2008)
7. Hikawa, H., Maeda, Y.: Improved learning performance of hardware self-organizing map using a novel neighborhood function (2015)
8. Lachmair, J., et al.: A reconfigurable neuroprocessor for self-organizing feature maps. Neurocomputing **112**, 189–199 (2013)
9. Aggarwal, C.C., et al.: On the surprising behavior of distance metrics in high dimensional space. Springer (2001)
10. Manolakos, I., Logaras, E.: High throughput systolic SOM IP core for FPGAs. In: IEEE International Conference on Acoustics, Speech and Signal Processing, 2007. ICASSP 2007, vol. 2, pp. 11–61. IEEE (2007)
11. Benini, L., De Micheli, G.: Networks on chips: a new soc paradigm. Computer **35**(1), 70–78 (2002)
12. Dally, W., Towles, B.: Route packets, not wires: on-chip interconnection networks. In: Design Automation Conference, 2001. Proceedings, pp. 684–689 (2001)

Local Models for Learning Inverse Kinematics of Redundant Robots: A Performance Comparison

Humberto I. Fontinele, Davyd B. Melo and Guilherme A. Barreto

Abstract In this paper we report the results of a comprehensive comparative analysis of the performances of six local models applied to the task of learning the inverse kinematics of a redundant robotic arm (Motoman HP6). The evaluated algorithm are the following ones: SOM-based Local Linear Mapping (LLM), Radial Basis Functions Network (RBFN), Local Model Network (LMN), Local Weighted Regression (LWR), Takagi-Sugeno-Kang Fuzzy Model (TSK) and Local Linear Mapping over K-winners (KSOM). Each algorithm is evaluated with respect to its accuracy in estimating the joint angles given the Cartesian coordinates along end-effector trajectories within the robot workspace. Also, a careful evaluation of the performances of the aforementioned algorithms is carried out based on correlation analysis of the residuals of the best model.

Keywords Local linear models · Inverse kinematics · Self-organizing maps · Locally weighted regression · Local model networks · Redundant Robots

1 Introduction

In Robotics, the forward kinematics function is a continuous mapping $\mathbf{f} : \mathcal{C} \subseteq \mathbf{\Theta}^n \rightarrow \mathcal{W} \subseteq \mathcal{X}^m$, which maps a set of n joint parameters from the configuration space, $\mathcal{C}$, to the m-dimensional task space, $\mathcal{W}$. If $m < n$, the robot has redundant degrees-of-freedom (dof's), i.e. it can achieve more postures than the strictly necessary to solve a

H.I. Fontinele (✉) · D.B. Melo (✉) · G.A. Barreto (✉)
Department of Teleinformatics Engineering, Center of Technology,
Federal University of Ceará (UFC), Campus of Pici, Fortaleza, Ceará, Brazil
e-mail: hicaropf@gmail.com

D.B. Melo
e-mail: davydmelo@gmail.com

G.A. Barreto
e-mail: gbarreto@ufc.br; guialenbar@gmail.com

© Springer International Publishing Switzerland 2016
E. Merényi et al. (eds.), *Advances in Self-Organizing Maps and Learning Vector Quantization*, Advances in Intelligent Systems and Computing 428,
DOI 10.1007/978-3-319-28518-4_15

given task. In general, control objectives such as the positioning and orienting of the end-effector are specified with respect to task space coordinates. However, a robot manipulator is typically controlled only in the configuration space (i.e. joint angles are sent to the robot controllers). Therefore, it is important to be able to find some $\theta \in \mathcal{C}$ such that $\mathbf{f}(\theta)$ is a particular target value $\mathbf{x} \in \mathcal{W}$. This is the inverse kinematics (IK) problem.

The IK of redundant robots is an *ill-posed* problem because there will be multiple configurations which result in the same task space location. Thus, computation of a direct inverse is problematic due to the many-to–one nature (and therefore non-invertibility) of the map $\mathbf{f}$. The IK problem can be solved in closed form only for certain manipulators, such as the Puma 560 and Motoman HP6 (see [2]). Solutions obtained by numerical methods can be an alternative to a closed form solution, usually either using the inverse of the Jacobian matrix of $\mathbf{f}$, or by using gradient-descent based methods. However, since these methods are iterative and require costly Jacobian or gradient computation at each step, they have limited use in real-time control scenarios.

Another alternative relies on machine learning algorithms, such as artificial neural networks (ANNs), which can be used to find an inverse mapping $\mathbf{f}^{-1}$ by implementing either direct inverse modeling (estimating $\mathbf{f}^{-1}$ explicitly) or differential kinematics methods. The main rationale behind the use of ANNs in robotics comes from the fact that such learning models can naturally deal with the non-linearity of the solution set. The multiplicity of solutions in inverse kinematics is usually dealt with by restriction to a single solution a priori. ANN models applied to IK learning have been mostly based on supervised learning architectures, specially the multilayer perceptron (MLP) network [5], but a great number of works has been based on the self-organizing map (SOM) [4] and extensions (see [1] and references therein).

Despite the existence of a number of successful applications of SOM-based techniques for robot modeling and control, a comprehensive performance comparison among them in learning the inverse kinematics of redundant robots is still missing. In order to fill this gap, in this paper we evaluate the performances of six local models on that complex robotic task, namely, the Local Linear Mapping (LLM) [11], the Local Linear Mapping over K-winners (KSOM) [9], the radial basis function network (RBFN) [7], the local model network (LMN) [6], the Takagi-Sugeno-Kang (TSK) fuzzy inference system [10] and the locally weighted regression (LWR) method [8]. Each local modeling algorithm will be evaluated with respect to its accuracy in estimating the joint angles given the Cartesian coordinates along end-effector trajectories within the robot workspace. A careful evaluation of the performances of the aforementioned algorithms is carried out based on correlation analysis of the residuals.

The remainder of the paper is organized as follows. The evaluated local models are described in Sect. 2. Then, the results of the computer experiments are presented and discussed in Sect. 3. Section 4 concludes the paper.

2 The Evaluated Models

In this section we briefly describe the six models to be evaluated in the task of learning the IK of a redundant robot.

2.1 The Local Linear Mapping (LLM)

The basic idea of the LLM [11] is to associate each neuron in the SOM with a linear mapping trained with a variant of the LMS rule. The SOM network is used to quantize the input space using a reduced number of prototype vectors, while the linear model associated with the winning neuron provides a local estimate of the output of the mapping being approximated.

For the IK learning task, vector quantization (VQ) of the input space is performed by the LLM model using training input samples $\{\mathbf{x}_k\}_{k=1}^{N}$ as in the usual SOM algorithm, with each neuron i owning a prototype vector $\mathbf{w}_i$. In addition, associated to each weight vector $\mathbf{w}_i$, there is a parameter matrix $\mathbf{M}_i$ used to generate the local estimate of the robot's joint angles as $\hat{\boldsymbol{\theta}}_i = \mathbf{M}_i \mathbf{x}_k$.

The adjustable parameters of the LLM model are the set of prototype vectors $\mathbf{w}_i$ and the corresponding parameter matrices $\mathbf{M}_i$, $i = 1, \ldots, Q$. For them, we need two learning rules. The rule for updating the prototype vectors $\mathbf{w}_i$ is exactly the one in the usual SOM network. The learning rule for the matrix $\mathbf{M}_i$ is a variant of the normalized LMS rule, that also takes into account the influence of the neighborhood function $h(i^*, i; t)$:

$$\mathbf{M}_i(k+1) = \mathbf{M}_i(k) + \alpha'(k)h(i^*, i; k)\Delta\mathbf{M}_i(k), \tag{1}$$

where $0 < \alpha'(k) < 1$ is the learning rate, and $\Delta\mathbf{M}_i(k) = [\boldsymbol{\theta}_k - \mathbf{M}_i(k)\mathbf{x}_k]\mathbf{x}_k^T / \|\mathbf{x}_k\|^2$, where $\boldsymbol{\theta}_k$ is the target output vector of the IK mapping being approximated. Training can demand cycling over the input-output data pairs for several epochs until convergence is observed. Once training is finished, the weight vectors $\mathbf{w}_i$ and the associated parameter matrices $\mathbf{M}_i$, $i = 1, 2, \ldots, Q$, remain unchanged for new input vectors.

2.2 Local Linear Mapping Over K-Prototypes (KSOM)

The KSOM [9] is a kind of lazy learning algorithm, whose main idea involves training firstly the SOM with a few prototypes (usually less than 100 units) in order to have a compact representation of the input-output mapping encoded in the weight vectors and build a local model only when required (i.e. whenever an input vector is presented). For that purpose, the weight vector of the i-th unit in the SOM has an increased dimension, i.e. $\mathbf{w}_i = [\mathbf{w}_i^{in} \ \mathbf{w}_i^{out}]^T \in \mathbb{R}^{m+n}$, where $\mathbf{w}_i^{in} \in \mathbb{R}^m$ is responsible for the vector quantization of the input samples $\mathbf{x}_k$, while $\mathbf{w}_i^{out} \in \mathbb{R}^n$ does the same for the output samples $\boldsymbol{\theta}_k$.

Once the SOM is trained, for each training input sample $\mathbf{x}_k$ we need to find the indices of the K first winning neurons, denoted by $\{i_1^*, i_2^*, \ldots, i_K^*\}$:

$$i_1^*(k) = \arg\min_{\forall i} \{\|\mathbf{x}_k - \mathbf{w}_i(k)\|\} \tag{2}$$

$$i_2^*(k) = \arg\min_{\forall i \neq i_1^*} \{\|\mathbf{x}_k - \mathbf{w}_i(k)\|\}$$

$$\vdots \qquad \vdots \qquad \vdots$$

$$i_K^*(k) = \arg\min_{\forall i \neq \{i_1^*, \ldots, i_{K-1}^*\}} \{\|\mathbf{x}_k - \mathbf{w}_i(k)\|\}$$

and use their weight vectors to build a local model for the current input. For this purpose, let the set of K winning weight vectors at iteration k to be denoted by $\{\mathbf{w}_{i_1^*}(k), \mathbf{w}_{i_2^*}(k), \ldots, \mathbf{w}_{i_K^*}(k)\}$.

Thus, we expect that the compact representation of the target input-output mapping $\boldsymbol{\theta}_k = \mathbf{f}^{-1}(\mathbf{x}_k)$ to be approximated locally over the K winning neurons by the following linear map: $\hat{\boldsymbol{\theta}}_k = \mathbf{M}(k)\mathbf{x}_k$, where $\mathbf{M}(k)$ is a matrix computed at iteration k. The idea behind the KSOM is that the matrix $\mathbf{M}(k)$ be constructed by means of the prototype vectors of the K winning neurons as

$$\mathbf{M}(k) = \mathbf{W}_{out}(k)\mathbf{W}_{in}^T(k)\left(\mathbf{W}_{in}(k)\mathbf{W}_{in}^T(k)\right)^{-1}, \tag{3}$$

where the matrices $\mathbf{W}_{in}(k) \in \mathbb{R}^3 \times \mathbb{R}^K$ and $\mathbf{W}_{out}(k) \in \mathbb{R}^m \times \mathbb{R}^K$ are defined as

$$\mathbf{W}_{in}(k) = [\mathbf{w}_{i_1^*}^{in}(k) \ \ \mathbf{w}_{i_2^*}^{in}(k) \ \ \cdots \ \ \mathbf{w}_{i_K^*}^{in}(k)], \tag{4}$$

$$\mathbf{W}_{out}(k) = [\mathbf{w}_{i_1^*}^{out}(k) \ \ \mathbf{w}_{i_2^*}^{out}(k) \ \ \cdots \ \ \mathbf{w}_{i_K^*}^{out}(k)]. \tag{5}$$

An important difference between LLM and KSOM is that while the LLM model has to store Q local matrices $\mathbf{M}_i$, $i = 1, \ldots, Q$ in memory, for posterior use, the KSOM, in its turn, builds a single local matrix $\mathbf{M}(k)$ every time an input vector is presented, without the need to store it in memory.

2.3 Radial Basis Functions Network (RBFN) and Local Model Network (LMN)

The RBFN is a classical feedforward one-hidden-layered neural network architecture, widely used for classification and regression [7]. In RBFNs, an Euclidean distance metric, $d_i(\mathbf{x}_k) = \|\mathbf{x}_k - \mathbf{c}_i\|$, and a Gaussian basis function, $z_i = \exp\{-d_i^2(\mathbf{x})/2\gamma^2\}$, are common choices. The parameter $\gamma > 0$ is the radius or width of the basis function and is assumed to be equal for all hidden units. In order to normalize the outputs of the basis functions to sum up to 1, we define

$$z_i(k) = \frac{\varphi(d_i(\mathbf{x}_k))}{\sum_{l=1}^{Q} \varphi(d_l(\mathbf{x}_k))} = \frac{\exp\{-d_i^2(\mathbf{x}_k)\}}{\sum_{l=1}^{Q} \exp\{-d_l^2(\mathbf{x}_k)\}} \tag{6}$$

Training of RBFNs usually requires 3 steps. The first step involves the positioning of the centers $\mathbf{c}_i$ in the input space, that can be done by a vector quantization algorithm, such as the SOM. In this case, we just set $\mathbf{c}_i = \mathbf{w}_i$, for $i = 1, 2, \ldots, Q$. The second step corresponds to the specification of the widths of the basis functions. In this paper, we use $\gamma = d_{max}(\mathbf{c}_i, \mathbf{c}_l)/\sqrt{2Q}$, $\forall i \neq l$, where $d_{max}(\mathbf{c}_i, \mathbf{c}_l) = \max_{\forall i \neq l}\{\|\mathbf{c}_i - \mathbf{c}_l\|\}$. The third step requires the computation of the weight matrix $\mathbf{M}$ that connects the hidden units to the output neurons. For this purpose, we build the matrices $\mathbf{Z}$ and $\boldsymbol{\Theta}$, such that

$$\mathbf{Z} = [\mathbf{z}_1 \mid \mathbf{z}_2 \mid \cdots \mid \mathbf{z}_N]_{(Q+1) \times N}, \quad \text{where} \quad \mathbf{z}_k = [1 \ z_1(k) \ z_2(k) \ \cdots \ z_Q(k)]^T, \tag{7}$$
$$\boldsymbol{\Theta} = [\boldsymbol{\theta}_1 \mid \boldsymbol{\theta}_2 \mid \cdots \mid \boldsymbol{\theta}_N]_{n \times N}, \tag{8}$$

which are used to compute the output weight matrix as $\mathbf{M} = \boldsymbol{\Theta}\mathbf{Z}^T \left(\mathbf{Z}\mathbf{Z}^T\right)^{-1}$. The output of the RBFN for the input vector $\mathbf{x}_k$ is then generated by a *linear combination* of Q nonlinear basis functions, i.e. $\hat{\boldsymbol{\theta}}_k = \sum_{i=1}^{Q} \mathbf{M}(:, i) z_i(k) = \mathbf{M}\mathbf{z}_k$, where $\mathbf{M}(:, i)$ is the i-th column of matrix $\mathbf{M}$ and $z_i(k)$ is computed as in Eq. (6).

The standard RBFN can be generalized to use not just a global weight matrix $\mathbf{M}$ operating at the current vector of outputs of the hidden neurons $\mathbf{z}_k$, but also a set of Q local matrices $\mathbf{M}_i$, $i = 1, \ldots, Q$, which operate on the current input vector $\mathbf{x}_k$. In this case, the response of the network should be rewritten in the form: $\hat{\boldsymbol{\theta}}_k = \sum_{i=1}^{Q} [\mathbf{M}_i \mathbf{x}_k] z_i(k) = \sum_{i=1}^{Q} \hat{\boldsymbol{\theta}}_i(k) z_i(k)$, where $\hat{\boldsymbol{\theta}}_i(k)$ is the estimate of the output vector by the local model associated with the i-th basis function.

This generalization of the RBF network is referred to as the Local Model Network (LMN) approach [6]. This approach can be viewed as implementing a decomposition of the complex, nonlinear system into a set of locally accurate submodels which are then smoothly integrated by associated basis functions. This means that a smaller number of local models can cover larger ares of the input space, when compared with the plain RBF network.

Estimation of the local matrices is carried out as $\mathbf{M}_i = \boldsymbol{\Theta}_i \mathbf{X}_i^T \left(\mathbf{X}_i \mathbf{X}_i^T\right)^{-1}$, where $\mathbf{X}_i = [\mathbf{x}_1^{(i)} \mid \mathbf{x}_2^{(i)} \mid \cdots \mid \mathbf{x}_{n_i}^{(i)}]$ and $\boldsymbol{\Theta}_i = [\boldsymbol{\theta}_1^{(i)} \mid \boldsymbol{\theta}_2^{(i)} \mid \cdots \mid \boldsymbol{\theta}_{n_i}^{(i)}]$. In words, the matrix $\mathbf{X}_i \in \mathbb{R}^m \times \mathbb{R}^{n_i}$ stores in its columns the n_i input vectors that are *relevant* for the i-th basis function, while the matrix $\boldsymbol{\Theta}_i \in \mathbb{R}^n \times \mathbb{R}^{n_i}$ in its turn stores the corresponding n_i target vectors. Note that $N = \sum_{i=1}^{Q} n_i$. The training input-output pair $(\mathbf{x}_k, \boldsymbol{\theta}_k)$, $k = 1, \ldots, N$, is relevant to the i-th basis function if $z_i(k) > z_j(k)$, $\forall j \neq i$.

2.4 Takagi-Sugeno-Kang (TSK) Model

The TSK fuzzy model is an important approach for local modeling of dynamical systems [10]. Thus, we consider a fuzzy model consisting of Q rules $R_i, i = 1, \ldots, Q$, that are of the following format:

$$R_i : \text{If } (x(k) \text{ is } A_x^i) \text{ and } (y(k) \text{ is } A_y^i) \text{ and } (z(k) \text{ is } A_z^i), \text{ Then } \hat{\boldsymbol{\theta}}_i(k) = \mathbf{M}_i \mathbf{x}_k, \quad (9)$$

characterized by "and" logical connectives operating on the premises of the antecedent of the rule and a local linear map in the consequent. The input variables $(x(k), y(k), z(k))$ comprise the current vector $\mathbf{x}_k$ of Cartesian coordinates of the robot end-effector, while $A_x^{(i)}$, $A_y^{(i)}$ and $A_z^{(i)}$ linguistic representations (e.g. small, large, etc.) of the fuzzy membership sets on the universe of discourse of x, y and z, respectively. The output of the TSK model is determined by a defuzzification process, implemented by means of the center-of-gravity formulation: $\hat{\boldsymbol{\theta}}_k = \sum_{i=1}^{Q} \mu_i \hat{\boldsymbol{\theta}}_i(k) / \sum_{i=1}^{Q} \mu_i$, where μ_i is the certainty of the premise of the i-th rule (e.g., the minimum of the certainties of each of the premise terms). Note that, regardless of inputs, we assume that $\sum_{i=1}^{Q} \mu_i \neq 0$.

Fuzzy membership functions for all variables are assumed to be Gaussian functions:

$$\mu_r(x) = \exp\left\{ -\frac{(x - c_{x,r})^2}{2\gamma_x^2} \right\}, \quad r = 1, \ldots, R, \quad (10)$$

where $c_{x,r}$ (γ_x) is the center (spread) of the fuzzy membership functions of the input variable x. Similar equations are used for the other two input variables (y and z). The centers of the membership functions of all variables can be estimated from data by applying a VQ algorithm, such as the K-means algorithm with R prototypes, over the training data. The components of the weight vector $\mathbf{w}_r, r = 1, \ldots, R$, being defined as $\mathbf{w}_r = [c_{x,r} \; c_{y,r} \; c_{z,r}]^T$.

A common value of the spread is shared by all the R membership functions of a given input variable, being computed as $\gamma_x = 2\lambda(x_{max} - x_{min})/(R - 1)$, where x_{max} and x_{min} are, respectively the maximum and minimum values over all the data for input variable x. The parameter λ is inserted as a degree of freedom to vary the spacing of the membership functions. Similar equations are used for the other two input variables (y and z). Finally, we can estimate the local parameter matrices $\mathbf{M}_i, i = 1, \ldots, Q$, one for each rule, using a procedure similar to that one used for LMN models. The training input-output pair $(\mathbf{x}_k, \boldsymbol{\theta}_k), k = 1, \ldots, N$, is relevant to the fuzzy Rule i if $\mu_i(k) > \mu_j(k), \forall j \neq i$.

After experimentation with the validation data, we finally set the number of fuzzy sets for each variable as $R = 7$. Since we have $m = 3$ input variables, we end up with a total of $R^m = 7^3 = 343$ rules. We also set $\lambda = 1$.

2.5   Local Weighted Regression (LWR)

For building the LWR model [8], let us consider that the set of N training samples $\{\mathbf{x}_k, \boldsymbol{\theta}_k\}_{k=1}^N \in \mathbb{R}^m \times \mathbb{R}^n$ is already stored in memory. Then, for every new Cartesian coordinate vector (a.k.a. *query point* in this context) $\mathbf{x}_q$, we need first to compute the $N \times N$ diagonal matrix $\mathbf{W} = [w_{kk}]$, where

$$w_{kk} = \exp\left\{ -\frac{1}{2}(\mathbf{x}_k - \mathbf{x}_q)^T \mathbf{D}(\mathbf{x}_k - \mathbf{x}_q) \right\}, \tag{11}$$

where the matrix $\mathbf{D}$ is usually assumed to be a global diagonal matrix $\mathbf{D} = \mathrm{diag}(1/\sigma_1^2, 1/\sigma_2^2, \ldots, 1/\sigma_m^2)$, with σ_i^2 denoting the variance of the i-th input variable. The next step requires building the matrices $\widetilde{\mathbf{X}}$ and $\boldsymbol{\Theta}$ as

$$\widetilde{\mathbf{X}} = [\widetilde{\mathbf{x}}_1 \mid \widetilde{\mathbf{x}}_2 \mid \cdots \mid \widetilde{\mathbf{x}}_N]_{(m+1)\times N}, \quad \text{where} \quad \widetilde{\mathbf{x}}_k = [1 \ (\mathbf{x}_k - \mathbf{x}_q)^T]^T, \tag{12}$$

$$\boldsymbol{\Theta} = [\boldsymbol{\theta}_1 \mid \boldsymbol{\theta}_2 \mid \cdots \mid \boldsymbol{\theta}_N]_{n\times N}, \tag{13}$$

which are used to compute the $n \times (m + 1)$ matrix of parameters $\mathbf{M}$ for the local linear model as $\mathbf{M} = \boldsymbol{\Theta}\widetilde{\mathbf{X}}^T(\widetilde{\mathbf{X}}\widetilde{\mathbf{X}}^T)^{-1}$. Finally, the predicted joint angles for the current query point $\mathbf{x}_q$ is given by $\hat{\boldsymbol{\theta}}_q = \mathbf{M}(:, 1)$, where $\mathbf{M}(:, 1)$ denotes the first column of the parameter matrix $\mathbf{M}$.

It is worth mentioning that the LWR is a type of lazy learning model just like the KSOM. In this regard, local model is built using the stored data only when a new input vector is available. However, the LWR stores all N training samples in memory, while the KSOM stores only the set of Q prototype vectors.

3   Experimental Results and Discussion

We report a number of experiments comparing the performances of the local models previously described in the task of IK learning for redundant robotic arms. Experiments were carried out for two robots, the PUMA560 and the Motoman HP6, but only the results for the latter one is reported. Firstly, we report the performances of all the models in terms of the mean squared error (MSE) for the joint angles over the testing set: $E_\theta = \frac{1}{N_T} \sum_{k=1}^{N_T} e_\theta^2(k) = \frac{1}{N_T} \sum_{k=1}^{N_T} \|\boldsymbol{\theta}_k - \hat{\boldsymbol{\theta}}_k\|^2$, where N_T is the number of testing samples. Secondly, an evaluation of the performances of the evaluated algorithms based on correlation analysis of the residuals is also carried out.

Dataset generation: For generating the input-output pairs $\{(\mathbf{x}_k, \boldsymbol{\theta}_k)\}_{k=1}^N$ for learning the Motoman HP6 robot , we used the *Robotics Toolbox* for Matlab [2]. Initially, we loaded the robot model (mdl_MotomanHP6) and generated a number of points (x, y, z) in Cartesian space inside a parallelepiped (with rectangular faces) defined within the robot's workspace. Then, for each coordinate vector $\mathbf{x}_k = (x(k), y(k), z(k))$, we used the IK command function ikine6s to get the

Table 1 Performances of the evaluated models in terms of MSE values

Model	Mean MSE	Standard dev.	Minimum MSE
LLM	2.2965e-04	2.9751e-05	1.8570e-04
KSOM	2.7024e-04	3.7857e-05	2.1869e-04
RBFN	**1.4724e-07**	**3.5469e-08**	**8.3337e-08**
LMN	**2.6949e-07**	**4.8677e-08**	**1.6837e-07**
LWR	5.4919e-05	4.0379e-06	4.7437e-05
TSK	6.3000e-03	8.3961e-04	4.9000e-03
ELM	**3.2265e-07**	**9.4905e-08**	**5.5536e-08**

corresponding values of the six joint angles $\boldsymbol{\theta}_k = [\theta_1(k) \; \theta_2(k) \; \cdots \; \theta_6(k)]^T$. If NaN (not-a-number) symbol was returned, then that Cartesian positioning was considered unreachable. At the end, we generated a total of $N = 1018$ input-output pairs. Two Cartesian trajectories, of circular and sinusoidal shapes, were generated inside the parallelepiped with the goal of evaluating the trained models in their capacity of estimating the joint angles for unseen Cartesian points. This set was further divided into training (60 %), validation (20 %) and testing (60 %) subsets. A total of 50 independent runs were executed. For each run the training/validation/testing samples were randomly selected.

Approximation Accuracy—For this experiment, the following specification for the best hyperparameters were selected for the evaluated models after extensive experimentation with the training/validation data sets. (i) **LLM**—The SOM is trained using the som_make function of the SOM toolbox[1] using $Q = 49$ neurons in a sheet-like 7×7 array, with hexagonal neighborhood topology, 100 initial training epochs and 1500 additional epochs for fine tuning. All other parameters of the som_make function use default values. (ii) **KSOM**—Same specification of the LLM model, but with $Q = 1024$ neurons in a sheet-like 32×32 array. (iii) **RBFN**—$Q = 97$ hidden Gaussian basis functions, whose centers are found via the K-means algorithm. Single Gaussian width parameter for all basis functions computed as in Sect. 2.3. (iv) **LMN**—Same specification of the RBFN model with $Q = 91$ hidden basis functions. (v) **LWR** - No parameter tuning is required. (vi) **TSK**—Number of membership functions $Q = 17$ and 5 membership functions per input variable. The algorithm K-means is used to find the centers of the Gaussian membership functions. The width parameters of all membership functions are set to 9.

The MSE results are shown in Table 1, where its is reported the MSE values averaged over 50 runs for each evaluated model, together with the standard deviation and the minimum values of the MSE. For the sake of completeness, we report also the performance of a feedforward global neural network model, the extreme learning

---

[1] Available for download from www.cis.hut.fi/somtoolbox/.

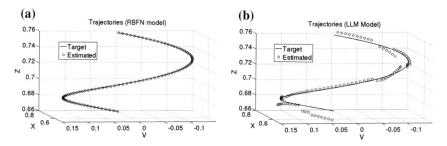

Fig. 1 Trajectories recovered by **a** the RBFN model and **b** the LLM model. *Solid line* Target trajectory. *Open circles* estimated points

machine (ELM) [3], on the same task. This network was trained with $Q = 110$ hidden neurons, all of them using the hyperbolic tangent activation function. In this table, the best performances are highlighted in boldface. Thus, one can easily note that the RBFN and the LMN models performed much better than the other local models, with performances equivalent to that of the ELM network.

A qualitative way of evaluating the accuracy of the models involves the control of the position of the robot's end-effector along a predefined trajectory. This task is implemented in the present context as follows. A certain number of Cartesian coordinates along a specific trajectory are defined in robot's workspace. Then, the corresponding joint angles (postures) are estimated by a given local/global model and used as inputs to the known forward kinematics of the Motoman HP6 robot.[2] By doing this, we assume that the control system is working perfectly. The results for the RBFN model (best local model, as inferred from Table 1) is shown in Fig. 1a. For the sake of comparison, we also show in Fig. 1b the resulting trajectory for the LLM model, where a common problem of this model (i.e. abrupt transitions between local models) is easily visualized. This problem does not occur with the other local models.

Residual Analysis—A common strategy to validate a regression model is via residual analysis. The rationale for this kind of analysis is that, for a well fitted model, the approximation errors should resemble white noise. The histograms of the residual errors produced by the RBFN model for the three first joint variables of the Motoman HP6 robot and the corresponding autocorrelation functions (ACFs) are shown in Fig. 2. As can be seen, the sequence of residuals for all three joint angles satisfy the whiteness test, i.e., they are all uncorrelated and follow approximately a bell-shaped distribution, confirming the goodness-of-fit of the RBFN model. A similar behavior was also observed for the sequence of residuals produced by the LMN model (second best one).

---

[2] Available in the *Robotics Toolbox* for Matlab [2].

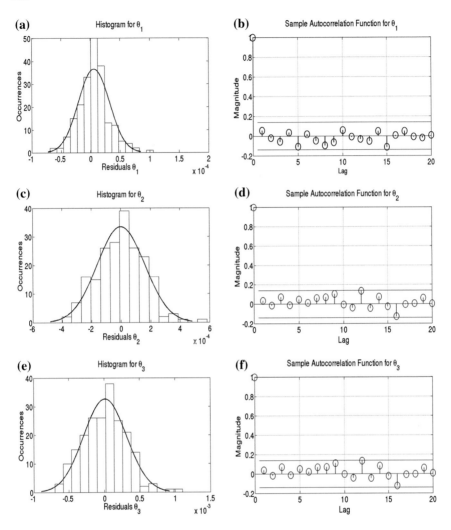

Fig. 2 Residual analysis of the RBFN model for the first three joints. **a, c, e**: Histograms of the residuals. **b, d, f**: Autocorrelation functions of the residuals

4   Conclusion

In this paper, a comprehensive performance evaluation of six local modeling methods applied to the inverse kinematics learning problem of redundant robots. All the models were evaluated with respect to the accuracy in estimating the joint angles given the Cartesian coordinates along end-effector trajectories within the robot workspace. The RBFN model achieved the best performance in this task, followed closely by the LMN model. Their performances were equivalent to that of the ELM network (a global model). Finally, an analysis of the residuals produced by the RBFN model is presented and discussed, confirming the quality of the fitted model.

Acknowledgments Authors thank CNPq (grant 309841/2012-7) and NUTEC for their financial support.

References

1. Barreto, G.A., Araújo, A.F.R., Ritter, H.J.: Self-organizing feature maps for modeling and control of robotic manipulators. J. Intell. Robot. Syst. **36**(4), 407–450 (2003)
2. Corke, P.: Robotics, Vision and Control: Fundamental Algorithms in MATLAB, vol. 73. Springer Science & Business Media (2011)
3. Huang, G.B., Zhu, Q.Y., Siew, C.K.: Extreme learning machine: theory and applications. Neurocomputing **70**(1), 489–501 (2006)
4. Kohonen, T.: Essentials of the self-organizing map. Neural Netw. **37**, 52–65 (2013)
5. Köker, R., Çakar, T., Sari, Y.: A neural-network committee machine approach to the inverse kinematics problem solution of robotic manipulators. Eng. Comput. **30**(4), 641–649 (2014)
6. Murray-Smith, R., Johansen, T.A.: Local learning in local model networks. In: Proceedings of the ICANN'95, pp. 40–46 (1995)
7. Poggio, T., Girosi, F.: Networks for approximation and learning. Proc. IEEE **78**(9), 1481–1497 (1990)
8. Schaal, S., Atkeson, C.G., Vijayakumar, S.: Scalable techniques from nonparametric statistics for real time robot learning. Appl. Intell. **17**(1), 49–60 (2002)
9. Souza, L.G.M., Barreto, G.A.: On building local models for inverse system identification with vector quantization algorithms. Neurocomputing **73**(10–12), 1993–2005 (2010)
10. Takagi, T., Sugeno, M.: Fuzzy identification of systems and its application to modeling and control. IEEE Trans. Syst. Man Cybern. **15**(1), 116–132 (1985)
11. Walter, J., Ritter, H., Schulten, K.: Non-linear prediction with self-organizing map. In: Proceedings of the IJCNN'90, pp. 587–592 (1990)

Part IV
Self-Organizing Maps in Neuroscience and Medical Applications

Using SOMs to Gain Insight into Human Language Processing

Risto Miikkulainen

Abstract While SOMs are commonly used as a tool for data visualization and data analysis, they can also serve as a model for cognitive functions in humans. Such functions include semantic and episodic memory, vision, and language. In this talk I will review how elements of sentence meaning can be laid out on a map, resulting in human-like graded semantic understanding instead of a single parse tree. I will also describe a model of the lexicon that can be fit to the individual patient with aphasia, and used to predict optical rehabilitation treatments.

Keywords Self-organizing maps · Cognitive models · Semantic understanding · Lexicon model

R. Miikkulainen (✉)
University of Texas at Austin, Austin, USA
e-mail: risto@cs.utexas.edu

© Springer International Publishing Switzerland 2016
E. Merényi et al. (eds.), *Advances in Self-Organizing Maps and Learning
Vector Quantization*, Advances in Intelligent Systems and Computing 428,
DOI 10.1007/978-3-319-28518-4_16

Prototype-Based Spatio-Temporal Probabilistic Modelling of fMRI Data

Nahed Alowadi, Yuan Shen and Peter Tiňo

Abstract Functional Magnetic Resonance Imaging (fMRI) is a powerful tool to study human brain activity in a non-invasive manner. It aims to detect brain activation areas in response to specific stimuli. A variety of spatio-temporal fMRI models have been developed to detect the link between cognitive tasks and haemodynamic responses. In conjunction with a parametric approach to HRF modeling, a prototype-based spatio-temporal fMRI model has been developed by (Shen et al. [1]) so as to significantly reduce model complexity, while sufficiently representing dominant spatio-temporal features of fMRI data. However, such a model poses a challenging non-linear parameter estimation task. We propose a modification of the parametric HRF model used in [1] so as to de-couple the HRF's response magnitude parameter from its shape parameters. This is justified through both algorithmical and conceptual arguments. We show in extensive numerical experiments that, compared with the original model, the model based on normalized HRF has much more favorable parameter estimation properties, although a careful data driven initialization can improve parameter estimation, especially in the original model employing un-normalized HRF.

Keywords Prototype based models · FMRI modeling · Normalized HRF

N. Alowadi · Y. Shen · P. Tiňo (✉)
School of Computer Science, The University of Birmingham,
Birmingham B15 2TT, UK
e-mail: pxt@cs.bham.ac.uk; P.Tino@cs.bham.ac.uk

N. Alowadi
e-mail: naa232@cs.bham.ac.uk

Y. Shen
e-mail: y.shen.2@cs.bham.ac.uk

© Springer International Publishing Switzerland 2016
E. Merényi et al. (eds.), *Advances in Self-Organizing Maps and Learning
Vector Quantization*, Advances in Intelligent Systems and Computing 428,
DOI 10.1007/978-3-319-28518-4_17

1 Introduction

With the growing interests in studying the working of the brain, several techniques have been developed to enable researchers to study brain activities, for example Positron Emission Tomography (PET), Electro Encephalo Graphy (EEG), or Functional Magnetic Resonance Imaging (fMRI). Each of these techniques has its own importance and application area. Due to its high spatial resolution, fMRI is particularly popular.

Various statistical methods have been suggested for analyzing fMRI data. In a broad sense, they can be categorized as model-driven (e.g. GLM) or data-driven (e.g. ICA) methods. Earlier approaches adopted a mass-univariate approach to fMRI data analysis and thus ignored the spatio-temporal nature of the brain activations. Over the past decade, a spatial-temporal approach to fMRI data analysis has been widely adopted by introducing spatial regularization into the existing mass-univariate approaches [2–4]. In [1] we proposed a prototype based model, SMM-HPM, constructed based on prototypical Hidden Process Models (HPM) [5]. In [5], a HPM is assigned to each of individual voxels in a region of interest (ROI). In [1], however, a small number of distinct HPMs (spatially organized temporal prototypes) are used to model fMRI time series across all voxels through a mixture of these HPM. The inter-voxel variability is accounted for by the voxel-varying mixture weights. For each HPM, the spatial distribution of its weight is modelled by a parametric function formulated as a mixture of multivariate Gaussians [1]. Such formulations are commonly referred to as a Spatial Mixture Model (SMM). Importantly, the number of model parameters in SMM-HPM does not grow with the number of voxels.

The aim of this work is to improve the learning of SMM-HPM and to pave the way for an extension to group SMM-HPM. In [1], Gamma function is adopted to model the shape of a Haemodynamic Response Function (HRF). However, variation of the two-dimensional parameter vector of Gamma function not only leads to the variation of its shape (i.e. both time-to-peak and peak width) but also causes the variation of peak height. This fact implies that the neural response level (NRL) in [1] depends on both the magnitude and shape parameters. On the other hand, HRF shape is subject dependent and task independent, while the opposite is true for NRL. Therefore, it conceptually makes sense to de-couple NRL from the shape parameter. In addition, such a de-coupling can improve SMM-HPM's parameter identifiability and thus make the learning of SMM-HPM more robust.

2 The Model

In this section, we first briefly review the SMM-HPM model [1] and then propose its modification based on normalized HRF.

2.1 SMM-HPM Overview

Let matrix $\mathbf{Y} \in \mathcal{R}^{V \times T}$ denote the fMRI data of V voxels and T volumes and vector $\mathbf{y}(v) \in \mathcal{R}^T$ is fMRI time series of T volume at voxel v, and scalar $y(v, t)$ a fMRI measurement at voxel v at volume t. The SMM-HPM model is a mixture of K spatio-temporal prototypes (indexed by $k = 1, 2, \ldots, K$) that account for K distinct neural activation patterns in an ROI. The model also contains a "null prototype" that accounts for the residuals in fMRI data ($k = 0$). Further, conditional on the model, fMRI measurements at different voxels and volumes are independent of each other. Thus, the key ingredient of SMM-HRF model is the likelihood of $y(v, t)$ written as

$$p\big(y(v, t); \Theta\big) = \sum_{k=0}^{K} p(k|v; \Theta^S) \cdot p\big(y(v, t)|k; \Theta_k^T\big) \tag{1}$$

where $p(k|v; \Theta^S)$ denotes the prior probability for the k-th prototype generating fMRI time series at voxel v and $p\big(y(v, t)|k; \Theta_k^T\big)$ the likelihood probability for $y(v, t)$ being generated by the k-th prototype. These probabilities are parameterized by spatial parameter Θ^S and the k-th HPM's temporal parameter Θ_k^T, respectively. Moreover, Θ denotes the set of all model parameters including spatial parameter Θ^S and temporal parameter $\Theta^T = \{\Theta_0^T, \ldots, \Theta_K^T\}$.

In the following we define K HPMs. Note that a HPM is independent of voxels. Also, all HPMs share a common model structure and differ only in their model parameters. Thus, to ease the presentation, the prototype index k is dropped in this paragraph. To start with, we first assume fMRI time series $y(t)$ to be a linear superposition of a signal component $x(t)$ and a noise component $\epsilon(t)$, i.e.

$$y(t) = x(t) + \epsilon(t). \tag{2}$$

The noise is further assumed to be i.i.d. white Gaussian, i.e. $\epsilon(t) \sim \mathcal{N}(0, \sigma^2)$. The key point of modelling $x(t)$ is that for each stimulus, its haemodynamic response is broken down to distinct constituents, which are the cognitive processes that evoked by that stimulus. Accordingly, $x(t)$ is given by:

$$x(t) = \sum_{s=1}^{S} \sum_{p=1}^{P} h_{p,s}(t), \tag{3}$$

where S is the number of stimuli, P is the number of cognitive processes and $h_{p,s}(t)$ is the haemodynamic response from the p-th cognitive process corresponding to the s-th stimulus:

$$h_{p,s}(t) = a_{p,s} \cdot \delta(t - (t_{p,s} + \tau_{p,s})) \bigotimes g_{p,s}(t), \tag{4}$$

where $a_{p,s}$ is the response magnitude, $t_{p,s}$ is the stimulus onset time, $\tau_{p,s}$ is the delay, and $g_{p,s}(t)$ is the shape function - in our case, gamma function

$$g_{p,s}(t) = g(t|\kappa_{p,s}, \theta_{p,s}) = \frac{t^{\kappa_{p,s}-1} \exp(-\frac{t}{\theta_{p,s}})}{(\theta_{p,s})^{\kappa_{p,s}} \Gamma(\kappa_{p,s})}, \tag{5}$$

where κ is the shape parameter and θ is the scale parameter.

The prior probability $p(k|v)$ varies within the ROI, but it is known that the haemo-dynamic response in the neighboring voxels is convergent. Thus, smoothness constraints are imposed to the spatial prior and so the prior probability is defined as:

$$p(k|v; \Theta^S) = \frac{p(v|k; \Theta_k^S)}{\sum_{k=0}^{K} p(v|k; \Theta_k^S)}, \tag{6}$$

where $p(v|k; \Theta_k^S)$ is the likelihood of the prototype k having voxel v in its region of influence. This likelihood is modelled by Gaussian distribution:

$$p(v|k) = \mathcal{N}(r_v|\mu_k, \Sigma_k), \tag{7}$$

where r is the location of the voxel v. The spatial parameter $\Theta_k^S = \{\mu_k, \Sigma_k\}$ includes the mean vector and the covariance matrix of the k-th Gaussian distribution, respectively. Ellipsoidal shape of the neighbourhood of the k-th prototype is characterized by the eigenvectors and eigenvalues of Σ_k.

2.2 HRF with Normalized Shape Function

Recall that the shape function of the p-th process evoked by stimulus s, i.e., $g_{p,s}(t)$ in Eq. 4 is a Gamma function. The peak of this shape function is located at $t_{max} = (\kappa_{p,s} - 1) \cdot \theta_{p,s}$. The height of this peak is given by $g_{p,s}(t_{max})$ which is not equal to 1. Thus, the neural response level (NRL) of the corresponding HRF is equal to $a_{p,s} \cdot g_{p,s}(t_{max})$. This gives rise to the dependence of NRL on HRF shape which could be a subject-specific HR property. On the other hand, NRL would vary across stimuli and is not related to any subject-specific HR property. Although this problem can be addressed e.g. by adopting a canonical HRF or by constraining HRFs, we propose a more direct and natural solution by employing a normalized HRF model with $g_{p,s}(t_{max}) = 1$.

The normalized HRF is thus obtained as $\tilde{g}_{p,s}(t) = \frac{g_{p,s}(t)}{g_{p,s}(t_{max})}$, which leads to

$$\tilde{g}_{p,s}(t) = \left(\frac{t}{t_{max}}\right)^{\kappa_{p,s}-1} \exp\left(-\frac{t - t_{max}}{\theta_{p,s}}\right), \quad \text{where} \quad t_{max} = (\kappa_{p,s} - 1)\theta_{p,s} \tag{8}$$

3 Learning Model Parameters

As in the original SMM-HPM (with unnormalized HRF), we learn the model parameters in Bayesian manner (MAP estimation) by maximizing the posterior

$$p(\Theta^S, \Theta^T | \mathbf{Y}) = p(\mathbf{Y} | \Theta^S, \Theta^T) \cdot p(\Theta^S, \Theta^T).$$

Part of the cost function to minimize is formed by the negative log model likelihood L

$$L = \sum_{v=1}^{V} \sum_{t=1}^{T} -\log \left(p\left(y(t, v) | \Theta^T, \Theta^S \right) \right). \tag{9}$$

For a particular subject and a particular ROI, it is reasonable to assume that the HRF shape is time constant across independent runs across a scan session or subsequent sessions across fMRI experiment [6]. When assuming that the shape and delay parameters are constant across all stimuli, there are two subsets of the parameters, temporal parameters

$$\Theta^T = \left\{ a_{p,s}^k, \theta_p^k, \kappa_p^k, \tau_p^k : p = 1, \ldots, P, \text{ and } s = 1, \ldots, S \right\}_{k=1}^{K}$$

and spatial parameters $\Theta^S = \{\mu_k, \Sigma_k\}_{k=1}^{K}$. These two subsets are optimized alliteratively using scaled conjugate-gradient optimization algorithms. As only HRF is modified in the new model, the gradients for optimizing the spatial parameters remain unchanged. For the HRF parameters, however, new formulae for the corresponding gradients need to be derived. Derivative of L, with respect to the temporal parameter $\Theta_k^T = \{\kappa_{k,p}, \theta_{k,p} \tau_{k,p}\}_{p=1}^{P}$ of the k-th prototype is given by

$$\nabla_{\Theta_k^T} L = \sum_{v=1}^{V} \sum_{t=1}^{T} (-) \cdot p(k | v; y_{vt}) \cdot \left(\frac{y_{vt} - x_k(t; \Theta_k^T)}{\sigma^2} \right) \cdot \left(\nabla_{\Theta_k^T} \mathbf{x_k}(t; \Theta_k^T) \right) \tag{10}$$

where $\nabla_{\{\kappa_{k,p}, \theta_{k,p} \tau_{k,p}\}} \mathbf{x_k}(t; \Theta_k^T)$ is equal to

$$\sum_{s=1}^{S} h_{p,s}^k \cdot \left\{ \frac{t}{\theta_{k,p}} - \kappa_{k,p} + 1, \frac{t}{(\kappa_{k,p} - 1)\theta_{k,p}}, \frac{1}{\theta_{k,p}} - \frac{\kappa_{k,p} - 1}{t - \tau_{k,p}} \right\} \tag{11}$$

4 Controlled Experiments

To examine the accuracy of parameters estimation when inferring the two spatio-temporal fMRI models (with un-Normalized and normalized HRF shape function) from fMRI data, a numerical experiment using synthetic data has been designed. The estimated model parameters will be compared to the ground truth as the estimation task is made progressively difficult by increasing the deviation (up to 30%) of the initial parameter setting from the ground truth.

Synthetic fMRI data have been generated for the purposes of controlled experiments. The size of these data are 1000 voxels organized on a 3-D lattice (i.e. $10 \times 10 \times 10$). Such voxel set size is comparable with the size of a large ROIs.[1] We assume that there are two sources of neural activation in the above voxel space. This means that there are two prototypes with distinct HPMs. The spatial prior of these two prototypes are computed using Eqs. 6 and 7 while $\{\mu_k\}_{k=1}^2$ and $\{\Sigma_k\}_{k=1}^2$ are predetermined as $\mu_1 = (3, 5, 5)^\mathsf{T}$, $\mu_2 = (7, 5, 5)^\mathsf{T}$, and $\Sigma_1 = \Sigma_2 = I_{3 \times 3}$ (identity matrix). The data was generated by a sequence of 50 stimuli with their Inter-Stimulus Interval (ISI) equal to 3.0 time units and regularly measured at a frequency of two volumes per time unit. This yields 300 fMRI volumes. Each stimulus evokes two artificial cognitive processes that are separated in time by 1.5 time units. The haemodynamic response of process p evoked by stimulus s is modeled as the product of the shape function g_p and the response magnitude $a_{p,s}$.

We assume that the shape function is time constant. Thus, its shape and delay parameters are the same for all stimuli. The variations in the haemodynamic response across prototypes, processes and stimuli come from the variations in the response magnitude. The response magnitude a as function of stimulus index s for process p in prototype k is modeled as a sine function

$$a_{p,s}^k = \sin\left(\frac{2\pi}{8} \cdot s \cdot \mathrm{ISI} + \delta_{k,p}\right),$$

where $s = 0, 1, \ldots, 49$, $\delta_{1,1} = 0$, $\delta_{1,2} = \frac{\pi}{2}$, $\delta_{2,1} = \frac{3\pi}{4}$, and $\delta_{2,2} = \frac{5\pi}{4}$.

All model parameters are initialized by assigning each parameter $\theta \in \{\Theta^S, \Theta^T\}$ a value θ^{init} that deviates from the corresponding ground truth θ^{tr} by a fixed relative difference η, that is, $\theta^{init} = \theta^{tr} \cdot (1 + \epsilon \cdot \eta)$, with $\epsilon \in \{-1, 1\}$.

where ϵ takes its value (i.e. $+1$ or -1) with equal probability. Note that as the number of model parameter (Θ^S, Θ^T) is very large, the above initialization method does yield a large pool of possible initializations for a given η. We experimented with various values of η, but in this study report results for $\eta = 30\%$.

4.1 Performance Measures

In order to measure the accuracy of parameter estimation, different summary statistics are used for each type of the model parameters. In the controlled experiments, the known ground truth values of the parameters were used as a reference for quantifying the estimation accuracy.

---

[1] The methodology developed here focuses on ROI-based analysis, rather than whole-brain analysis.

Spatial Prior: For prototype k, we denote its ground-truth spatial prior by $\mathcal{N}_g^k(\mu_g^k, \Sigma_g^k)$ and the estimated one by $\mathcal{N}_e^k(\mu_e^k, \Sigma_e^k)$. The accuracy of (μ_e^k, Σ_e^k), denoted by $A_{\Theta_k^S}$, is quantified by the symmetrized Kullback–Leibler divergence between these two multivariate Gaussian distributions:

$$
\begin{aligned}
A^{S_k} &= \frac{D(\mathcal{N}_g^k \parallel \mathcal{N}_e^k) + D(\mathcal{N}_e^k \parallel \mathcal{N}_g^k)}{2} \\
&= \mathrm{Tr}\left(\frac{\left(\Sigma_g^k\right)^{-1} \Sigma_e^k + \left(\Sigma_e^k\right)^{-1} \Sigma_g^k}{2} \right) \\
&\quad + (\mu_e^k - \mu_g^k)^{\mathsf{T}} \frac{\left(\Sigma_g^k\right)^{-1} + \left(\Sigma_g^k\right)^{-1}}{2} (\mu_e^k - \mu_g^k) - 3
\end{aligned}
\tag{12}
$$

The overall accuracy of the spatial prior estimation is then $A^S = \frac{1}{K}\sum_k A^{S_k}$.

HRF Shape Function: To measure the accuracy of the haemodynamic response shape parameters κ_p^k and θ_p^k for prototype k and process p, the difference between the ground truth HRF and the estimated HRF has been computed as their L_1 distance: $A_{\tilde{g}_{k,p}} = \frac{1}{n}\sum_{i=1}^n \left| \tilde{g}_{k,p}^g(i\,\Delta t) - \tilde{g}_{k,p}^e(i\,\Delta t) \right|$,, where n is the number of sample points ($n = 2000$) and $\Delta t = 0.01$. The overall accuracy of shape function estimation is given by $A_{\tilde{g}} = \frac{1}{P \cdot K}\sum_k \sum_p A_{\tilde{g}_{k,p}}$.

HRF Response Magnitudes: The accuracy of the haemodynamic response magnitude ($a_{p,s}^k$) estimation was measured by two summary statistics:

(i) integral L_1 difference between the ground truth and the estimated response magnitudes as follows

$$
A_{a_{k,p}} = \frac{1}{S}\sum_{s=1}^S \left| a_{k,p,s}^g - a_{k,p,s}^e \right|,
$$

where n is the number of stimuli. The overall accuracy of response magnitude estimation is given by $A_a = \frac{1}{P \cdot K}\sum_k \sum_p A_{a_{k,p}}$;

(ii) zero-lag cross correlation between the ground truth and the estimated time series of HRF response magnitudes, denoted by C_0^2. Note that due to the way the synthetic data is generated, the ground truth value of C_0^2 is -1.

4.2   Results

The performance of parameter estimation for the model with the normalized HRF is reported in Table 1. It shows that the results remain stable with initialisation using varying degree of deviation from the ground truth (up to 40 %).

Results of the experiments reporting the accuracy in recovering the parameters are shown in Table 2 for models involving both the un-normalized and normalized HRF when the initialisation based on 30 % deviation from the groud truth is used.

Table 1 Parameter estimation results from experiments with initialization using different degree of deviation from the ground truth: Spatial prior estimation accuracy A_S, HRF shape function estimation accuracy A_g, HRF magnitude estimation accuracy A_a and Zero-lag cross-correlation coefficient C_0^2

Degree of deviation (%)	Statistics	A_S	A_g	A_a	C_0^2
15	Mean	1.97E-4	0.012	0.041	−0.998
	StDev	0	0.003	0.007	0.003
20	Mean	2.04E-4	0.014	0.064	−0.981
	StDev	0	0.006	0.006	0.044
25	Mean	1.97E-4	0.011	0.049	−0.997
	StDev	0	0.003	0.003	0.001
30	Mean	1.98E-4	0.011	0.050	−0.996
	StDev	0	0.002	0.015	0.003
40	Mean	1.98E-4	0.015	0.079	−0.998
	StDev	0	0.006	0.021	0.003

Table 2 Parameter estimation results from experiments with initialization using 30 % deviation from the ground truth: Spatial prior estimation accuracy A_S, HRF shape function estimation accuracy A_g, HRF magnitude estimation accuracy A_a and Zero-lag cross-correlation coefficient C_0^2

HRF setting	Statistics	A_S	A_g	A_a	C_0^2
Un-norm HRF	Worst	28.06	0.042	2.566	−0.831
	Best	2E-4	0.004	0.435	−0.998
	Mean	5.02	0.017	1.226	−0.935
	StDev	8.82	0.014	0.792	0.058
Norm HRF	Worst	2E-4	0.016	0.083	−0.991
	Best	2E-4	0.008	0.023	−0.999
	Mean	2E-4	0.011	0.050	−0.996
	StDev	0	0.002	0.015	0.003

Compared to the un-normalized HRF shape function (u-HRF), the learning of SMM-HPM with the normalized one (n-HRF) is clearly much more robust. Both the bias and variance of the n-HRF based estimation are lower. Detailed analysis of the results (not reported in detail due to lack of space) revealed that in the case of u-HRF based model, 5 out of 10 random initializations lead to poor parameter estimates (Group1). Interestingly enough, the remaining 5 random initializations lead to comparable results with n-HRF (Group 2). Particularly, the estimated spatial priors of n-HRF were all very close to the ground truth whereas those of u-HRF in Group 1 significantly deviate from the ground truth. The same was observed for HRF-related estimates in Group 1. For Group 2, although the error in HRF magnitude estimates of u-HRF are still considerably large (when compared to those of n-HRF), the corresponding zero-lag cross-correlation estimates are very close to the ground truth. This indicates that the estimates of HRF response magnitudes are nearly a rescaling of the ground truth. This is a result of interaction between HRF shape and magnitude parameters. The effect is further amplified by the mis-matching between model and data when the data generated by the model with the normalised HRF were fitted to the model with the unnormalised HRF.

While this analysis revealed estimation stability of the two approaches around the ground truth parameter values, In practice, the ground truth model parameters are unknown. In particular, even a good guess of HRF response parameters is not available. Instead, an appropriate data-driven parameter initialization based on the observed fMRI signal is used for a full-blown model learning as described in Sect. 3. Details of such a parameter initialization will be presented in the next section.

5   Data-Driven Parameter Initialization

In the proposed data-driven parameter initialization, we initialize each prototype individually. First, the spatial prior of a prototype (i.e. its μ and Σ) are initialized by (functional) clustering fMRI time series [7] (**Step 1**). Then, this Gaussian distribution is used to select a subset of voxels that most represent that prototype (**Step 2**). Following this, fMRI data on these voxels are used to initialize the corresponding HPM model. In theory, we could estimate HRF shape parameter and magnitude parameter iteratively by minimizing L in the same way as for the full model. An initialization of HRF shape parameter is still needed and thus we randomly sample this parameter vector from its permissible range (**Step 3a**). Given this setting of HRF shape parameter, HRF magnitude parameter can be initialized by applying the conventional GLM method (**Step 3b**). Note that at Step 3, the model we fit to the data is not a mixture model but a HPM. It makes the initializing and learning much simpler (when compared to the full model), which in turn allows us to use a large number of random initialization for HRF shape parameter and find the best solution by choosing the one with the least L. Below we summarize technical details of each initialization step.

Step 1: To cluster fMRI time series, we employ the functional K means method. The functional (signal based) clustering distance D between voxel v_1 and v_2 is defined as

$$D(v_1, v_2) = d(v_1, v_2) - \lambda \cdot C_0^2\big(\mathbf{y}(v_1), \mathbf{y}(v_2)\big)$$

where $d(\cdot, \cdot)$ denotes the Euclidean distance in the voxel space, $C_0^2(\cdot, \cdot)$ denotes zero-lag cross-correlation, and λ is a tuning parameter. After performing the clustering, for each cluster, we fit a three-dimensional Gaussian distribution to the location of all voxels in this cluster and use its μ and Σ to initialize the spatial prior of the corresponding prototype;

Step 2: To determine the "most representative voxels for prototype k", we rank all voxels by $p(v|k)$ (see Eq. 7) and take the first n voxels in the rank with $\sum_{i=1}^{n} p(v_i|k) = 20\%$;

Step 3a: A HRF shape parameter (θ, κ) is permissible if the corresponding time-to-peak T^p and peak width W are both within their permissible ranges. Note that T^p is the mode of a Gamma function (i.e. $T^p = (\kappa - 1)\theta$) and W is the square root of its variance (i.e. $W = \sqrt{2\ln 2} \cdot \sqrt{\kappa}\theta$). Their permissible ranges are given by $[W_{min} = 3s, W_{max} = 6s]$ and $[T_{min}^p = 3s, T_{max}^p = 7s]$, respectively;

Step 3b: To perform a GLM based analysis, we define a regressor in the design matrix X for each pair of stimulus and process. Thus, the resulting X is a matrix of size $T \times P \cdot S$. Each column of X corresponds to a regressor and the one corresponding to stimulus s and process p is given by $h_{p,s}(j \cdot \text{TR})$ (see Eq. 4). The regression coefficient vector β contains all HRF magnitude parameters. A (least-squares) estimate of β is given by $\hat{\beta} = (X^T X)^{-1} X^T Y$ where Y denotes the fMRI data.

Results of the experiments involving data-driven parameter initialization (on the same set of synthetic fMRI data used in Sect. 4.2) are shown in Table 3. The data-driven initialization indeed massively improves u-HRF based estimation of spatial priors of the HPM prototypes. The corresponding estimation of the response amplitudes is also improved. Data-driven initialization with the normalized HRF yields a smooth time course of response magnitudes deviating from the ground truth to a larger degree than in the deviation based initialization. In contrast, response amplitude time courses obtained by the data-driven initialization with the unnormalised HRF are typically much less "smooth". However, subsequent full-blown learning typically smooths the response amplitude estimates. Interestingly, the estimation error A_a mainly arises from a phase shift between the estimated and ground truth time courses (preserving the main shape of the response amplitude time course).

Table 3 Parameter estimation results for experiments with data-driven parameter initialization: Spatial prior estimation accuracy A_S, HRF shape function estimation accuracy A_g, HRF magnitude estimation accuracy A_a and Zero-lag cross-correlation coefficient C_0^2

HRF setting	Statistics	A_S	A_g	A_a	C_0^2
Un-norm HRF	Mean	1.3E-4	0.016	0.583	−0.998
	StDev	0	0.011	0.001	0.000
Norm HRF	Mean	2.0E-4	0.009	0.757	−0.999
	StDev	0	0.002	0.020	0.002

6 Conclusion

We have empirically investigated the bias and variance of parameter estimation of a probabilistic prototype based model of fMRI data [1]. In particular, we suggested a modification of the original model with a normalized HRF. Such a modification not only constitutes a more natural model formulation, but also stabilizes the parameter estimation (when compared with the original un-normalized HRF case). Moreover, normalized HRF formulation is a necessary ingredient for building population based models. Investigation of possibilities for building such group-level SMM-HPM models is a matter for future research.

References

1. Shen, Y., Mayhew, S., Kourtzi, Z., Tino, P.: Spatialtemporal modelling of fMRI data through spatially regularized mixture of hidden process models. NeuroImage **84**, 657–671 (2013)
2. Woolrich, M.W., Jenkinson, M., Brady, J.M., Smith, S.: Full bayesian spatio-temporal modeling of FMRI data. IEEE Trans. Med. Imaging **23**, 213–231 (2004)
3. Penny, W., Flandin, G., Trujillo-Barreto, N.J.: Spatio-temporal models for fMRI. Stat. Parametr. Mapp. Model. Brain Imaging **12**, 313–322 (2006)
4. Ciuciu, P., Vincent, T., Risser, L., Donnet, S.: A joint detection-estimation framework for analysis within-subject fMRI data. Journal de la Societe Francaise de Statistique **151**, 58–89 (2010)
5. Hutchinson, R.A., Niculescu, R.S., Keller, T.A., Rustandi, I., Mitchell, T.M.: Modeling fMRI data generated by overlapping cognitive processes with unknown onsets using hidden process models. NeuroImage **46**, 87–104 (2009)
6. Boynton, G.M., Engel, S.A., Glover, G.H., Heeger, D.J.: Linear system analysis of functional magnetic resonance imaging in human V1. NeuroImage **8**, 360369 (2006)
7. Heller, R., Stanley, D., Yekutieli, D., Rubin, N., Benjamini, Y.: Cluster-based analysis of FMRI data. NeuroImage **33**, 599–608 (2006)

LVQ and SVM Classification of FDG-PET Brain Data

Deborah Mudali, Michael Biehl, Klaus L. Leenders
and Jos B. T. M. Roerdink

Abstract We apply Generalized Matrix Learning Vector Quantization (GMLVQ) and Support Vector Machine (SVM) classifiers to fluorodeoxyglucose positron emission tomography (FDG-PET) brain data in the hope to achieve better classification accuracies for parkinsonian syndromes as compared to the decision tree method which was used in previous studies. The classifiers are validated using the leave-one-out method. The obtained results show that GMLVQ performs better than the previously studied decision tree (DT) method in the binary classification of group comparisons. Additionally, GMLVQ achieves a superior performance over the DT method regarding multi-class classification. The performance of the considered SVM classifier is comparable with that of GMLVQ. However, in the binary classification, GMLVQ performs better in the separation of Parkinson's disease subjects from healthy controls. On the other hand, SVM achieves higher accuracy than the GMLVQ method in the binary classification of the other parkinsonian syndromes.

Keywords Learning Vector Quantization · Support Vector Machine · Parkinsonian syndromes · Classification

D. Mudali (✉) · M. Biehl (✉) · J.B.T.M. Roerdink (✉)
Johann Bernoulli Institute for Mathematics and Computer Science,
University of Groningen, Groningen, The Netherlands
e-mail: d.mudali@rug.nl
URL: http://www.cs.rug.nl/svcg

M. Biehl
e-mail: m.biehl@rug.nl

J.B.T.M. Roerdink
e-mail: j.b.t.m.roerdink@rug.nl

K.L. Leenders
Department of Neurology, University Medical Center Groningen,
University of Groningen, Groningen, The Netherlands
e-mail: k.l.leenders@umcg.nl

J.B.T.M. Roerdink
University of Groningen, University Medical Center Groningen, Neuroimaging
Center, Groningen, The Netherlands

© Springer International Publishing Switzerland 2016
E. Merényi et al. (eds.), *Advances in Self-Organizing Maps and Learning
Vector Quantization*, Advances in Intelligent Systems and Computing 428,
DOI 10.1007/978-3-319-28518-4_18

205

1 Introduction

Diagnosis of neurodegenerative diseases (NDs), especially at an early stage, is very important to affect proper treatment [1], but it is still a challenge [19]. Nevertheless, some studies report considerable success in differentiating between some of these diseases [23]. In fact, promising classification performances were obtained for the multiple system atrophy (MSA) and progressive supranuclear palsy (PSP) groups versus the healthy control group in the study [16] where the decision tree (DT) method was used. The same study showed that discriminating the Parkinson's disease (PD) group from healthy controls (HC) on the basis of PET brain scan imaging data remains a challenge. Therefore, in this paper other classification methods are applied in the hope to improve classification of parkinsonian syndromes, in particular PD, MSA, and PSP. The classification methods used in this study are Generalized Matrix Learning Vector Quantization (GMLVQ) and Support Vector Machine (SVM).

LVQ is a method which uses prototypes assigned to each class. A new case is classified as belonging to the class of the closest prototype [12]. In the training phase, a set of appropriately chosen prototypes is computed from a given set of labeled example data. This training process can be based on a suitable cost function, as for instance in the so-called Generalized LVQ (GLVQ) introduced in [17]. The conceptional extension to matrix-based relevance learning was introduced in [18]; simpler feature weighting schemes had been considered earlier in [10]. Relevance learning provides insight into the data in terms of weighting features and combinations of features in the adaptive distance measure. Moreover, GMLVQ allows for the implementation of multi-class classification in a straightforward way.

The Support Vector Machine is a supervised learning method for classifying data by maximizing the margin between the defined classes, see for instance [4, 7]. The aim of SVM training is to minimize the classification error while maximizing the gap or margin between the classes by computing an optimally separating hyperplane. The training data points that lie closest to the hyperplane define the so-called support vectors [6, 25]. This method was originally designed for binary classification but has been extended to multi-class classification, see for instance [11] and references therein. Moreover, several studies including [9, 13] have used SVM to classify neurodegenerative diseases with high accuracy. Other examples of SVM applications like biological data mining are described in [7].

2 Method

The data used in this study is described in [22]. The brain data were obtained from 18 healthy controls (HC), 20 Parkinson's Disease (PD), 17 progressive supranuclear palsy (PSP) and 21 multi system atrophy (MSA) cases. We apply the scaled subprofile model with principal component analysis (SSM/PCA), based on the methods by Spetsieris et al. [21], to the datasets to extract features. The method was implemented in Matlab R2014a. The SSM/PCA method [14, 15, 20] starts by double centering the data matrix and then extracts metabolic brain patterns in the form of principal component images, also known as *group invariant subprofiles*. The original images are projected onto the extracted patterns to determine their weights, which are called *subject scores*. The subject scores then form the features that are input to the classifiers to classify the subject brain images. Because of the application of the PCA method, the computed subject scores are dependent on the whole input dataset, an unusual circumstance in the standard situation. This makes the number of features extracted equal to the number of samples in the dataset.

A leave-one-out cross validation (LOOCV) of the classifiers is performed to predict their performance on new subject cases. For each run, a subject (test sample) is left out, then the SSM/PCA process is performed on the rest of the subjects (training set) to obtain their scores on the principal components. These subject scores are then used to train the GMLVQ and the SVM classifiers. The test subject is projected onto the invariant profiles to obtain its scores on the extracted profiles. Then the test subject scores are used to evaluate the trained classifier. The sensitivity (true positive rate), specificity (true negative rate) and classifier accuracy are determined. Note that the test subject is removed *before* the SSM/PCA process in order to deal with dependencies of the extracted features on both the training and test sets. In addition, the test set receiver operating characteristic (ROC) curve and Nearest Prototype Classifier (NPC) confusion matrix are computed for all the left-out subjects. The area under curve (AUC) of the ROC curve is a measure of the ability of the features (i.e., subject scores on the principal components) to separate the groups.

For both the SVM and GMLVQ classifiers, we do binary and multi-class classification. The binary classification involves comparing the distinct disease groups (PD, PSP, and MSA) with the healthy control group. The multi-class classification concerns the comparison of all the groups, i.e., HC versus PD versus PSP versus MSA (a total of 76 subjects), as well as only the disease groups, i.e., PD versus PSP versus MSA (a total of 58 subjects). The goal is to determine the class membership (healthy or diseased) of a new subject of unknown diagnosis and also determine the type of parkinsonian syndrome.

For SVM training and testing, we use the Matlab R2014a functions "fitcsvm" and "predict", respectively, with default parameters and a linear kernel, representing a large margin linear separation in the original feature space. Also, all features are centered at their mean in the dataset and scaled to have unit standard deviation. The "fitcsvm" returns an SVM classifier which can be used for classification of new data samples. It also provides class likelihoods which can be thresholded for an ROC analysis. For the SVM multi-class classification we use the LIBSVM library [5] with the one-against-one method, since the previously mentioned Matlab functions support only binary classification. The one-against-one method has a shorter training time than the one-against-all, as reported in [11].

As for GMLVQ, we employ it in its simplest setting with one prototype w_k per class. A global quadratic distance measure of the form $d(w_k, x) = (x - w_k)^T \Lambda (x - w_k)$ is used to quantify the dissimilarity of an input vector x and the prototypes. The measure is parameterized in terms of the positive semi-definite relevance matrix Λ [18]. Both, prototypes and relevance matrix are optimized in the training process which is guided by a suitable cost function [18]. We employed the gmlvq-toolbox [2], which performs a batch gradient descent minimization with automated step size control, see [2] for details. All the results presented here were obtained using the default parameter settings of [2]. After 100 gradient steps, the training errors and cost function appeared to have converged in all considered classification problems.

3 Results

3.1 Generalized Matrix Relevance LVQ (GMLVQ)

As mentioned earlier, in order to validate the classifiers the training process is repeated with one test subject removed from the training set before applying the SSM/PCA process. This section presents the LOOCV results for the distinct disease groups versus the healthy control group in the binary and multi-class classification. Important to note is that all the features (100 %) as extracted from the brain image data using the SSM/PCA method are provided to the GMLVQ classifier. In the tables, sensitivity (%) is the percentage of correctly classified patients, specificity (%) the percentage of correctly classified healthy controls, and AUC is the area under the ROC curve. In addition, the corresponding results are visualized in terms of projections on the leading two eigenvectors of the relevance matrix. This exploits the fact that GMLVQ displays a tendency to yield low-rank matrices which correspond to an intrinsically low-dimensional representation of the feature space [3, 18]. Additionally, we include the corresponding plots showing diagonal and off-diagonal matrix elements for one LOOCV iteration as an example illustration.

3.1.1 Binary Classification

The objective here is to separate the individual disease groups from the healthy control group. The GMLVQ results are shown in Table 1.

The results in Table 1 are much better than those of the decision tree as reported in [16]. In fact a tremendous improvement can be seen in the PD vs HC group, whose LOOCV performance has increased from 63.2 % (decision trees) to 81.6 %

Table 1 GMLVQ Classifier performance in LOOCV for the different data sets (patients versuss healthy controls, number of cases in brackets)

Feature set (size)	Perf. (%)	Sensitivity (%)	Specificity (%)	AUC
PD-HC (38)	81.6	75	88.9	0.84
MSA-HC (39)	92.3	90.5	94.4	0.99
PSP-HC (35)	88.6	82.4	94.4	0.97

The column Perf.(%) indicates the percentage of subject cases correctly classified per group. Perf. as well as Sensitivity and Specificity correspond to the Nearest Prototype Classifier (NPC)

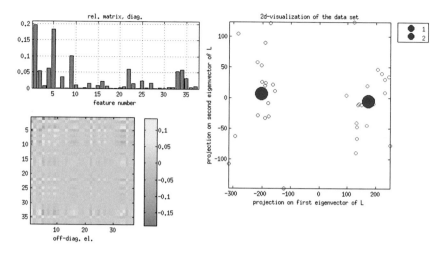

Fig. 1 Illustrations of the results of a single GMLVQ training process in the LOOCV of the PD vs HC two class-problem, 1 = HC, 2 = disease group. Graphs show diagonal relevances (*upper left*), and off-diagonal relevance matrix elements (*lower left*). The visualization of the training data in terms of their projection on the two leading eigenvectors of the relevance matrix is displayed on the right

(GMLVQ). The use of the relevance matrix to weight features according to their relevance appears to boost performance. An illustration is shown in Fig. 1 where the training data points are displayed in a feature space of the two leading eigenvectors of the relevance matrix. Observe that the subject scores do not overlap after the GMLVQ classifier training phase, which corresponds to error-free classification of the training set. Further, the resulting AUC measures (for the different groups) are relatively high. This means that the GMLVQ weighted features are very suitable for separating the groups.

As observed in Fig. 1, the PD vs HC comparison shows a clear separation between the PD group and the healthy group. Apart from a few outliers, most of the data points cluster around the specific prototypes, i.e., the two bigger circles that each represent a class. Further, the relevance matrix histogram shows the features and their diagonal weights as used in the classification process. For example, in the PD vs HC group feature 1 was weighted the highest, implying that feature 1 carries relevant information required to separate the two groups. As a matter of fact, the highly weighted feature should be given more attention, i.e., critically analyze the principal component image corresponding to this feature to gain insights from the clinical perspective.

3.1.2 Multi-class Classification

Here we show the results for the LOOCV of the GMLVQ classifier on the multi-class datasets, i.e., the classification of all the four classes, and the three disease

Table 2 Four-class problem

GMLVQ classification	HC	PD	PSP	MSA
HC (18)	**14**	3	1	0
PD (20)	5	**13**	1	1
PSP (17)	2	2	**11**	2
MSA (21)	0	1	4	**16**
Class accuracy (%)	77.8	65	64.7	76.2
Overall performance (%)	71.1			

The table shows the number of subject images correctly classified for each class in bold and the overall performance in percentage as obtained in the LOOCV

Table 3 Three-class problem

GMLVQ classification	PD	PSP	MSA
PD (20))	**19**	0	1
PSP (17)	2	**12**	3
MSA (21)	2	3	**16**
Class accuracy (%)	95	70.6	76.2
Overall performance (%)	81.03		

The table shows the number of subject images correctly classified for each class in bold with the overall LOOCV performance in percentage

classes, respectively. The latter is considered separately, because the main task in clinical practice is to distinguish the three parkinsonian syndromes. Additionally, for the four-class comparison, we include the HC group because we want to build a classifier which can also distinguish a healthy subject from the parkinsonian groups. The results are shown in Tables 2 and 3 for four-class comparison and three disease groups, respectively. Also included are the scatter plots showing the distribution of training data points in the two-dimensional projection of the feature space in a single run of the training process.

Four-Class Comparison

From the results in Table 2, we notice that most of the misclassified HC subjects are classified as PD and *vice versa*. As already observed in [16], the PD and HC subjects have a closely related metabolic pattern. Likewise, the PSP and MSA groups display a similarity, in view of the fact that four (majority of the misclassification) MSA subjects are misclassified as PSP.

Three-Class Comparison

The classifier results show that the PD group is clearly separable from the other two disease groups. On the other hand, the PSP and MSA groups seem to overlap more

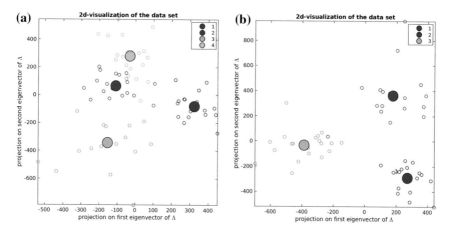

Fig. 2 The visualization of the training data with respect to their projections on the two leading eigenvectors of the relevance matrix as observed in a single run of GMLVQ training. **a** Four class problem; 1-HC, 2-PD, 3-PSP, 4-MSA. **b** Three class problem; 1-PD, 2-PSP, 3-MSA, 4-MSA

strongly. We observe that the majority of the misclassification for both the PSP and MSA belong to either classes, which shows that these two groups are quite similar. In fact, it is known that PSP and MSA are hard to distinguish because the patients with either disorders show similar reduction in striatal and brain stem volumes [8].

Visualization of the Data Points

The scatter plots show the training data points with respect to their projections on the two leading eigenvectors of the relevance matrix. It can be observed in Fig. 2a that the PSP and healthy groups are clearly separable from the rest of the groups. But a small overlap exists between the PD and MSA groups even in the training set. Meanwhile, the three-class comparison in Fig. 2b shows a clear separation among the disease groups. This is encouraging since we are generally interested in distinguishing between the parkinsonian syndromes.

3.2 Support Vector Machine (SVM)

Next we show the results of the leave-one-out cross validation of the SVM classifier for the different groups, both in a binary and multi-class comparison. Note that, as before, a subject is left out before the SSM/PCA process.

3.2.1 Binary Classification

Here, the classifier was used to separate each disease group from the healthy control group to determine its classification performance. As seen in Table 4, apart from the

Table 4 SVM classifier LOOCV performance for the different data sets (patients versus healthy controls, number of cases in brackets)

Feature set(size)	Perf. (%)	Sensitivity (%)	Specificity (%)	AUC
PD-HC (38)	76.3	75	77.8	0.84
MSA-HC (39)	94.9	90.5	100	0.97
PSP-HC (35)	91.4	88.2	94.4	0.92

The column Perf. (%) indicates the percentage of subject cases correctly classified per group, Sensitivity (%) the percentage of correctly classified patients, and Specificity (%) the percentage of correctly classified healthy controls

Table 5 Four-class problem

SVM classification	HC	PD	PSP	MSA
HC (18)	12	3	2	0
PD (20)	4	12	1	3
PSP (17)	1	2	9	5
MSA (21)	0	2	2	17
Class accuracy (%)	66.7	60	52.9	81.0
Overall performance (%)	65.8			

The confusion matrix and the overall performance of the SVM in the LOOCV scheme

PD vs HC comparison, the other groups' performances improve in comparison to GMLVQ (cf. Table 1). However, the AUC measures for MSA and PSP are lower than those of GMLVQ, indicating that it outperforms the SVM when choosing an appropriate class bias to modify the nearest prototype classification. In comparison to the linear SVM in [16], the results differ because different features have been used. Furthermore, here the LOOCV is done correctly by removing the test subject from the training set before applying the SSM/PCA method, whereas in [16] the SSM/PCA method was applied to all subjects to obtain the scores before the LOOCV was performed.

3.2.2 Multi-class Classification

We also applied SVM to the multi-class datasets to determine its performance on larger datasets.

Four-Class Comparison

This involved the comparison of all the four groups, i.e., HC, PD, PSP, and MSA. In Table 5, the SVM four-group classification accuracy is slightly above chance level and lower than that of GMLVQ (see Table 2). But the classifier can separate the MSA group from the rest of the groups with an accuracy of 81 %.

Table 6 Three-class problem

SVM classification	PD	PSP	MSA
PD (20))	**17**	1	2
PSP (17)	2	**10**	5
MSA (21)	3	2	**16**
Class accuracy (%)	85	58.8	76.2
Overall performance (%)	74.1		

The table shows the confusion matrix with the number of subject images correctly classified by the SVM for each class in bold and the overall LOOCV performance in percentage

Three Disease Groups

This involved the comparison of only the disease groups, i.e., PD, PSP and MSA without the healthy group. The separation of the disease groups using SVM yields a better performance accuracy than the separation of the four groups (including the healthy group). Also, as in the GMLVQ classification, the PD group appears to be well separated from PSP and MSA (Table 6).

4 Discussion and Conclusion

Both GMLVQ and SVM were studied and tested for the binary and multi-class problems. In the binary classification, GMLVQ performs better than SVM in the PD vs HC comparison (performance of 81.6 %), but both achieve the same sensitivity of 75 %. However, SVM performs better in the MSA vs HC and PSP vs HC comparisons. For the two-class problems we also considered the Area under Curve (AUC) of the ROC, as it does not depend on the choice of a particular working point (threshold, class bias) in the classifier. In terms of the AUC, GMLVQ was seen to outperform or equal the performance of the SVM classifier. Additionally, in the multi-class problems, GMLVQ achieves a better accuracy than SVM.

The GMLVQ relevance matrix, which makes use of an adaptive weighting of features according to their discriminative power, displayed overall superior classification performance. In particular, for the PD vs HC comparison which has been challenging to discriminate using decision trees, GMLVQ was able to separate PD from HC with an accuracy of 81.6 %, better than SVM by a margin of 5.3 %. Although SVM classification performance for the MSA vs HC and PSP vs HC comparisons is better than GMLVQ, the AUC measures show that GMLVQ achieves superior binary classification of the distinct groups. Overall, GMLVQ also achieves a better accuracy for the multi-class classification. In addition, when it comes to explaining the results to the physicians, GMLVQ is more intuitive than SVM. The analysis of the resulting relevance matrix allows for the identification of particularly relevant features and combinations of features. These results should trigger further investigations from the clinical perspective.

Clearly, the number of cases in the available data set is fairly small and our findings could be partly skewed by the small sample size. For instance, leave-one-out validation schemes are known to frequently yield unreliable estimates of performance. It is also possible that the performance of decision trees in [16], which was found inferior to GMLVQ and SVM, might improve significantly for larger data sets (see comparable work in [24]). We intend to extend our work in this direction as more data become available in the future. Moreover, variants of the considered classifiers could be considered, e.g., SVM with more powerful kernels or LVQ systems with several prototypes per class or local distance matrices [18].

References

1. Appel, L., Jonasson, M., Danfors, T., Nyholm, D., Askmark, H., Lubberink, M., Sörensen, J.: Use of 11C-PE2I PET in differential diagnosis of parkinsonian disorders. J. Nucl. Med. **56**(2), 234–242 (2015). http://jnm.snmjournals.org/content/56/2/234.abstract
2. Biehl, M.: A no-nonsense Matlab (TM) toolbox for GMLVQ (2015). http://www.cs.rug.nl/biehl/gmlvq.html
3. Bunte, K., Schneider, P., Hammer, B., Schleif, F.M., Villmann, T., Biehl, M.: Limited rank matrix learning, discriminative dimension reduction and visualization. Neural Netw. **26**, 159–173 (2012)
4. Burges, C.J.: A tutorial on support vector machines for pattern recognition. Data Min. Knowl. Disc. **2**(2), 121–167 (1998)
5. Chang, C.C., Lin, C.J.: LIBSVM: A library for support vector machines. ACM Trans. Intell. Syst. Technol. **2**, 27:1–27:27 (2011). Software available at http://www.csie.ntu.edu.tw/cjlin/libsvm
6. Cortes, C., Vapnik, V.: Support-vector networks. Mach. Learn. **20**(3), 273–297 (1995). http://dx.doi.org/10.1007/BF00994018
7. Cristianini, N., Shawe-Taylor, J.: An Introduction to Support Vector Machines and Other Kernel-based Learning Methods. Cambridge University Press (2000)
8. Eckert, T., Sailer, M., Kaufmann, J., Schrader, C., Peschel, T., Bodammer, N., Heinze, H.J., Schoenfeld, M.A.: Differentiation of idiopathic Parkinson's disease, multiple system atrophy, progressive supranuclear palsy, and healthy controls using magnetization transfer imaging. Neuroimage **21**(1), 229–235 (2004)
9. Haller, S., Badoud, S., Nguyen, D., Garibotto, V., Lovblad, K., Burkhard, P.: Individual detection of patients with Parkinson disease using support vector machine analysis of diffusion tensor imaging data: initial results. Am. J. Neuroradiol. **33**(11), 2123–2128 (2012)
10. Hammer, B., Villmann, T.: Generalized relevance learning vector quantization. Neural Netw. **15**(8–9), 1059–1068 (2002)
11. Hsu, C.W., Lin, C.J.: A comparison of methods for multiclass support vector machines. IEEE Trans. Neural Netw. **13**(2), 415–425 (2002)
12. Kohonen, T.: The self-organizing map. Neurocomputing **21**(1), 1–6 (1998)
13. Magnin, B., Mesrob, L., Kinkingnéhun, S., Pélégrini-Issac, M., Colliot, O., Sarazin, M., Dubois, B., Lehéricy, S., Benali, H.: Support vector machine-based classification of Alzheimers disease from whole-brain anatomical MRI. Neuroradiology **51**(2), 73–83 (2009)
14. Moeller, J.R., Strother, S.C.: A regional covariance approach to the analysis of functional patterns in positron emission tomographic data. J. Cereb. Blood Flow Metab. **11**(2), A121–135 (1991)
15. Moeller, J.R., Strother, S.C., Sidtis, J.J., Rottenberg, D.A.: Scaled subprofile model: a statistical approach to the analysis of functional patterns in positron emission tomographic data. J. Cereb. Blood Flow Metab. **7**(5), 58–649 (1987)

16. Mudali, D., Teune, L.K., Renken, R.J., Leenders, K.L., Roerdink, J.B.T.M.: Classification of Parkinsonian syndromes from FDG-PET brain data using decision trees with SSM/PCA features. Comput. Math. Methods Med. Artic. ID **136921**, 1–10 (2015)
17. Sato, A., Yamada, K.: Generalized learning vector quantization. Adv. Neural Inf. Process. Syst. 423–429 (1996)
18. Schneider, P., Biehl, M., Hammer, B.: Adaptive relevance matrices in learning vector quantization. Neural Comput. **21**(12), 3532–3561 (2009)
19. Silverman, D.H.: Brain 18F-FDG PET in the diagnosis of neurodegenerative dementias: comparison with perfusion SPECT and with clinical evaluations lacking nuclear imaging. J. Nucl. Med. **45**(4), 594–607 (2004)
20. Spetsieris, P.G., Eidelberg, D.: Scaled subprofile modeling of resting state imaging data in Parkinson's disease: methodological issues. NeuroImage **54**(4), 2899–2914 (2011). http://www.sciencedirect.com/science/article/pii/S1053811910013170
21. Spetsieris, P.G., Ma, Y., Dhawan, V., Eidelberg, D.: Differential diagnosis of parkinsonian syndromes using PCA-based functional imaging features. NeuroImage **45**(4), 1241–1252 (2009). http://www.sciencedirect.com/science/article/pii/S1053811908013335
22. Teune, L.K., Bartels, A.L., de Jong, B.M., Willemsen, A.T., Eshuis, S.A., de Vries, J.J., van Oostrom, J.C., Leenders, K.L.: Typical cerebral metabolic patterns in neurodegenerative brain diseases. Mov. Disord. **25**(14), 2395–2404 (2010)
23. Van Laere, K., Casteels, C., De Ceuninck, L., Vanbilloen, B., Maes, A., Mortelmans, L., Vandenberghe, W., Verbruggen, A., Dom, R.: Dual-tracer dopamine transporter and perfusion SPECT in differential diagnosis of parkinsonism using template-based discriminant analysis. J. Nucl. Med. **47**(3), 384–392 (2006)
24. Westenberg, M.A., Roerdink, J.B.T.M.: Mixed-method identifications. In: Du Buf, J.M.H., Bayer, M.M. (eds.) Automatic Diatom Identification, Series in Machine Perception and Artificial Intelligence, vol. 51, chap. 12, pp. 245–257. World Scientific Publishing Co., Singapore (2002)
25. Zhang, X.: Using class-center vectors to build support vector machines. In: Neural Networks for Signal Processing IX, 1999. Proceedings of the 1999 IEEE Signal Processing Society Workshop, pp. 3–11, Aug 1999

Mutual Connectivity Analysis (MCA) for Nonlinear Functional Connectivity Network Recovery in the Human Brain Using Convergent Cross-Mapping and Non-metric Clustering

Axel Wismüller, Anas Z. Abidin, Adora M. DSouza and Mahesh B. Nagarajan

Abstract We explore a computational framework for functional connectivity analysis in resting-state functional MRI (fMRI) data acquired from the human brain for recovering the underlying network structure and understanding causality between network components. Termed mutual connectivity analysis (MCA), this framework involves two steps, the first of which is to evaluate the pair-wise cross-prediction performance between fMRI pixel time series within the brain. Here, we use a Generalized Radial Basis Functions (GRBF) neural network as a nonlinear time series predictor. In a second step, the underlying network structure is subsequently recovered from the affinity matrix using non-metric network clustering approaches, such as the so-called Louvain method. Finally, we use convergent cross-mapping (CCM) to study causality between different network components. We demonstrate our MCA framework in the problem of recovering the motor cortex network associated with hand movement from resting state fMRI data. Results are compared with a ground truth of active motor cortex regions as identified by a task-based fMRI sequence involving a finger-tapping stimulation experiment. Our results on whole-slice fMRI analysis demonstrate that MCA-based model-free recovery of regions associated with the primary motor cortex and supplementary motor area are in close agreement with localization of similar regions achieved with a task-based fMRI acquisition.

A. Wismüller (✉) · A.Z. Abidin · M.B. Nagarajan
Department of Imaging Sciences, University of Rochester Medical Center, NY, USA
e-mail: axel_wismueller@urmc.rochester.edu; axel.wismueller@gmail.com

A. Wismüller · A.Z. Abidin
Department of Biomedical Engineering, University of Rochester, NY, USA

A. Wismüller · A.M. DSouza
Department of Electrical Engineering, University of Rochester, NY, USA

A. Wismüller
Department of Clinical Radiology, Ludwig Maximilian University, Munich, Germany

© Springer International Publishing Switzerland 2016
E. Merényi et al. (eds.), *Advances in Self-Organizing Maps and Learning Vector Quantization*, Advances in Intelligent Systems and Computing 428,
DOI 10.1007/978-3-319-28518-4_19

217

Keywords Resting-state fMRI · Functional connectivity · Mutual connectivity analysis · Convergent cross-mapping · Non-metric clustering · Louvain method

1 Introduction

There has been significant growth in research aimed at exploring structural and functional connectivity in the human brain [1]. Of particular interest is the analysis of functional connectivity at fine-grained spatial and temporal resolution scales, based on the acquisition capabilities provided by advanced in vivo neuro-imaging techniques, such as state-of-the-art fMRI. Here, several contemporary analytic techniques such as seed-based functional connectivity analysis [2], independent component analysis [3], Granger causality [4], etc., imply inherent simplifications, such as assuming linearity or implicit time series separability, which can obscure the characteristics of the complex system being investigated. Another drawback of such approaches is that they transform the original high-dimensional imaging data into simpler low-dimensional representations, which discards valuable information and thus limits the interpretability of brain connectivity analysis.

Our primary goal with this contribution is to introduce a computational framework for analyzing functional network connectivity between pixel time series in the human brain, while simultaneously avoiding some of the information loss induced by the previously mentioned techniques. To this end, we present a mutual connectivity analysis (MCA) approach for non-linear functional connectivity analysis in large time series ensembles obtained from resting state fMRI data. Our approach involves connectivity characterization through large scale non-linear mutual time series cross-prediction [5] followed by functional network identification by partitioning the resulting affinity (or dissimilarity matrix) through non-metric clustering approaches, such as the Louvain method [6]. Subsequently, causality analysis (which is used to study directional influence between time series) is performed using a convergent cross-mapping (CCM) framework [7] on identified network components.

We demonstrate the applicability of our MCA framework to identifying and visualizing the motor cortex through analysis of resting-state fMRI data. It has been previously shown that frequency fluctuations (<0.1 Hz) from regions of the motor cortex associated with hand movement are strongly correlated both within and across hemispheres [2]. We explore non-linear connectivity and causality between time series ensembles from different regions of the motor cortex associated with hand movement, as discussed in the following sections.

2 Data

Functional MRI images were acquired from a healthy male volunteer (age 25 years) with a 1.5T GE SIGNA^{TM} whole-body MRI scanner (GE, Milwaukee, WI, USA). Two image sequences were acquired; the first was under resting state conditions while

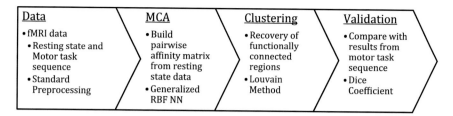

Fig. 1 Flow chart of the analysis steps followed in this study. Causality analysis is an extension to these steps

the second one involved a finger-tapping task stimulus to localize the left motor cortex (LMC), right motor cortex (RMC), and supplementary motor area (SMA) regions for establishing ground truth (example shown in Fig. 1). During the resting-state scan, the subject was instructed to stay still and keep eyes closed. The fMRI sequences were performed with the following parameters—echo time (TE)—40 ms, echo-repetition time (TR)—500 ms, and flip angle (FA)—90°. 512 fMRI scans were acquired from two slice locations that corresponded to the motor cortex; each image had a slice thickness of 10 mm and an in-plane pixel resolution of 3.75×3.75 mm. The first 24 time points of fMRI data were discarded to avoid any impact on the data analysis by initial saturation effects.

3 Methods

3.1 Pre-processing

Motion artifacts were compensated by automatic image alignment and signal drifts were corrected with linear de-trending. In addition, resting state fMRI time series were subject to low pass filtering with a cut-off frequency of 0.08 Hz for minimizing the influence of respiratory and cardio-vascular oscillations while preserving the frequency spectrum pertaining to functional connectivity [2].

Finally, the time series were further normalized to zero mean and unit standard deviation to focus on signal dynamics rather than amplitude [9]. Data from the two slices (~600 pixels and 488 time points each) were analyzed independently.

3.2 MCA—Pair-Wise Affinity Evaluation

Our first step is to build an affinity/similarity matrix $\mathbf{A}$, for all time series of brain pixels on a single fMRI slice, populated based on cross-prediction performance of each pair of time series. Given two pixel time series $\mathbf{X}$ and $\mathbf{Y}$, (where $\mathbf{X}, \mathbf{Y} \in \mathbf{T} = \{\mathbf{T}_f,$

$f = 1, \ldots, n\}$ and where $\mathbf{T}$ represents the set of all n pixel fMRI time series under consideration), we aim to describe the degree of their dynamic coupling as a measure of their cross-prediction performance, which is stored as matrix element $(\mathbf{A})_{\mathbf{X},\mathbf{Y}}$. This is accomplished in the following way: We break down time series $\mathbf{X}$ of length l into a set of vectors $\mathbf{x}_t, t \in \{1, 2, \ldots, l - d + 1\}$ of dimension d, which can be interpreted as a sliding window of length d moving along $\mathbf{X}$. The corresponding prediction target vectors for $\mathbf{x}_t$ are vectors $\mathbf{y}_t$ of dimension $e$extracted from $\mathbf{Y}$. In this study, the parameters d and e were chosen $d = 10$ and $e = 1$. Here, $\mathbf{x}_t$ is mapped to the corresponding future $\mathbf{y}_t$, e.g. the vector $\mathbf{x}_t$ that comprises of the first 10 time points of $\mathbf{X}$ is mapped to $\mathbf{y}_t$, which corresponds to the 11th time point of $\mathbf{Y}$.

Generalized Radial Basis Function (GRBF) Network: Here, the set of $\mathbf{x}_t$ and their corresponding $\mathbf{y}_t$ are randomly split into a training (Tr) (70 % of available $\mathbf{x}_t$ vectors) and test (Te) set. The training set is then used to create a non-linear mapping f, i.e., $\mathbf{y}_t^{\mathrm{Tr}} = f(\mathbf{x}_t^{\mathrm{Tr}})$. Once defined, this mapping is subsequently used to compute target vector estimates $\hat{\mathbf{y}}_t^{\mathrm{Te}}$ for $\mathbf{x}_t^{\mathrm{Te}}$, i.e.,

$$\hat{\mathbf{y}}_t^{\mathrm{Te}} = f(\mathbf{x}_t^{\mathrm{Te}}), \tag{1}$$

from which $\hat{\mathbf{y}}$ can be constructed.

For defining the approximating function f, we use a GRBF neural network with three layers, i.e. the input, hidden and output layers. The activation pattern of the input layer with d neurons is represented by d-dimensional vector $\mathbf{x}_t$. In the training phase, this activity $(\mathbf{x}_t^{\mathrm{Tr}})$ is propagated to m neurons of the hidden layer through directed connections with prototypical weight vectors $\mathbf{w}_j \in \mathbb{R}^d, j = 1, 2, \ldots, m$. These weight vectors are representations of the training set $\mathbf{x}_t^{\mathrm{Tr}}$ and are computed using an unsupervised clustering approach; fuzzy C-means is used in this study [8], although a Self-Organizing Map (SOM) might be used as well. Given previous work on the comparison of choosing different clustering methods for the performance of GRBF networks, e.g. [22], we would not expect significantly different results when using SOMs in this step. The activity a_j of neurons in the hidden layer is defined as

$$a_j(\mathbf{x}_t^{\mathrm{Tr}}) = \frac{e^{-(\mathbf{x}_t^{\mathrm{Tr}} - \mathbf{w}_j)^2/2\rho^2}}{\sum_{i=1}^m e^{-(\mathbf{x}_t^{\mathrm{Tr}} - \mathbf{w}_i)^2/2\rho^2}}, \tag{2}$$

i.e., generalized (given the normalization in the denominator of Eq. (2)) radial basis functions [15, 21], where the ρ parameter controls the width of the radial basis function kernel and defines the neighborhood of vectors that contributes to the computation of f [10]. In the final step of the training phase, the activity of the output layer $\hat{\mathbf{y}}_t^{\mathrm{Tr}}$ is computed as a weighted sum of the hidden layer activations a_j, i.e.,

$$\hat{\mathbf{y}}_t^{\mathrm{Tr}} = \sum_{j=1}^m a_j(\mathbf{x}_t^{\mathrm{Tr}} \cdot \mathbf{s}_j), \tag{3}$$

where $\mathbf{s}_j$ are the output weights obtained through minimization of the cost function $E = \|\hat{\mathbf{y}}_t^{\mathrm{Tr}} - \mathbf{y}_t^{\mathrm{Tr}}\|^2$. After the training phase is completed, f is subsequently used to process the test set $\hat{\mathbf{y}}$ and is constructed from target vector estimates $\hat{\mathbf{y}}_t^{\text{Te}}$. Further details concerning our GRBF neural network approach can be found in [15]. We haven chosen the value of ρ as 0.5 (which is not updated further during the training phase) and used 20 hidden layer neurons based on initial experiments.

Once $\hat{\mathbf{y}}$ is constructed using the approach described above, its similarity to $\mathbf{y}$ is measured using cross correlation. In this manner, the pair-wise affinity/similarity matrix $\mathbf{A}$ is computed for all the time series under investigation.

3.3 MCA—Non-metric Clustering

From the affinity matrix $\mathbf{A}$, we use the Louvain method [6] to recover the underlying network structure through non-metric clustering. In network science such a matrix is referred to as a network graph, with the rows and columns representing different *nodes* (various pixels in this case) and the entries representing *edges/links* or degree of coupling (non-linear predictability in our study). The Louvain method aims to find high modularity clusters in such network graphs, where modularity is defined as the ratio of the density of intra-community node linkage to the density of inter-community node linkage [11]. Modularity Q is mathematically represented as

$$Q = \frac{1}{2m} \sum_{i,j} \left[A_{ij} - \frac{k_i k_j}{2m} \right] \delta(C_i, C_j) \tag{4}$$

where $\mathbf{A}_{ij}$ represents the affinity between nodes i and j, $k_i = \sum_j A_{ij}$ is the sum of affinities of nodes attached to i, C_i is the community to which node i is assigned, $\delta(u, v) = 1$ if $u = v$, and 0 otherwise, and

$$m = \frac{1}{2} \sum_{ij} A_{ij}. \tag{5}$$

Thus, a complex network is decomposed into clusters with strong intra-community links and weak inter-community links. The algorithm involves an iterative process during which different nodes of the network are merged into larger communities, if the modularity is improved as a consequence. The process is discontinued when no further improvement in modularity can be achieved. Further details pertaining to this clustering approach can be found in [6].

In order to avoid the creation of large super-communities that encompass smaller and more interesting clusters, we also pursue an approach frequently applied in spectral clustering to make the affinity matrix sparser [12]. Specifically, we only consider the k most similar nodes for any given node i. Additionally, only mutual k most-similar nodes are considered, i.e., bi-directional links in the k most-similar nodes. In this study, $k = 100$ is chosen empirically from preliminary analyses; this corresponds to approximately 20 % of the nodes in the network. Similarity between clustering results and the ground truth was evaluated using the Dice coefficient [13].

3.4 MCA—Causality Analysis

We further extend MCA to include convergent cross-mapping (CCM) [7] for investigating causality between different regions of the primary motor cortex network. CCM explores the phenomenon of causation (cause-effect relationship of two time series) in non-linear systems, where the ability of time series X to better predict (or "cross-map") Y with increasing time series length L is investigated. The length L can be modified by using various percentages of the time series as training set with a GRBF neural network. Thus, according to [7], observing the degree to which X and Y are cross-mapped over increasing L enables one to establish grounds for causation.

In this study, we restrict the examination of causation to specific regions of the motor cortex, as identified using the ground truth. Thus, MCA is used to build a smaller affinity matrix involving pixels time series only from the LMC, RMC and SMA. From the collection of vectors $\mathbf{x}_i$ from time series $\mathbf{X}$, a randomly chosen subset (of 10–80%) can have different variations. So we compute an affinity matrix for 20 different variations of subsets of $\mathbf{x}_i$, and use their average for CCM analysis.

For interpretation of results achieved with CCM causality analysis, we present a pair-wise regional visualization of presumed causal influences between the LMC, RMC and SMA. Thus, when comparing any two regions in the motor cortex network, each pixel of a specific region is assigned an influence score based on its cross-prediction from MCA with respect to all pixels in the other region under consideration.

All procedures were implemented using MATLAB 8.1 (MathWorks Inc., Natick, MA, 2013). The Louvain method implementation was taken from [14].

4 Results

4.1 Network Recovery

Figure 3 shows the results of recovering communities associated with the motor cortex from a single resting state fMRI slice through non-metric clustering of the MCA affinity matrix using the Louvain method [6]. As seen here, MCA with non-metric clustering is able to recover the community structure of bilateral primary motor cortices and the supplementary motor areas. The overlap between the recovered regions of the primary motor cortex and ground truth, measured using the Dice coefficient, was 0.51, indicating a fair, but not perfect agreement between the two. Network recovery using MCA can be seen to slightly over-estimate the regions of the motor cortex. Here, it should be noted that the physiological processes pertaining to the resting state may be different from the brain activity related to a targeted finger-tapping motor task. Therefore, the task-related activation pattern in Fig. 2 may serve

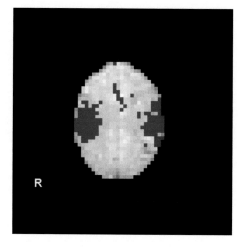

Fig. 2 Ground truth example for the subject. The identified primary motor cortex (left and right motor cortex) and pixels corresponding to the supplementary motor area (shown in *red*) are superimposed on the original slice

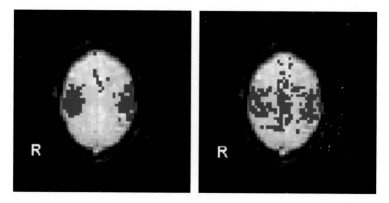

Fig. 3 (*Left*) Ground truth for primary motor cortex regions (LMC, RMC and SMA). (*Right*) Motor cortex regions recovered from our MCA framework. Note the similarity of the identified brain networks revealing bilateral primary motor cortices and supplementary motor areas. Dice coefficient between the ground truth and our MCA network analysis results is 0.51

as a 'localization aid' for the motor cortex rather than the 'ground truth' for its exact delineation. However, in the absence of any better reference criterion for defining the motor cortex, the above result can serve as an indicator for the usability of our method.

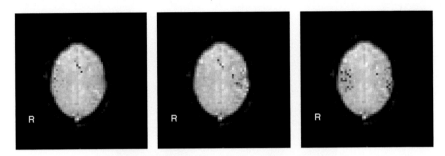

Fig. 4 Pair-wise regional causality analysis performed on a pixel-wise basis. From *left* to *right*, the figures show interactions between pairs of regions *i.e.* RMC and SMA, LMC and SMA, and the LMC and RMC respectively. The influence score of each pixel, as described in Sect. 3.4 is color-coded; *red pixels* are "influencers" while *blue pixels* are "influencees". We see here that the SMA is generally influenced by the RMC and LMC. Whereas between the LMC and RMC there isn't a directional interaction seen (all pixels are 'influencees')

4.2 Causality Analysis

Figure 4 shows a visualization of the results of pair-wise causality analysis between the LMC, RMC and SMA for the same image slice. A specific direction of causation is noted between the different regions, i.e., both LMC and RMC appear to influence the SMA, though further experiments are needed to validate these findings more extensively.

5 Discussion

We present a computational framework for analysis of non-linear functional connectivity in the brain from resting state fMRI data for purposes of recovering the underlying network structure and establishing connectivity. While other methodologies for assessing functional connectivity through fMRI exist, such as seed-based approaches [2], ICA [3] and others, our framework avoids certain shortcomings, such as assumptions of linearity, time series separability, etc. We instead propose to use non-linear mutual connectivity analysis (MCA) to evaluate the pair-wise cross-prediction quality between resting state fMRI time series acquired from the human brain. Our results, as seen in Figs. 3 and 4, suggest that such pair-wise affinity matrices can reveal valuable information concerning the underlying network structure between functionally connected brain regions. The agreement of our results based on resting state time series with the motor task sequence highlights the usability of our approach. Over estimation of the motor cortex as seen, could be attributed to the fact that we only recover regions responsible for hand movements in the finger tapping sequence whereas MCA may be able to provide a better representation. This will be explored in further studies with analysis of the recovery of other functionally connected regions as reported in fMRI literature [3]. Although some directional connections are seen in Fig. 4, further analysis is required to establish and confirm causal interactions within the motor cortex. For future outlook, one may also use other methods of non-metric clustering, such as agglomerative clustering [16], pair-wise clustering through deterministic annealing [17], SOM-related techniques, such as topographic mapping of proximity data (TMP) [18], spectral clustering [19] and non-distance based CONNvis [20] in place of the Louvain method.

6 Conclusion

We present a mutual connectivity analysis (MCA) framework for analysis of functional connectivity and causality in the brain from resting state fMRI data, which combines local non-linear time series prediction, such as by using a GRBF neural network, with non-metric clustering, such as the Louvain method, for recovering the underlying functional brain network structure. By successfully recovering the network structure of the motor cortex, the results observed in our study demonstrate the applicability of our method to exploring connectivity in the human brain.

Acknowledgments This research was funded by the National Institutes of Health (NIH) Award R01-DA-034977. This work was conducted as a Practice Quality Improvement (PQI) project related to American Board of Radiology (ABR) Maintenance of Certificate (MOC) for Prof. Dr. Dr. Axel Wismüller. The authors would like to thank Prof. Dr. Dorothee Auer at the Institute of Neuroscience, University of Nottingham, UK, for her assistance with the fMRI data acquisition process. The authors would also like to thank Prof. Dr. Herbert Witte and Dr. Lutz Leistritz, Institute of Medical Statistics, Computer Sciences, and Documentation, Jena University Hospital, Friedrich Schiller University Jena, Germany, Dr. Oliver Lange and Prof. Dr. Dr. h.c. Maximilian F. Reiser, FACR, FRCR, of the Institute of Clinical Radiology, Ludwig Maximilian University, Munich, Germany for their support.

References

1. Margulies, D.S., Böttger, J., Long, X., Lv, Y., Kelly, C., Schäfer, A., Goldhahn, D., Abbushi, A., Milham, M.P., Lohmann, G., Villringer, A.: Resting developments: a review of fMRI post-processing methodologies for spontaneous brain activity. Magn. Reson. Mater. Phys. Biol. Med. **23**(5–6), 289–307 (2010)
2. Biswal, B., Yetkin, F.Z., Haughton, V.M., Hyde, J.S.: Functional connectivity in the motor cortex of resting human brain using echo-planar MRI. Magn. Reson. Med. **34**, 537–541 (1995)
3. Beckmann, C.F., DeLuca, M., Devlin, J.T., Smith, S.M.: Investigations into resting-state connectivity using independent component analysis. Philos. Translations R. S. B Biol. Sci. **360**, 1001–1013 (2005)
4. Zhou, Z., Ding, M., Chen, Y., Wright, P., Lu, Z., Liu, Y.: Detecting directional influence in fMRI connectivity analysis using PCA based Granger causality. Brain Res. **1289**, 22–29 (2009)
5. Wismüller, A., Lange, O., Auer, D.P., Leinsinger, G.: Model-free functional MRI analysis for detecting low-frequency functional connectivity in the human brain. In:Proceedings of SPIE Medical Imaging 7624: 1M1-8 (2010)

6. Blondel, V.D., Guillame, J.-L., Lambiotte, R., Lefebvre, E.: Fast unfolding of communities in large networks. J. Stat. Mech. Theory Exp. P10008 (2008)
7. Sugihara, G., May, R., Ye, H., Hsieh, C.H., Deyle, E.R., Fogarty, M., Munch, S.: Detecting causality in complex ecosystems. Science **338**, 496–500 (2012)
8. Bezdek, J.C.: Pattern Recognition with Fuzzy Objective Function Algorithms. Springer, US (1981)
9. Wismüller, A., Lange, O., Dersch, D.R., Leinsinger, G.L., Hahn, K., Pütz, B., Auer, D.: Cluster analysis of biomedical image time-series. Int. J. Comput. Vision **46**, 103–128 (2002)
10. Moody, J., Darken, C.J.: Fast learning in networks of locally-tuned processing units. Neural Comput. **1**, 281–294 (1989)
11. Newman, M.E.J.: Analysis of weighted networks. Phys. Rev. E **70**, 056131 (2004)
12. von Luxburg, U.: A tutorial on spectral clustering. Technical Report TR-149, Max Planck Institute for Biological Cybernetics (2006)
13. Dice, L.R.: Measures of the amount of ecologic association between species. Ecology **26**(3), 297–302 (1945)
14. Scherrer, A.: Community detection algorithm based on louvain method [software]. http://perso.uclouvain.be/vincent.blondel/research/Community_BGLL_Matlab.zip
15. Wismüller, A., Vietze, F., Dersch, D. R.: Segmentation with neural networks. In: Handbook of Medical Imaging, pp. 107–126 (2000)
16. Duda, R.O., Hart, P.E., Storck, D.G.: Pattern Classification, 2nd edn. Wiley (2001)
17. Hofmann, T., Buhmann, J.: Pairwise data clustering by deterministic annealing. IEEE Trans. Pattern Anal. Mach. Intell. **19**, 1–14 (1997)
18. Graepel, T., OberMayer, K.: A stochastic self-organizing map for proximity data. Neurocomputing **11**, 139–155 (1999)
19. Shi, J., Malik, J.: Normalized cuts and image segmentation. IEEE Trans. Pattern Anal. Mach. Intell. **22**, 888–905 (2000)
20. Taşdemir, K., Merényi, E.: Exploiting the data topology in visualizing and clustering of self-organizing maps. IEEE Trans. Neural Netw. **20**(4), 549–562 (2009)
21. Moody, John, Darken, Christian J.: Fast learning in networks of locally-tuned processing units. Neural Comput. **1**(2), 281–294 (1989)
22. Lange, O.: MRT-bildverarbeitung durch intelligente mustererkennungsalgorithmen: Zeitreihenanalyse durch selbstorganisierende Clustersegmentierung. Dissertation, LMU München (2004)

SOM and LVQ Classification of Endovascular Surgeons Using Motion-Based Metrics

Benjamin D. Kramer, Dylan P. Losey and Marcia K. O'Malley

Abstract An increase in the prevalence of endovascular surgery requires a growing number of proficient surgeons. Current endovascular surgeon evaluation techniques are subjective and time-consuming; as a result, there is a demand for an objective and automated evaluation procedure. Leveraging reliable movement metrics and tool-tip data acquisition, we here use neural network techniques such as LVQs and SOMs to identify the mapping between surgeons' motion data and imposed rating scales. Using LVQs, only 50 % testing accuracy was achieved. SOM visualization of this inadequate generalization, however, highlights limitations of the present rating scale and sheds light upon the differences between traditional skill groupings and neural network clusters. In particular, our SOM clustering both exhibits more truthful segmentation and demonstrates which metrics are most indicative of surgeon ability, providing an outline for more rigorous evaluation strategies.

Keywords SOM · LVQ · Skill assessment · Surgical training

1 Introduction

Medical advancements in recent years have increased the popularity of endovascular surgery as an alternative to more traditional surgical methods [1]. In the most basic sense, endovascular surgery is a form of minimally invasive surgery (MIS) which allows access to various parts of the body through blood vessels and the endovascular

B.D. Kramer (✉) · D.P. Losey · M.K. O'Malley
Rice University, Houston, Tx 77005, USA
e-mail: bkramer@rice.edu; bk18@rice.edu

D.P. Losey
e-mail: dlosey@rice.edu

M.K. O'Malley
e-mail: omalleym@rice.edu

© Springer International Publishing Switzerland 2016
E. Merényi et al. (eds.), *Advances in Self-Organizing Maps and Learning Vector Quantization*, Advances in Intelligent Systems and Computing 428,
DOI 10.1007/978-3-319-28518-4_20

227

system. The surgeon introduces a catheter into the vasculature of the patient, typically via the femoral artery, and from there navigates the catheter to the desired location so as to perform some type of procedure. During these procedures, surgeons must rely on fluoroscopy and other forms of medical imaging in order to determine tool position. This imaging is often limited, and complications may go unnoticed until they become too serious; therefore, it is imperative that surgeons be proficient at endovascular techniques. Aside from the risk of possible complications, surgeon skill level significantly affects clinical outcomes after successful surgeries [2].

1.1 Previous Work

As a result, there is medical interest in understanding an effective means to determine a surgeon's skill [3]. There are presently two preeminent methods for assessing a surgeon. The most common involves an expert observing task completion by a novice, which is entirely subjective and vulnerable to significant amounts of variability [4]. The second method is simply a measurement of the number of cases performed by the surgeon; although it stands to reason that an individual with more practice will likely be better, it is also likely that individual surgeons will improve at different rates. Either method is insufficient, and therefore a primary goal of the endovascular community is the development of an objective assessment technique [5, 6].

In an effort to more objectively study surgeons, sensors have been used to record the tool tip trajectory [7]. The results are then processed to calculate a variety of motion-based metrics; the most indicative of these metrics are correlated to user smoothness, such as minimum jerk [8] and spectral arc length [9]. An alternative, yet similarly-minded, method is the extraction of submovement number and duration from a larger task [10]. To date, researchers have attempted to show that there exist correlations between these movement metrics and the standard methods of skill evaluation. Surgeon force and motion signatures have been leveraged to objectively assess performance; hidden Markov models were then used to learn the nonlinear mapping between performance data and skill [11]. Lin et al. demonstrated the ability to decompose a surgical procedure into a series of sub-tasks by parsing raw motion data in order to provide on-line training feedback [12]. Estrada et al. specifically quantified the correlation between various metrics and the standard methods of surgeon evaluation on both manual and robotic platforms [13].

1.2 Motivation/Objective

Successfully mapping metrics to skill may improve training procedures, reduce the amount of oversight required, and ultimately automate this task. While the statistically significant correlation between various objective metrics and current subjective assessments is an important initial finding, it fails to provide a holistic approach to skill classification. Hence, the motivation for our work is to understand the mapping between movement metrics and surgeon proficiency, which we will reveal through neural networks. We will first train an LVQ to classify surgeons using standardized novice, intermediate, or expert labels, and then study the LVQ's accuracy using testing data. Next we will utilize SOMs to examine the underlying clusters; by comparing these SOM clusters with pre-labeled classes, we can evaluate the veracity of the medically imposed class labels. We hypothesize that the traditional "novice, intermediate, and expert" labeling—while commonly assumed to be correct—does not actually reflect the motion data, and, as such, more sophisticated classification is recommended. Our secondary goal is to identify which motion patterns contribute most to the surgeon's classification; this knowledge may improve the feedback which can be provided during and after the surgeon's training.

2 Methods

2.1 Input Data and Class Labels

The data used in this paper was collected during a previous study [13]. Actual and virtual tool-tip trajectories were recorded for fifteen surgeons over three sessions while completing four separate tasks. Five of the subjects (i.e., surgeons) were deemed "novices," six were labeled "intermediates," and the remaining four were regarded as "experts." The two platforms used during experimentation can be seen in Fig. 1, along with sample input vectors. For the purposes of our research, we did not differentiate between the platforms, sessions, or tasks, yielding a total of 120 separate trials. The motion metrics associated with each trial—described in more detail below— were then utilized as a unique input vector; hence, our results were obtained using 120 input vectors.

The input vectors for our LVQ and SOM neural networks were constructed from previously calculated motion metrics. These metrics were all computed from the three-dimensional catheter position data, which was collected at 30 Hz frequency. Based upon the findings of Estrada et al. [13], we selected motion metrics which were shown to individually correlate with traditional skill labels. Eleven metrics (listed below) were chosen, and each comprised an element of the eleven-dimensional input vectors. Although the units for the various metrics are not detailed here, it should be noted that they were kept consistent throughout our work. We found that our best results occurred with the inclusion of (1) Spectral Arc Length, (2, 3) Average Sub-movement Duration (LGNB and MinJerk Profiles), (4, 5) Number of Submovements (LGNB and MinJerk Profiles), (6) Normalized Velocity, (7) Mean Arrest Period Ratio (10 % was used for this study), (8) Completion Time, (9, 10) Submovement Overlap (LGNB and MinJerk Profiles), and (11) Average Frequency. Example input vectors can be seen in Fig. 1. Note that these values are all well defined over a continuous

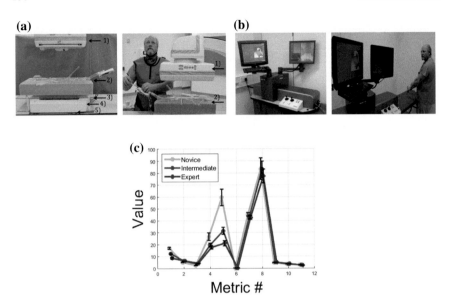

Fig. 1 A comparison of the manual and virtual simulators which were navigated during the various tasks. In both platforms the surgeon is operating a catheter—in (**a**) the tip position is tracked using a magnet at the tool-tip, while (**b**) offers a platform leveraging teleoperation. In (**c**) each of the plotted points corresponds to one of the eleven motion metrics derived from the surgeon's trajectory. Average input vectors associated with a novice, intermediate, and expert surgeon are shown (standard error bars included). **a**. Manual Simulator [13] **b**. Virtual simulator [13] **c**. Example Input Vector

range, and that the chosen metrics mitigated statistical outliers which may skew results of our input-space neural networks.

Each of these 120 input vectors was associated with a class label corresponding to the surgeon's proficiency; the three classes consisted of either "novice," "intermediate," or "expert." Forty input vectors were labeled novice, forty-eight input vectors were denoted intermediate, and thirty-two input vectors were termed expert. When performing supervised learning, stratified four-fold cross-validation was leveraged to select exclusive sets of ninety input vectors for training and thirty input vectors for testing.

2.2 Classification with LVQs

In order to determine the mapping from input data to desired classification, we used an LVQ with supervised learning [14]. More specifically, we used an LVQ2 with stratified four-fold cross-validation; the LVQ was initialized with 120 prototypes, as this was found to provide the best classification accuracy, where forty prototypes were allocated to novices, forty-eight were allocated to intermediates, and the remaining thirty-two were allocated for experts. Thus, this prototype allocation was done in proportion to class size. LVQ neurons were randomly initialized and scaled to the range of the input data. Ideally, the trained LVQ would adapt to the externally imposed classification structure, and, as such, would serve as an autonomous means towards identifying the class of the surgeon's skill—novice, intermediate, or expert—based solely on motion metrics. On the other hand, the reliability of the traditionally imposed class labels may be questionable [15]. These labels are based on the number of cases performed; however, it is conceivable that a surgeon could perform a large number of cases with improper technique, and therefore be labeled an "expert" by this traditional evaluation while actually maintaining a "novice" level of ability. In order to examine the performance of our classification with LVQs, we will show confusion matrix data and statistics across all four folds, as well as a visualization of our best results.

2.3   Clustering with SOMs

As we will demonstrate, the best classification accuracies obtained with LVQs were unsatisfactory, suggesting that more analysis into the label veracity is needed for effective machine learning. Further analysis of the input data—and, in particular, clusters present in the input data—was performed and visualized through the use of SOMs [16]. We leveraged forty-nine prototypes for the SOM, which were arranged into a seven-by-seven rectangular grid in the lattice space. A Gaussian neighborhood function was used while updating the prototypes, and mU-matrix visualization was employed to visualize clusters. Our rationale for using an SOM was to capitalize upon the strengths of unsupervised learning; we sought to obtain an objective view of the data structure without needing potentially erroneous labels. Therefore, we had two primary goals behind this SOM application. First, we wanted to validate or disprove the classification labels (novice, intermediate, and expert) previously used for our LVQ training. By superimposing these labels over the SOM lattice while visualizing SOM clusters, we could test label veracity and hopefully understand why the LVQ machine learning underperformed. Second, we wanted to identify clusters within the data in order to determine the relative importance of surgeon attributes and motion methods when distinguishing between skilled and unskilled surgeons. By comparing the input vectors associated with different clusters, we can better understand which motion metrics were consistent and which varied amongst clusters. These insights may enable more efficient evaluation of surgeons and more directed training strategies. SOM clustering will be revealed through plots of the lattice space.

3 Results and Discussion

3.1 LVQ Classification Results

The results obtained by implementing an LVQ were reasonable, but did not provide sufficiently accurate classification for the purposes of automated evaluation. Our best results were obtained with an LVQ2 using a learning rate of 0.001 and 10,000 on-line learning steps, although other learning rates and learning step counts were tested. Both the training and testing accuracy were plotted as a function of learning steps to ensure that overtraining did not occur. To summarize, we consistently found that we were able to differentiate the skill groups and correctly classify surgeons within the novice, intermediate, and expert labels 80 % of the time for training data and 50 % of the time for testing data. In particular, the LVQ struggled to distinguish "intermediate" from "expert" surgeons, logically suggesting a larger skill gap from novice to intermediate than from intermediate to expert. This disparity is depicted in Fig. 2. We also note that, while LVQ1, LVQ2, and LVQ3 were tested, there was not significant variation among the performance of these algorithms.

By inspecting the confusion matrices, summarized in Fig. 3, we can further verify that novices were reasonably distinguished from intermediates and experts, but intermediates and experts were largely lumped together. We hypothesize that this stems from at least partially inaccurate training labels; the imposed classifications may not truly identify the skill level of each surgeon, since intermediate surgeons, despite having performed fewer cases than experts, may be more proficient than their caseload suggests. Moreover, the use of only three classes is likely insufficient to accurately capture the gradient in surgeon skill, and perhaps more nuanced labels would better reflect our motion data. The overall statistics show that the LVQ procedure netted consistent and accurate training classification, but the testing accuracy and hence machine learning was unacceptable. We conclude that the LVQ was unable to generalize for the given data, and suggest that this inability stems from the lack

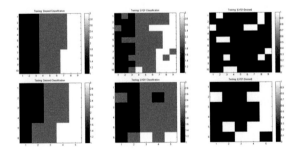

Fig. 2 Sample LVQ results. These plots are from one fold of the four-fold stratified cross-validation procedure: training classification top; testing classification bottom. The *black* pixels represent novice surgeons, the *grey* pixels represent intermediate surgeons, and the *white* pixels represent expert surgeons

		Training Results			Testing Results		
		Novice	Intermediate	Expert	Novice	Intermediate	Expert
Labels	Novice	87.5% (26.25)	10% (3)	2.5% (0.75)	70% (7)	15% (1.5)	15% (1.5)
	Intermediate	6.25% (2.25)	81.9% (29.5)	11.8% (4.25)	20.8% (2.5)	52.1% (6.25)	27.1% (3.25)
	Expert	2.1% (0.5)	27.1% (6.5)	70.8% (17)	9.4% (0.75)	65.6% (5.25)	25% (2)

Classification Summary					
	Training	Testing		Training	Testing
Mean	81% (291)	51% (61)	Std.	4.2% (15)	9.5% (11)

Fig. 3 Average confusion matrix over the four folds. Data is given in the form % of hits (number of hits). Diagonal elements represent correctly classified data, while off-diagonal elements show incorrect classifications. The mean and standard deviation for training and testing accuracy are also shown. Poor results likely stem from incorrect class labels, particularly between intermediates and experts

of labeling precision and correctness for intermediate and expert surgeons. To verify this claim, we will subsequently explore SOM clusters in the data space.

3.2 SOM Clustering Results

Following the failure of LVQs to successfully identify this mapping, SOMs were applied to both test our concerns with the imposed classification labels and help us further explore nuances within the data. The best results presented in this paper were obtained using a seven-by-seven rectangular SOM grid in lattice space, where the forty-nine prototypes were initialized randomly over the input space. The learning rate α started at 0.005 and reached 0.001 following a linear decrease across 100,000 learning steps; similarly, the Gaussian neighborhood width σ started at 4 and linearly decreased to 2 over the same number of learning steps. We experimentally observed the SOM training to converge after around 80,000 to 90,000 on-line learning steps, at which point no changes occurred in the mapping. The results shown below were found to be repeatable and superior to those identified using different parameters, which gives us confidence in the subsequent conclusions.

Selecting the learning parameters as described above while observing the system visualizations depicted in Fig. 4, we repeatedly converged to a similar, if not the same, solution each time we trained the SOM. Instances in which we did not converge to the results outlined in Fig. 4 involved some type of rotation of the lattice—however, this did not alter the SOM clustering. Using U-Matrix techniques, we readily discerned some distinct clusters which were identified by the SOM; we then checked these locations with superimposed novice, intermediate, and expert labels in the lattice space, and determined whether there existed agreement between medically defined clusters and clusters identified by the SOM.

From the modified U-Matrix density map and the projection of classifications into lattice space, we can deduce (a) that there exist some SOM clusters which roughly correspond with traditional groups, but (b) other SOM clusters disagree with the

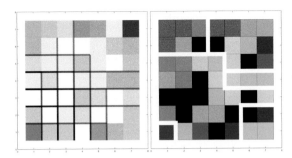

Fig. 4 SOM final results. The left visualization is a modified U-Matrix [17]; *red-scale* represents the number of mappings (i.e., relative density), while the *gray-scale bars* signify the distance between prototypes in the data space. The redder the neuron, the more input vectors are contained within its Voronoi cell; likewise, the darker the bar separating neurons, the greater the difference between their weight vectors. The right visualization shows the known surgeon classifications projected onto the SOM lattice—here red signifies novice, *green* represents intermediate, and *blue* indicates expert surgeons, with color intensity representing the number of mappings (more intensity again means increased density). *Black* neurons indicate that no input vectors are mapped to a particular node. The clusters found in the mU-matrix are identified using white lines in the right visualization. We can quickly observe that while novices (*red*) are primarily separated, clustering in the upper and lower left, intermediates (*green*) and experts (*blue*) are largely intermingled, clustering along the right side, a result which supports our LVQ findings

medical consensus. We have marked these SOM identified clusters in Fig. 5. For instance, the bottom left section of the SOM lattice clearly clusters several surgeons who performed poorly, and are correctly labeled as novices. Likewise, the top left SOM cluster corresponds to another group of novice surgeons, which again matches the medical labeling. Moving to the right side of the SOM lattice, however, we can see two regions: in the upper right, there exists a mixed cluster—some experts, intermediates, and novices are included here, suggesting labeling inaccuracy. Finally, in the bottom right of the SOM lattice we find a cluster of increasing ability, with intermediates and experts grouped together; perhaps these surgeons are closer in ability than their classification would suggest. By applying SOMs to the input space of motion metrics, we were therefore able to demonstrate that a surgeon's experience is not sufficient when attempting to classify that surgeon's skill. Although there are some similarities between the medical labels and SOM clusters, there is also sufficient disparity to suggest that perhaps more precise skill assessment is required. These findings also explain the inability of our LVQs to distinguish "intermediate" and "expert" surgeons, as SOM clusters revealed overlaps between these classifications.

In order to further investigate clustering and the distinctions between various groups, it was instructive to look at the weight vector within these individual clusters, as illustrated in Fig. 5. There are a few hypotheses which can be formed from visualizing these prototypes and clusters. First, completion time is not necessarily an accurate measure of skill. In fact, completion time appears to be somewhat counterintuitive; experts often take longer than less successful intermediates and novices, perhaps because they are utilizing slower and more deliberate movements. A quick

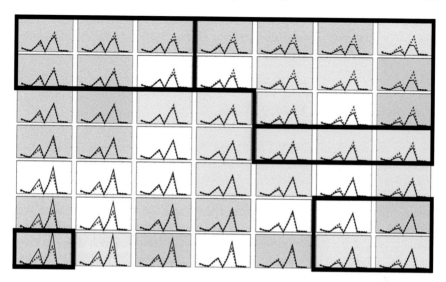

Fig. 5 SOM weight vector plotted in the grid cells. This figure shows both the final results of our SOM grid with the known classifications projected onto the lattice, as well as the weight vector of each PE with respect to the average weight vector across all nodes. The weight vector of a given PE is shown in *black*, while the average weight vector across all nodes is plotted in a *dotted magenta line*. The color coding of the prototypes is the same as before, with a slight fading of the colors in order to better visualize the weight vectors. *Black boxes* were used to mark the SOM cluster boundaries identified in Fig. 4. By comparing the differences in weight vectors between members of different clusters, we can visualize which metrics most impact distinctions in surgeon skill. With respect to the average weight vector, novices appear to complete the task in less time but require an increased number of motions; on the other hand, proficient surgeons move slowly but smoothly, reducing submovement duration and number. The combination of SOM clustering and neuron weight vectors reveals errors within traditional labeling and provides insight into important motion attributes. The existing labeling of novice, intermediate, and expert does not agree with knowledge gained through motion metrics (as shown by differences in clustering), and the contribution of various metrics can be analyzed to yield better categorization (as shown by comparing weight vectors)

procedure is ideal, but not if it comes at the cost of deliberate, precise movements. Second, some metrics may provide redundant differentiation, therefore requiring the use of fewer metrics—and other metrics may be entirely irrelevant for classification purposes. Finally, the number of submovements appears to be particularly useful when distinguishing surgeons; we observed that the most proficient cluster employed substantially smoother motions than did novice or mixed clusters.

With these ideas in mind, we can describe the five classes of surgeons from SOM clustering (Fig. 5). Class one (lower left) will perform the task slowly and with very little smoothness; likely true beginners. Class two (upper left) will perform the surgery quickly with little smoothness; likely novice surgeons. Class three (upper right) will perform the surgery quickly at the expense of some smoothness metrics; likely competent surgeons primarily concerned with completion time. Class four (middle right) will perform the surgery above average in terms of time and smoothness; likely experienced surgeons. Class five (lower right) will perform the surgery at an average pace with exceptional dexterity; likely skilled, precise surgeons.

4 Conclusions and Future Work

Based on the results of our LVQ and SOM, there does appear to be some consistent mapping between motion metrics and desired classification; using the LVQ we achieved around 50 % testing accuracy. We hypothesized that this poor LVQ machine learning, particularly when discerning between intermediate and expert surgeons, stemmed from inaccurate class labeling. Using the SOM approach, we were able to identify some clusters which roughly corresponded to the known classification groups; however, we also discovered that several clusters disagreed with the given labels. Indeed, from Fig. 4 we were able to conclude that the traditional labeling based on surgeon experience disagreed with SOM clustering in the motion metrics. We were further able to suggest which metrics may best be able to indicate ability, as can be seen in Fig. 5. By replacing the subjective medical grouping with the actual measured features, we may be able to improve on skill assessment for endovascular surgeons. Similar to the work by Cotin et al. [5], we suggest that it may be better to first identify statistics which are significant to expert clusters, and then create a scoring system which classifies users based on their accordance with those statistics. Summarily, SOM clustering, as seen in Fig. 5, helps accomplish our goals of both disproving classical labels and suggesting improved alternatives.

References

1. Schanzer, A., Steppacher, R., Eslami, M., Arous, E., Messina, L., Belkin, M.: Vascular surgery training trends from 2001–2007: a substantial increase in total procedure volume is driven by escalating endovascular procedure volume and stable open procedure. J. Vasc. Surg. **49**(5), 1344–1399 (2009)
2. Cox, M., Irby, D.M., Reznick, R.K., MacRae, H.: Teaching surgical skills–changes in the wind. N. Engl. J. Med. **355**(25), 2664–2669 (2006)
3. Reiley, C.E., Lin, H.C., Yuh, D.D., Hager, G.D.: Review of methods for objective surgical skill evaluation. Surg. Endosc. **25**(2), 356–366 (2011)
4. Darzi, A., Mackay, S.: Assessment of surgical competence. Qual. Health Care 10(suppl 2), ii64–ii69 (2001)
5. Cotin, S., Stylopoulos, N., Ottensmeyer, M., Neumann, P., Rattner, D., Dawson, S.: Metrics for laparoscopic skills trainers: the weakest link! In: Medical Image Computing and Computer-Assisted Intervention, pp. 35–43. Springer, Berlin (2002)
6. van Hove, P.D., Tuijthof, G.J.M., Verdaasdonk, E.G.G., Stassen, L.P.S., Dankelman, J.: Objective assessment of technical surgical skills. Br. J. Surg. **97**(7), 972–987 (2010)
7. Kuipers, J.: Object tracking and determining orientation of object using coordinate transformation means, system and process (1975)

8. Hogan, N., Sternad, D.: Sensitivity of smoothness measures to movement duration, amplitude, and arrests. J. Mot. Behav. **41**(6), 529–534 (2009)
9. Balasubramanian, S., Melendez-Calderon, A., Burdet, E.: A robust and sensitive metric for quantifying movement smoothness. IEEE Trans. Biomed. Eng. **59**(8), 2126–2136 (2012)
10. Rohrer, B., Hogan, N.: Avoiding spurious submovement decompositions: a scattershot algorithm. Biol. Cybern. **94**(5), 409–414 (2006)
11. Rafii-Tari, H., Payne, C.J., Liu, J., Riga, C., Bicknell, C., Yang, G.Z.: Towards automated surgical skill evaluation of endovascular catheterization tasks based on force and motion signatures. In: Proceeding of IEEE International Conference on Robotics and Automation, pp. 1789–1794 (2015)
12. Lin, H.C., Shafran, I., Yuh, D., Hager, G.D.: Towards automatic skill evaluation: detection and segmentation of robot-assisted surgical motions. Comput. Aided Surg. **11**(5), 220–230 (2006)
13. Estrada, S., O'Malley, M.K., Duran, C., Schulz, D., Bismuth, J.: On the development of objective metrics for surgical skills evaluation based on tool motion. In: 2014 IEEE International Conference on Systems, Man and Cybernetics (SMC), pp. 3144–3149. IEEE (2014)
14. Kohonen, T.: Learning vector quantization. In: Self-Organizing Maps. Springer Series in Information Sciences, vol. 30, pp. 175–189. Springer, Berlin (1995)
15. Bismuth, J., Donovan, M.A., O'Malley, M.K., El Sayed, H.F., Naoum, J.J., Peden, E.K., Davies, M.G., Lumsden, A.B.: Incorporating simulation in vascular surgery education. J. Vasc. Surg. **52**(4), 1072–1080 (2010)
16. Kohonen, T.: The self-organizing map. Neurocomputing **21**(1), 1–6 (1998)
17. Merényi, E., Tasdemir, K., Zhang, L.: Learning highly structured manifolds: harnessing the power of SOMs. In: Similarity-based clustering. Lecture Notes in Computer Science, pp. 138–168. Springer, Berlin (2009)

Visualization and Practical Use of Clinical Survey Medical Examination Results

Masaaki Ohkita, Heizo Tokutaka, Nobuhiko Kasezawa and Eikou Gonda

Abstract At present, the Metabolic Syndrome (MS) is judged by checking whether a number of reference values are exceeded. It is regarded as non-metabolic if those values are not exceeded. Using our previously established non-ill area for these values, we evaluate their applicability in medical practice. We start with a clinical survey of medical checkup data and apply our methodology to six items of the checkup sheet: glucose metabolism, liver, diabetes, kidney, blood-general and inflamed-immunity. We outline our methodology, called Dr. Ningendock, and evaluate it on the results of the liver function.

Keywords Clinical survey medical examination · Metabolic syndrome judgment · Self-Organizing Map (SOM) · Dr. Ningendock

1 Introduction

An evaluation method for the Metabolic Syndrome (MS) was developed that is part of the early lifestyle-related disease prevention program [1] in our country. The diagnostic criteria of MS were recommended by the Japanese Society of Internal Medicine and related societies and are now widely enforced in Japan. At present, according to its diagnostic criteria, the "extraordinary" condition refers to only the case of slightly exceeded reference values for each fat, the blood pressure, and blood sugar. Whether a condition is considered "abnormal" depends on the number of exceeded reference values. However, when observing medical checkup data in practice, as a time series at the individual level, there are many cases for which the values sometimes come

M. Ohkita (✉) · H. Tokutaka
SOM Japan Inc., Tottori, Japan
e-mail: mohkita111@yahoo.co.jp

N. Kasezawa
Fuji Iki-iki Hospital, Health Support Center, Shizuoka, Japan

E. Gonda (✉)
Yonago National College of Technology, Yonago, Japan
e-mail: gonda@yonago-k.ac.jp

© Springer International Publishing Switzerland 2016
E. Merényi et al. (eds.), *Advances in Self-Organizing Maps and Learning Vector Quantization*, Advances in Intelligent Systems and Computing 428,
DOI 10.1007/978-3-319-28518-4_21

239

close to the border of the reference values so that it is difficult to judge a condition as extraordinary. Therefore, the MS judgment should be carefully monitored for slight deviations. In order to address this issue, we introduced the concept of "non-ill area" [2], defined as a gray zone between the upper limit of the normal values and the high abnormal ones (i.e., the critical region). The distribution of each MS item recorded in the general medical checkup was observed and categorized into three conditions: "normal", "the non-ill area" and "the critical region". Then, the checkup data categorized in this way (i.e., according to our method) was applied to a Self- Organizing Map (SOM) [3–6] for calculating the MS score and for observing trends in the MS condition of the examinee.

In our previous paper, we proposed the MS judging tool for check-up data as described above [4]. In the present paper, the MS judging tool is applied to medical examination data for the following 6 lifestyle related functions: sugar metabolism, liver function, renal function, blood (blood-general), and inflammation-immunity. The results are then displayed into a hexagonal coordinate system. Among others, an indication for the upcoming year can be given by using our slider-based functionality. Therefore, we are able to focus on an item that displays a potential problem, and then proceed to a detailed judgment.

2 Visualization of Medical Examination Results with Self-Organizing Maps

2.1 Clinical Survey of Medical Examination

Upon receipt of the complete clinical survey (Ningen dock in Japanese), a table of the physical examination medical checkup can be compiled for each patient as shown in Fig. 1.

This table is only one example as its format differs between medical facilities. The table includes measurements of height, weight, BP, BMI, results of blood tests, abdominal ultrasonography such as liver, pancreas, gallbladder, kidney, and so on. Their reference values also included. Mostly they are computer-generated numerical tables. Then, the outcome for each item is judged by the degree by which it differs from the reference value:

A non-abnormal, when inside the reference value

B slightly abnormal, but there is no problem for everyday life

C follow-up

D treatment or detailed re-examination is required

Here, we propose the following method which attempts to more legibly re-organize and utilize the data. Firstly, in the table of Fig. 1, the data enclosed in the green frame is extracted and compared with the officially approved (conventional) diagnostic criteria. Related data (blood general, liver function, renal function, inflammation-immunity) are shown in the red frames, two of which are shown in Fig. 1 (there are 4 of them in total, [1–4]).

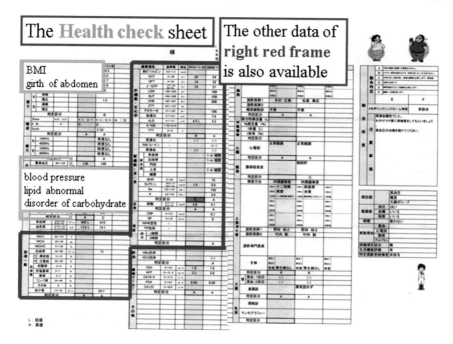

Fig. 1 Table with a patient's medical checkup data. After consultation, it is mailed after about one month

2.2 Visualization of Score (points) Map and Components Map

The degree of MS was evaluated using the previously developed Dr. Metabo tool (metabolic syndrome (MS) judgment tool [3], See 2.4 Calculation of MS(score) in [3]).

The Metabo tool describes 4 regions in the MS score map that are marked by 4 different gray scales: Region I: Non MS region (0 < score < 20) labeled "DM-Normal", Region II: MS boundary (20 <= score < 40) labeled "DM boundary", Region III: MS corresponding (40 <= score < 60) labeled "DM-abnormal, and Region IV: MS critical(60 <= score <= 100) labeled "DM-critical".

According to the MS evaluation method of the Japanese Society of Internal Medicine, a 0–19 score is in the normal region, and a score above 20 is marked as MS. In our tool, we consider two additional regions II and III as "non-ill" ones. These additional regions, between non-MS and MS critical, provide a way to monitor the deterioration of the patient's condition, thus, as a follow up on the conventional MS judgment method. Indeed, with the proposed method, the data of the examinee are judged step by step in relation to the non-ill region. With the conventional method, as soon as the examinee's value exceeds even once a reference value, he/she is immediately judged as MS. The result of the proposed method is displayed in the following

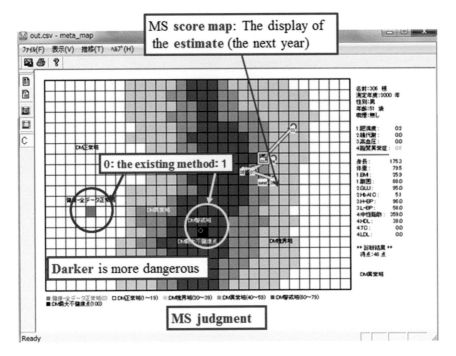

Fig. 2 The conventional method divides the map into metabolic (1) and non-metabolic (0). However, with the proposed method, it is clear that the examinee's condition gradually moves in the direction of the *black region* (metabolic syndrome (MS))

map (Fig. 2) obtained by mapping the above four regions. Also, the relation between the components that make up the map of Fig. 2 is clarified in Fig. 3.

For the visualisation, we considered a torus type of SOM [7] because of its improved learning accuracy and clustering. We trained the SOM with the tool described in [3] considering the following parameters: For the SOM we take a square map sized 30 by 20 (i.e, the number of columns and row). We start from randomized initial weights and further take a Gaussian neighborhood function with initial radius of 30, an initial learning rate of 0.1, and 100,000 learning steps (iterations).

2.3 Construction of a Visualization Judgment Tool of Clinical Survey Data

In the present paper, digital data, except for the images, but including the part in the green and red frames in Fig. 1, were collected in the following 6 fields and already analyzed using the metabolic analysis method described in [1–3]: 1. metabolic judg-

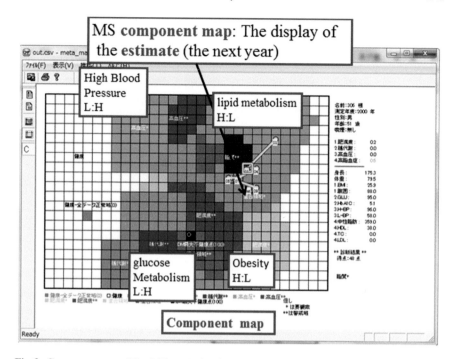

Fig. 3 Component map of Fig. 2. The examinee's condition gradually moves into the obese direction towards early lipid abnormality. In the *red colored rectangular* boxes, for L:H of High Blood Pressure, L shows the low score in the left side and H the high score in the right side of the High Blood Pressure zone. Similarly for the other items

ment (described above), 2. glucose metabolism, 3. liver function, 4. renal (kidney) function, 5. Blood general, and 6. Immunology and inflammation. The input data format for the analysis is shown in Fig. 4.

The SOM-based analytical method previously used for metabolic analysis was then applied to the analysis of the upper 6 fields. The result of the analysis can be displayed visually and organ-specific anomalies and anomalies in each function systematically grasped. In addition, these six areas are almost always part of the normal medical checkup except for the image data. Hence, when storing individual test data on a portable computer, a face-to-face health support can be provided to the patient during the checkup at the health center. Continuous and detailed health evaluation of medical examination data can be established. All this enhances the patient's motivation to change his/her lifestyle and increases the individual's health instruction effect even more. Figure 4 shows the input data used for the diagnosis of metabolic health in columns F to Q. For the R to AK columns, the test values of each item of liver function, kidney function, blood general, and inflammation & immunity are listed per check-up day.

In "Doctor metabolic syndrome tool", the MS degree was evaluated by the score. The same idea was also applied to judgment of these six areas. The results of the score

	A	B	C	D	E	F	G	H	I	J	K	L	M	N	O	P	Q
1						Drメタボ, Dr糖代謝						Drメタボ					
2	受診日	名前	年齢	性	喫煙	身長	体重	BMI	腹囲	血糖	HbA1c	最高血圧	最低血圧	TG	HDL-C	TC	LDL-C
3	20050326	1SK-ab	53	1	0	161.8	49.9	19.1	72	87	5.3	124	83	144	52	264	187
4	20060311	1SK-ab	54	1	0	161.9	50.6	19.3	73	86	5.1	134	89	163	42	251	171
5	20080315	1SK-ab	56	1	0	160.8	51.4	19.9	74.1	99	5.2	138	87	148	48	248	176
6	20090321	1SK-ab	57	1	0	160.8	51.4	19.9	74.8	89	5.2	126	88	191	48	258	185
7	20100327	1SK-ab	58	1	0	161.2	50.8	19.5	75	94	5.3	121	83	238	48	261	177
8						1	2	3	4	5	6	7	8	9	10	11	12
10			男1	喫1		**Diabetes**							**Diabetes**				
11			女0	無0									**Metabo**	**From F to Q**			
12										合わせてメ...				**Column are MS**			

R	S	T	U	V	W	X	Y	Z	AA	AB	AC	AD	AE	AF	AG	AH	AI	AJ	AK
					Dr腎機能					Dr血液一般					Dr炎症・免疫				
LDH	γ-GTP	AST	ALT	PLT	年齢	尿蛋白	CRE	BUN	尿酸	RBC	PLT	HGB	HCT	WBC	WBC	CRP	好中球	リンパ球	LDL-C/HDL-C
222	22	35	59	31.1	53	-	0.8	14.8	5.8	539	31.1	16.6	50.6	8.96	8.96	0.09	58.2	27.4	3.6
185	21	25	41	32.3	54	-	0.9	15.9	6.2	510	32.3	16.3	47.6	8.4	8.4	0.05	55.7	31.4	4.07
199	23	29	57	34.6	56	-	0.9	14.3	6.2	491	34.6	15.3	44.8	8.39	8.39	0.02	61.3	27.1	3.67
205	24	35	76	33.2	57	-	0.9	13.3	6.5	490	33.2	15.6	45.2	7.31	7.31	0.03	57.1	28.5	3.85
193	20	30	57	33.2	58	-	0.8	14.8	6.6	502	33.2	15.5	45.7	7.18	7.18	0.03	53.2	32.3	3.69
13	14	15	16	17	18	19	20	21	22	23	24	25	26	27	28	29	30	31	32

Liver func.   **Kidney func.**   **Blood func.**   **Inflamed**

Fig. 4 Input data format example for 6 field decision tool analyzing lifestyle-related diseases

are summarized in the hexagonal coordinate system of Fig. 5. The center of gravity is also displayed. The center of gravity is intended to monitor the evolution of the 6 fields as a change in the annual 6 field medical checkup results is otherwise difficult

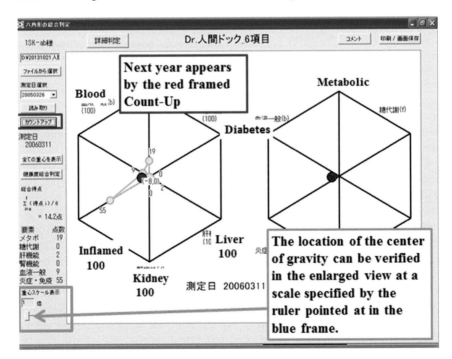

Fig. 5 Lifestyle-disease related items are judged by the location of the score in a *hexagonal* coordinate system. The average score of all six fields is shown in the lower *left corner* in the figure. The *red filled circle* shows the center of gravity

to detect. For a long-term clinical survey spanning 10 to 20 years, it is considered useful to accurately grasp the trend. The traditional medical examination system does not consider such a long time span.

In Fig. 5, output levels are plotted on the hexagonal axes. The unhealthy marks (scores) are shown at the end of each axis and correspond to 100 points. The center is at 0 point (where all data are in the normal range). The result for year 2006 of a given examinee is shown in Fig. 5. Clockwise the shown scores are 19, 0, 4, 0, 55, and 9 points. The center of gravity of these six points is shown by the red filled circle. The location of the center of gravity can be verified in the enlarged view.

The 6 fields of the medical examination for a given year can be easily understood when plotted in a hexagonal coordinate system. However, it is difficult to display subsequent changes in position especially when they are close to one another. The temporal change of each position along its hexagonal coordinate was also displayed on a vertical axis. This is shown in Fig. 6. In this way, the chronological order of each field can be judged. In the example shown in Fig. 6, abnormal values are assigned

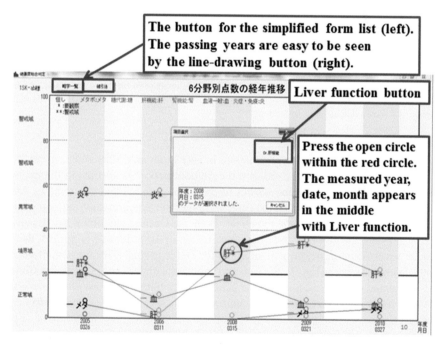

Fig. 6 Graph of the general evolution of the itemized score over five years of a subject. Items of which all data is in the normal range are not displayed in the graph. A * or ** mark is attached when the score of an item surpasses 20 score points. Also, when clicking the open circle within the red circle, the liver function data of 2008 appears, 0315. The shown menu for a detailed analysis appears when clicking the Dr. Liver function icon in the *red circle*. In addition, in this figure, the region of more than 20 points above the borderline was colored thinly *pink*. Abbreviations: MS = Metabolic syndrome, Lf = liver function, BG = blood (general), Ii = inflammation-immunity. Note that GM = glucose metabolism, and Rf =renal function (Kidney) are not shown here

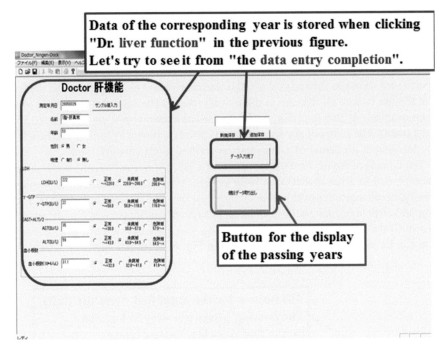

Fig. 7 The subject's data of 15 March 2008 (20080315) appears in the large *red frame*. After clicking the right upper small *red frame* of the figure, the result score and component maps will appear (not shown due to space limitations)

to Ii, MS, Lf, and BG by the abbreviation. In case of an unhealthy score, the mark (*) appears in the range of 20–60, and the caution mark (**) appears in the range of more than 60. So, when one wishes to focus on an individual abnormality, for example, if you want to examine the contents of the liver function, then the examiner can click on the black dot in the red circle. Then, the Dr liver function window pops up. When subsequently the "liver function" button in the red frame is pushed, then the Dr. Liver function screen will appear (Fig. 7), for example, for the data of year 2008. With this screen (Fig. 7), the year and temporal variation can be analyzed from the score- and component maps of the patient's liver function. Thus, after entering the screen of Fig. 6, the examiner can immediately assess the corresponding data in detail.

The score- and component maps of 15 March 2008 (20080315) are displayed by pushing the small red frame in Fig. 7. Also, the score- and component maps for Liver function from years 2005 to 2010 can be called (Fig. 8) by pushing the small blue frame in Fig. 7.

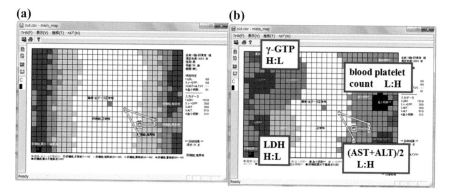

Fig. 8 The temporal variation of **a** the score map and **b** the component map from 2005 till 2010 corresponding to Fig. 6

3 Entering Comments into the Component Map

The judgments can be inserted as comment lines at particular positions of the component map. Examples are shown in Fig. 9.

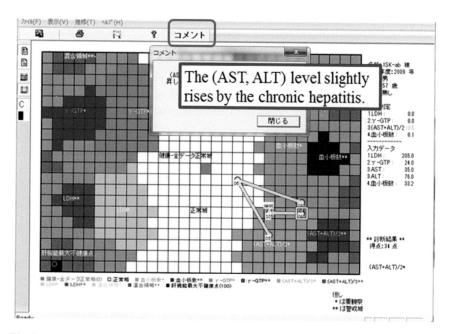

Fig. 9 When clicking "the comment" in the upper section (*blue frame*), a comment equivalent to input data and judgment data of the *yellow frame* on the map appears in the *rectangular* comment box on the top

4 Summary

It has been verified that our MS decision tool can avoid the "over-estimation" and "overlooking" of a patient's MS condition the currently used recommendation method is prone to. Building on this advantage, the non-ill level outside the reference range was defined. A method for evaluating the degree of abnormality was devised in terms of the extent to which the data exceeds the upper normal limit. The tool has been expanded to include items of six areas of health screening and medical examination in general. The result of the six fields can be judged when displayed on a regular hexagonal coordinate system. In this way, the result can be grasped. In addition, the temporal change in the center of gravity of the six fields' scores can be visualized. Also, a comprehensive view on the patient's health condition can be displayed by a bar graph which also shows the temporal evolution of the general medical examination result. As a result, the examiner no longer needs to solely rely on clinical survey- and medical examination results expressed as lists of numerical values, but instead can use a visualized and personalized health display. This display in turn can serve as an incentive for the patient to change his/her lifestyle and thus act as a health promotion tool. The temporal evolution of the center of gravity generates a comprehensive and long-term estimate of one's health dynamics. It could also be useful in non-ill prevention. Finally, it is believed that the display and evaluation of medical examination data over a long time span is an effective way of utilizing medical examination data not seen in conventional medical examination tools.

Finally, with the proposed tool, the search for trends in medical examination data and general judgment can run on a personal computer. Furthermore, it is possible to check abnormality by switching to the corresponding screen. Thus, in this way, a continuous and detailed health evaluation of medical examination data is supported. In addition, since in a face-to-face consultation the computer displays of our tool can be shared with the patient, it can be expected that those displays will support the health advice given to the patient.

Acknowledgments The authors are very obliged to Prof. M. Van Hulle of KU Leuven, Belgium, for kindly reading and correcting our manuscript.

References

1. The metabolic syndrome diagnosis standard exploratory committee: The definition and the diagnosis standard of the metabolic syndrome. J. Jpn. Soc. Intern. Med. (J. Jpn. Soc. Int. Med.) **94**, 188–213 (2005)
2. Tokutaka, H., Maniwa, Y., Kihato, P.K., Fujimura, K., Ohkita, M.: Application of SOM in a health evaluation system. In: WSOM2007, Bielefeld, Germany, 3–6 Sept (2007)
3. Tokutaka, H., Ohkita, M., Kasezawa, N., Ohki, M.: Verification of metabolic syndrome checkup data with a self-organizing map (SOM): towards a simple judging too. In: WSOM2012, Chile (2012)

4. Tokutaka, H., Kasezawa, N.: Verification of distinction ability by self organizing maps (SOM) about metabolic syndrome data in health evaluation and promotion. Health Eval. Promot. **37**(3):389–397 (2010) (in Japanese)
5. Kasezawa, N., Toyama, K., Nakano, M., Hirota, K., Morishita, T., Tokutaka, H.: Usefulness of computer-based support tool based on self-organizing maps (SOM) for metabolic syndrome checkup and aftercare in health evaluation and promotion. Health Eval. Promot. **38**(5):574–583 (2011) (in Japanese)
6. http://www.somj.com
7. Ohkita, M., Tokutaka, H., Fujimura, K., Gonda, E. (eds): The Self-Organizing Maps and the tools. Springer-Japan Inc., (2008) (in Japanese)

The Effect of SOM Size and Similarity Measure on Identification of Functional and Anatomical Regions in fMRI Data

Patrick O'Driscoll, Erzsébet Merényi, Christof Karmonik
and Robert Grossman

Abstract We demonstrate the advantage of larger SOMs than those typically used in the literature for clustering functional magnetic resonance images (fMRI). We also show the advantage of a connectivity similarity measure over distance measures for cluster discovery and extraction. We illustrate these points through maps generated from a multiple-subject investigation of the genesis of willed movement, where clusters of the fMRI time-courses signify functional (or anatomical) regions, and where accurate delineation of many clusters is critical for tracking the relationships of neural activities across space and time. While we do not provide an automated optimization of the SOM size it is clear that for this study increasing it up to 40 × 40 facilitates clearer discovery of more relevant clusters than from a 10 × 10 SOM (a size frequently used in the literature), and further increase has no benefits in our case despite using large data sets (all data from whole-brain scans). We offer insight through data characteristics and some objective justification.

This work was partially supported by the Program for Mind and Brain, Department of Neurosurgery, Houston Methodist Hospital. Figures are in color, request a color copy by email: patrick.odriscoll@rice.edu, erzsebet@rice.edu

P. O'Driscoll (✉)
Applied Physics Program, Rice University, Houston, TX, USA
e-mail: patrick.odriscoll@rice.edu; po2@rice.edu

E. Merényi
Department of Statistics and Department of Electrical & Computer Engineering,
Rice University, Houston, TX, USA
e-mail: erzsebet@rice.edu

C. Karmonik · R. Grossman
Department of Neurosurgery, Houston Methodist Neurological Institute,
Houston Methodist Research Institute, Houston, TX, USA
e-mail: CKarmonik@houstonmethodist.org

R. Grossman
e-mail: RGrossman@houstonmethodist.org

C. Karmonik
Weill Medical College of Cornell University, New York, NY, USA

© Springer International Publishing Switzerland 2016
E. Merényi et al. (eds.), *Advances in Self-Organizing Maps and Learning Vector Quantization*, Advances in Intelligent Systems and Computing 428,
DOI 10.1007/978-3-319-28518-4_22

251

Keywords Conscience self-organizing map · CONNvis · Cluster extraction · Functional magnetic resonance imaging · Willed movement · Data-driven model

1 Background and Motivation

In this paper we aim to demonstrate that SOM size significantly influences cluster identification. We also aim to demonstrate the benefits of a connectivity based (rather than distance based) measure for cluster extraction from a converged SOM. To do this we analyze full brain functional magnetic resonance imaging (fMRI) data of humans generating willed movement initiated from a visual stimulus. fMRI is an accepted method to non-invasively infer real-time neural activity from a hemodynamic response known as the blood oxygen level dependence (BOLD) signal. fMRI data comprises time-courses, or time-series, of the BOLD signal at each voxel in a regular three-dimensional grid over a brain volume. Traditionally, a map reflecting neural activity level is constructed by computing the statistical likelihood of each voxel's fit to a given model of the BOLD signal. Activity maps, however, only provide a comparison of the activation strengths of various regions, but do not reveal the functional relationships of the activation patterns (time-courses).

Voxels clustered based on the similarity of their time-courses can be used to identify functional regions of the brain, in a model-free (data-driven) approach. Various techniques including graph based, statistical, and artificial neural network methods have been applied for this purpose. Kohonen SOMs [1] in particular, have been successful in either outperforming other methods or providing deeper insights (e.g., [2–6]). While it is widely known that too small SOMs can be suboptimal for cluster extraction, fMRI studies tend to use small SOMs ranging from 3 × 3 to 12 × 12 neurons, often trained only on selected subsets of the available data. Such small SOMs can work for specific goals as in the examples we review below. We will argue, however, that larger SOMs could allow more detailed discoveries or more comprehensive analyses of the whole brain.

Authors of [2–4] use the whole brain (or substantial portion) but constrain their focus to relatively few functional regions. The interest in [2] is to capture 4–5 functional regions from each of the resting and a goal directed state. After experimenting with SOM sizes ranging from 4 × 4 to 12 × 12 neurons the authors conclude that a 10 × 10 SOM suffices for finding the targeted functional regions. A 10 × 10 SOM is used by [3] to examine the effects of age on autism, by capturing 16 clusters that represent a handful of expected active areas of the rest state and the default mode network (DMN). Similarly, 10–20 clusters are extracted from a 11 × 11 SOM in [4], delineating expected regions mainly in the motor cortex. Other studies limit the amount (and complexity) of the data by processing only selected parts. Both [5, 6] take the widely used approach of excluding voxels that fall below some activation level. [5] uses unsupervised SOM, [6] uses supervised SOMs to obtain a small number of clusters/classes (3–8) of very small numbers of voxels (few hundred to a few thousands), and evaluate clustering quality or classification accuracy as a function of the number of voxels processed. The SOMs are small (6 × 4 in [5], undeclared size in [6] but look no larger than ~10 × 10.) [5] concludes that keeping only active voxels with increasing ROI specificity (smaller and smaller sets of voxels) improves results. [6] shows that increasing the ratio of active voxels to inactive ones improves classification, albeit the accuracies are rather low ($\lesssim 0.5$ for real data). However, neither paper investigates how a larger SOM would facilitate better results by coping with more voxels or providing more resolution for cluster separation. For clustering the SOM, typically ℓ_2-distance based measures are applied although some works use more sophisticated clustering methods than others. Visualization, where used, is most often the plotting of prototype vectors into their SOM grid locations.

In this work we show the benefits of using larger SOMs than those typically found in fMRI literature, and we also show the advantage of using a non-distance-based metric to extract clusters from converged SOMs. We demonstrate these points on *whole brain* fMRI data.

2 Data Collection, Acquisition, and Pre-processing

Here we describe the experiment performed for our data collection, the acquisition parameters and resulting dataset, and the pre-processing of that data.

Experiment A series of ten human faces (five pleasant and five unpleasant) are presented to subjects in a random order, generally with a 50s rest period. Each face is shown for 10s, and judged by the subject to be pleasant or unpleasant. The subject is instructed to squeeze a ball placed in his/her right hand if the face is judged to be unpleasant, until the face goes away. If the subject finds the face pleasant, he/she does nothing. Figure 1 shows part of the experiment with expected BOLD signals, generated in the left motor cortex, as a result of the subject's reaction to unpleasant faces. When the subject sees an unpleasant face, he/she makes a willed movement, thereby generating a series of neural activities that travel through both time and space in the brain. The activity originates in the visual cortex upon perceiving the face, then travels to other parts of the brain, and finally reaches the left sensory-motor cortex when the subject squeezes the ball. We are investigating the spatial and temporal relationships between the areas of the brain that participate in this process. In this paper we concentrate on describing the methods used to extract this information by clustering.

Data Acquisition and Pre-processing The data of six subjects from a larger study under an IRB approved protocol are analyzed. The fMRI data is collected using a Siemens Vario 3 Tesla scanner. Each subject sees each face for 10s. The duration of the rest period is generally 50s, long enough to allow the expected BOLD signal to completely subside before the next face presentation. The voxel size ranges between $2.750 \times 2.750 \times 5.000$ mm^3 and $3.594 \times 3.594 \times 5.000$ mm^3, the temporal resolution varies from 1.0 to 1.5s per brain scan across subjects, yielding data cubes of ~$64 \times 64 \times 24 \times 460$ (i.e., approx. 100 k time-courses each with approx. 460 samples). Pre-processing follows that in [7], which performed well in our experiments: motion correction, high- and low-pass filtering (which removes signal outside the

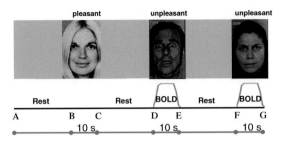

Fig. 1 Sample experiment consisting of showing three faces (one pleasant, and two unpleasant). Time windows A to B, C to D, and E to F are rest periods, B to C is a pleasant face presentation, and D to E and F to G are unpleasant face presentations. The expected BOLD signal (in the left motor cortex) is shown for the two unpleasant face presentations, our windows of interest

0.008–100.0 Hz frequency range), and each time-course is scaled by its ℓ_2-norm. Areas outside the brain are masked (excluded) from processing. All these steps are carried out using AFNI [8], an open source data visualization and processing software. To concentrate on relevant information in the time-courses, the windows of interests—such as the windows of face showing—may be extracted and concatenated to form the input vectors for clustering. We follow another approach using a single window. Since data from the first unpleasant face presentation is most likely to be free of irrecoverable artifacts in all subjects we use an interval of 36 points (40–50s) encompassing the entire ramp up and down of the BOLD signal generated by this event.

3 Analysis Methods

We use a SOM with conscience learning, or Conscience SOM (CSOM) [9], for maximum entropy (equiprobabilistic) mapping, thus potentially more faithful matching of the *pdf* of the data by the SOM prototypes. Compared to the Kohonen SOM algorithm, this is achieved by the use of a *bias* at winner selection, thereby discouraging frequently winning nodes from winning and encouraging infrequent winners to win more:

$$c(\mathbf{x}) = argmin_i (||\mathbf{x} - \mathbf{w}_i|| - bias_i), i = 1, ..., N \tag{1}$$

Here N is the number of SOM prototypes $\mathbf{w}_i \in R^n$, $\mathbf{x}$ is a point in the data manifold $M \subset R^n$, and c indexes the winning prototype $\mathbf{w}_c$. The bias for prototype $\mathbf{w}_i$ is computed as in Eq. (2) where γ is a user-controlled parameter, and F_i is the winning frequency of $\mathbf{w}_i$, updated after each learning step. The weight update rule remains the same as for the KSOM (Eq. 3).

$$bias_i = \gamma (1/N - F_i) \tag{2}$$

$$\mathbf{w}_i(t+1) = \mathbf{w}_i(t) + \alpha(t)h_{c,i}(t)(\mathbf{x} - \mathbf{w}_i(t)) \tag{3}$$

The CSOM neighborhood function $h_{c,i}$ can have a constant small radius r (of 1 or 2) throughout the learning process because the "conscience" ensures the propagation of collaboration among prototypes. We use $r = 1$ or $r = \sqrt{2}$, (updating the 4 or 8 immediate neighbors in diamond-shaped or square neighborhoods, respectively), in a rectangular lattice. This significantly reduces computational cost. The equiprobabilistic mapping property of the CSOM was shown in [9] for 1-dimensional data, and demonstrated for higher-dimensional data in [10, 11].

Cluster Extraction For capturing clusters of fMRI time-courses from converged SOMs we compare the relative merits of two frequently used inexpensive visualizations, mU-matrix [10] and the plot of prototype vectors at their SOM grid locations, with CONNvis [12] (Fig. 2). We note that visualizations such as U, P, AU*, AP, matrices [13] (and references therein) – while attractive and effective when used for an emergent SOM – are not applicable in our case. They require the number of prototypes to be close to the number of data points, which is not practical for our large data size. Just as importantly, large number of prototypes does not help clustering of our fMRI data, as we will see.

The mU-matrix [10] is a refinement of the classic U-matrix [14]. It represents the Euclidean distance of a prototype to each of its eight lattice neighbors. The distances are visualized as thin gray-scale "fences" between adjacent SOM grid cells (instead of shading each grid cell to the average value of the distances). Dark fence means small distance, bright fence means strong separation and therefore may indicate cluster boundary. The mU-matrix also encodes the mapping density by the brightness of a monochrome cell color (red in Fig. 2a) which is proportional to the number of data points mapped to the cell. An example can be seen in Fig. 2a. We also plot the prototypes at their lattice locations as it is a customary way to show the learned SOM in fMRI studies, and it provides a direct visual assessment of the pattern differences (Fig. 2c).

The CONNvis is a visualization of the CONN similarity measure, which expresses *connectivity* rather than distances. The connectivity, $CONN(i,j)$, of two prototypes $\mathbf{w}_i$, $\mathbf{w}_j$, is the number of times $\mathbf{w}_i$ and $\mathbf{w}_j$ are selected as a pair of best matching unit (BMU) and second BMU for any data point. $CONN(i,j) > 0$ means that $\mathbf{w}_i$, $\mathbf{w}_j$ are Voronoi neighbors in M. The visualization shows the connectivity for every pair of prototypes (black points in Fig. 2b) by a connecting line where the line width is proportional to the (normalized) $CONN$ value. For visualization purposes the line widths are also binned to help the human eye. The binning, described in detail in [12], is non-linear and governed by the data statistics. Discontinuities or weakly connected regions of the manifold emerge where no or very thin connections are drawn. The connections of a prototype to its Voronoi neighbors are ranked by their relative strengths and the ranking is indicated by colors: red line connects to the most important Voronoi neighbor, followed by blue, green, yellow, and gray shades. The ranking expresses local manifold relations and provides finer details for the identification of cluster boundaries. As an additional benefit CONNvis shows topology violations: prototypes connected with line segments longer that one lattice unit violate topology preservation. The line width indicates the severity of the violation. A procedure for

(a)   **(b)**   **(c)**   **(d)**

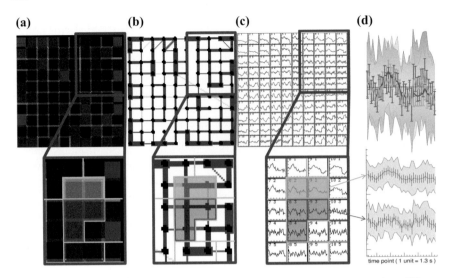

Fig. 2 Example of extracting two clusters belonging to the visual cortex from three different visualizations of a 10 x 10 CSOM. These two clusters are indicated by the *light green* and *dark green* outlines, highlights and lines. **a** mU-matrix, **b** CONNvis, **c** prototypes plotted at their SOM grid cells, and **d** *top*: average time-courses of the two *green* clusters vertically exaggerated and overlain for comparison, with standard deviations (*vertical bars*), and ranges shown; bottom: the same two average time-courses shown separately. Other clusters found in the boxed SOM area (some also related to the visual cortex) are outlined in *orange*. The mU-matrix representation, which expresses clusters well for many other types of data, seems insensitive to the small differences in prototype distances that appear to characterize fMRI data. Owing to the connectivity measure, the CONNvis shows clearer clusters despite their high degree of similarity. The shapes of the prototypes are consistent with the extracted clusters

cluster extraction based on CONNvis is also outlined in [12]. Figure 2 shows an example of extracting two clusters, indicated in green boxes, from a 10×10 CSOM. In Fig. 2b these are defined by groups of prototypes with strong connections to each other (thick red lines) while each group's connection to another group of prototypes is less strong (blue lines). The two clusters highlighted in green primarily make up the visual cortex. Their close relationship is expressed by the strong ranking (blue) of their interconnections in the CONNvis representation. Figure 2c provides evidence for this grouping. Other clusters in this inset are indicated in orange boxes but not discussed here.

Data Post-Processing For the purpose of tracking the generation of the willed movement, we filter the extracted SOM clusters for displays of brain maps showing associations with the visual stimulus and the clenching of the right fist. The filtered clusters are those whose average time-courses correlate relatively strongly with the mean of the cluster identified as the visual cortex. Other clusters are assumed to represent the rest state or other involvement. The correlation threshold, in this case 0.5, is empirically determined and can vary for different data and tasks. Our discussion of the clustering quality as a function of SOM size, however, includes all clusters we delineate, not only the filtered ones.

4 Effects and Evaluation of SOM Size

All clusters extracted from the 10×10 SOM in Fig. 2, and from a 40×40 SOM (18 and 29, respectively) can be seen in Fig. 3. Filtered clusters mapped back to two selected brain slices are shown in Fig. 4. The quality of the extracted clusters can greatly differ depending on the SOM size. By allocating more prototypes to high-density areas, the 40×40 SOM facilitates separation of groups of similar fMRI time-courses with small but consistent differences. This translates to finer spatial resolution and delineation of more, functionally distinct, areas in the brain than from the 10×10 SOM. An example can be seen by the comparisons made in Fig. 4. While clusters belonging to the superior frontal and medial frontal gyri (the magenta clusters) are detected from both the 10×10 and 40×40 SOMs, the 40×40 SOM also allows to fully resolve the sensory-motor area (dark red cluster), and the detection of the cerebellum (dark blue cluster). These regions cannot be mapped from the 10×10 SOM without including large swaths of other brain areas. The visual cortex is

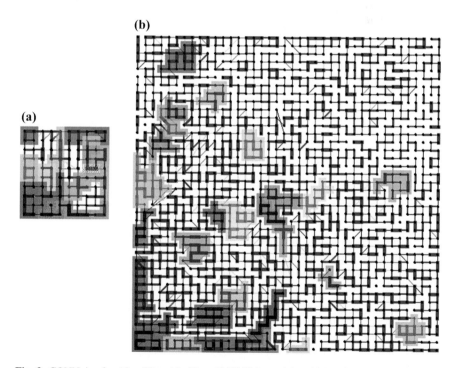

Fig. 3 CONNvis of **a** 10×10 and **b** 40×40 CSOM, overlain with extracted clusters (colored groups of prototypes). The color coding of clusters belonging to the same functional regions in the brain is as similar as possible in the two SOMs, but cannot be made identical due to more resolved clusters in the 40×40 SOM. Unclustered areas of the 40×40 SOM contain prototype groups that map to spatially incoherent sets of voxels or unimportant features in the brain (such as spinal fluid). Data: Subject 2

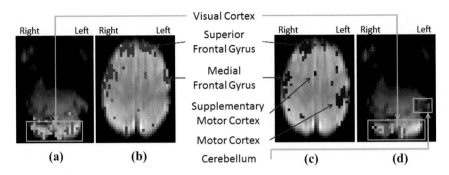

Fig. 4 Comparison of clusters extracted from the 10×10 and 40×40 CSOMs in Fig. 3, filtered and mapped to the brain. The two selected axial slices display clusters associated with the visual, motor, and cognitive functions. **a**, **b**: Clusters identified from the SOM in Fig. 3a. **c**, **d**: Clusters identified from the SOM in Fig. 3b

resolved in both SOMs (Figs. 4a and d). Both these clusterings as well as one from a 20×20 SOM were validated and compared by neuroscientist experts, judging the 40×40 clustering as significantly better than the others.

The advantage of the larger SOM size can also be measured objectively using cluster validity indices. There exist many indices, and some are better suited for high-dimensional data with complex cluster structure than others. We give here measurements by four indices, listed in columns 3–6 of Table 1. Two of them, the classic Davies-Bouldin Index (*DBI*, [15]), and the newer Pakhira-Bandyopadhyay-Maulik-index (*PBM*) favors spherical clusters when ℓ_2 distances are used. *PBM* strongly favors a small number of clusters (penalizes the number of clusters quadratically). Composed density between and within clusters (*CDbw*) rewards clusters with homogeneous density. *CONNindex* [16] is a recent one developed to address difficulties caused by irregular clusters and complicated cluster structure. We sketch the essence of *DBI* and *CONNindex* below. Due to space constraints please see formulae and references for *PBM* and *CDbw* in [16].

DBI is defined as the average, over all clusters, of the maximum ratio of the average intra-cluster scatter (standard deviation in this case) to the inter-cluster separation. The inter-cluster separation is the distance between cluster centers. *CONNindex* relies on the *CONN* connectivity measure [12]. As defined in [16], $CONNindex = Intra\_Conn \times (1 - Inter\_Conn)$ where *Intra_Conn* is the average *intra-cluster connectivity*, and *Inter_Conn* is the average of the maximum *inter-cluster connectivities* where averaging is over all clusters C_k. The intra-cluster connectivity of a cluster C_k is the proportion of connections between prototypes that reside inside C_k, to all connections that the prototypes of C_k have to any other prototypes. The inter-cluster connectivity of two clusters C_k, C_l is the proportion of connections between prototypes of C_k and C_l (in either direction), to all connections (to any cluster) of those prototypes in C_k which have at least one connection to C_l. Both *Intra_Conn* and $1 - Inter\_Conn$ are 1 when all clusters are completely separated. The value ranges of these measures are shown in Table 1, along with arrows pointing from worst to

Table 1 Quality measures (explained in the text) for clusterings of the same (Subject 2) fMRI data from three different SOMs, with best in bold face and worst in italics

SOM size	Nr clusters	DBI $0 \leftarrow \infty$	PBM $0 \rightarrow \infty$	CDbw $0 \rightarrow \infty$	CONNind $0 \rightarrow 1$	Intra_Conn $0 \rightarrow 1$	Sep_Conn $0 \rightarrow 1$
10×10	18	2.854	**0.0008**	*0.023*	*0.364*	*0.535*	*0.681*
20×20	25	*2.934*	0.0007	**0.027**	0.408	0.550	0.741
40×40	29	**2.761**	*0.0006*	0.025	**0.572**	**0.716**	**0.799**
60×60	—	—	—	—	—	—	—

Value ranges and arrows pointing from worst to best are under the respective measures

best value. *Sep_Conn* stands for $1 - Inter\_Conn$. While it is hard to compare open-ended indexes, it is helpful to know that *DBI* values tend to be below 10, and $DBI > 1$ indicates overlaps but $DBI < 1$ does not necessarily mean separated clusters. *CDbw* values can be much larger. *PBM* is scaled by $\frac{1}{K^2}$ where K is the number of clusters, which can make its values magnitudes smaller compared to *DBI*.

Quality measures for clusterings of the same fMRI data from SOMs of three different sizes are summarized in Table 1. Both *DBI* and *CDbw* assign very similar scores to all SOMs although the 40×40 SOM is slightly better by the *DBI* and the 20×20 SOM by the *CDbw*. However, given the typical value ranges of these indices all scores are poor, and the differences are negligible. A reasonable explanation is the model-dependence of these indices. *DBI* misjudges clusterings with non-spherical and unevenly sized clusters. *CDbw* is likely failing because of possibly heterogeneous densities. If we ignored the quadratic penalty by *PBM* (scaled it back by K^2, i.e., 324, 625, and 841, respectively) it would indicate substantial differences, progressively to the advantage of the larger SOM. While the 40×40 SOM is confirmed by experts as the best, the *DBI*, *CDbw*, and *PBM* have difficulty correctly judging the highly irregular fMRI clusters. *CONNindex*, in contrast, handles irregular clusters and shows significant increase, given its range, in quality from 10×10 to 40×40 SOM size. Examining the components of *CONNindex*, the 40×40 SOM preforms significantly better in both metrics. It is noteworthy though that the larger increase is in the intra-cluster connectivity term, indicating more self-contained clusters. This is due to a sufficient number of prototypes for accurate mapping of the manifold structure, increasing the proportion of connections inside clusters regardless of their shapes. The connectivity measure senses this improvement correctly. No sensible cluster extraction could be done from a 60×60 SOM, which we attribute to the highly mixed an noisy signals (discussed below) in fMRI voxels. The 60×60 SOM has enough prototypes to begin to model the structure of the noise rather than the characteristics of the functional regions we aim to capture.

fMRI data is highly complex, partly because the voxels are large compared to the spatial extent of distinct neuronal signals and the variations of tissue types. This results in heavily mixed signals (time-courses) of tissue types and functional regions within a voxel. Exacerbating this mixing is the nature of the BOLD signal, which is not always constant within the same functional region. It reflects overlapping spatial and temporal influences, potentially from many voxels depending on the functional region, subject and other factors. The result is a large degree of overall mixing that dilutes the discriminating characteristics of distinct functional regions. Figure 2d is an illustration of the level of similarity.

While formal optimization of SOM size is beyond the scope of this paper, we can also draw approximate justification for the 40×40 SOM from a Growing SOM (GSOM, [17]), which returns a $7 \times 6 \times 4 \times 4 \times 3 \times 2 \times 2$ SOM. With the last two dimensions close to vanishing the rest of this SOM comprises 2016 neurons, a number close to the 1600 neurons in the 40×40 SOM we use, and much larger than the number of neurons in a 10×10 or 20×20 SOM.

5   Results from Multiple Subjects

Figure 5 shows the localization of filtered clusters extracted from 40×40 SOMs and mapped back to the three-dimensional brains for each of the six subjects. The presented clusters belong to brain regions involved in the visual processing and motor response, and show commonality of the activated areas across subjects. Representative slices are chosen to exhibit the visual, motor, and supplementary motor cortex. Not all extracted clusters can be displayed in each of the three slices. For example in subject 2 the activation in the visual cortex is shown in the coronal slice, but not in the more laterally located sagittal slice. The visual cortex and cuneus (the group of green clusters in the coronal slices, and at the bottom of the sagittal slices) are activated by the visual stimulus. The left motor cortex and sensory cortex (red clusters at right in the axial, and at top in the sagittal slices) are active, consistent with squeezing the ball with right hand. Subjects 1, 2, and 6 exhibit some bi-lateral activity of the motor areas, with the larger response in the left brain (corresponding to the movement in the right hand). The supplementary motor area, also used in the generation of movement, is activated in each subject (red clusters at the center of the axial slices) with subjects 1 and 3 generating the largest and most coherent response areas. A number of clusters also appear, consistently across subjects, in other functional regions such as the superior and medial frontal gyri (magenta colors). While those, and several more that map to other brain slices (e.g., cerebellum, thalamus, cyngulate gyrus, precuneus, and caudate nucleus, not shown here) may correlate with the visual cortex to lesser extent, their common activation in all (or most) subjects calls attention to relationships worth investigating, and may hold keys to new discoveries of neuronal processes.

We note that, since clusters reflect similarity of time-courses, the same cluster may occur in multiple areas. For example, in the axial slice of subject 4, the green clusters cover parts of the sensory cortex (adjacent to the red motor cortex cluster at center right) and a section of the precuneus (the green cluster at the bottom of the slice). This means in subject 4 these areas are highly correlated, likely a result of

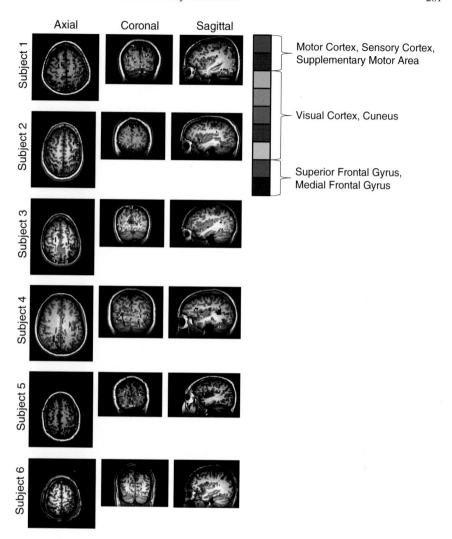

Fig. 5 Filtered clusters shown for all six subjects, in selected axial, coronal, and sagittal slices on the anatomical substrate. Here we only show clusters which occur in all subjects in the motor cortex, supplementary motor area and visual cortex (where activation is expected during our experiment), and in the cuneus, superior frontal gyrus, and medial frontal gyrus. The color wedge codes clusters which are present in these slices. Cluster colors are grouped into three hues that signify closely related functional/anatomical regions. The slices shown are selected to display the same functional regions in each subject. (Geometric co-registration remains a follow-up task at this time.)

a slightly different neural pathway that subject 4 uses to complete the task. Slight deviations of the pathways are expected in each subject. Thus, the same level of correlation between the same regions is not common in all subjects.

6 Conclusions

Our objective is to call attention to the untapped potentials of larger SOMs than those ($\sim 10 \times 10$) typically employed in fMRI analyses; to CSOM; and to connectivity (non-distance-based) measures, for better SOM manifold learning and cluster extraction. To that end we demonstrate, through real, full-brain fMRI data that increasing the SOM size up to a point (40×40 lattice in our case) facilitates cleaner capture of more relevant clusters than small SOMs. Importantly, further increase of the SOM size is detrimental to the clustering. We provide justification that this is due to the highly mixed and noisy time-course signals in fMRI data. Clusters in functional regions relevant to the generation of willed movement (the goal-oriented task we analyze), as well as others, are consistently identified from 40×40 SOMs across six subjects. This in turn supports more detailed elucidation of the functional relationships of brain regions and potentially allows discoveries of more nuanced neuronal activities related to the goal-oriented task. Follow-up work will strive for more comprehensive computational experiments and more formal investigation of the dependence of SOM sizes on the data characteristics.

References

1. Kohonen, T.: Self-Organizing Maps, 2nd edn. Springer, Berlin (1997)
2. Peltier, S.J., Polk, T.A., Noll, D.C.: Detecting low-frequency functional connectivity in fMRI using a Self-Organizing Map (SOM) algorithm. Hum. Brain Mapp. **20**, 220–226 (2003)
3. Wiggins, J.L., Peltier, S.J., Ashinoff, S., Weng, S.-J., Carrasco, M., Welsh, R.C., Lord, C., Monk, C.S.: Using a Self-Organizing Map algorithm to detect age-related changes in functional connectivity during rest in autism spectrum disorders. Brain Res **1380**, 187–197 (2011). Mar
4. Fischer, H., Hennig, J.: Neural network-based analysis of MR time series. Magn. Reson. Med. **41**(1), 124–131 (1999)
5. Erberich, S.G., Willmes, K., Thron, A., Oberschelp, W., Huang, H.: Knowledge-based approach for functional MRI analysis by SOM neural network using prior labels from talairach stereotaxic space. Med. Imag. pp. 363–373 (2002)
6. Hausfeld, L., Valente, G., Formisano, E.: Multiclass fMRI data decoding and visualization using supervised Self-Organizing Maps. NeuroImage **96**, 54–66 (2014)
7. O'Driscoll, P.: Using Self-Organizing Maps to discover functional relationships of brain areas from fMRI images. Masters thesis, Rice University (2014)
8. Cox, R.W.: AFNI: Software for analysis and visualization of functional magnetic resonance neuroimages. Comput. Biomed. Res. **29**(0014), 162–173 (1996)
9. DeSieno, D. Adding a conscience to competitive learning. In: Proceeding of ICNN, July 1988 (New York) vol. I, pp. 117–124, (1988)
10. Merényi, E., Jain, A., Villmann, T.: Explicit magnification control of Self-Organizing Maps for "forbidden" data. IEEE Trans. Neural Netw. **18**, 786–797 (2007). May
11. Merényi, E.: Precision mining of high-dimensional patterns with Self-Organizing Maps: interpretation of hyperspectral images. In: Sinčak, P., Vaščak J. (eds.) Quo Vadis Computational Intelligence, Vol 54. Physica-Verlag, (2000)
12. Taşdemir, K., Merényi, E.: Exploiting data topology in visualization and clustering of Self-Organizing Maps. IEEE Trans. Neural Netw. **20**(4), 549–562 (2009)
13. Lötsch, J., Ultsch, A.: Exploiting the structures of the U-matrix. In: Advances in Self-Organizing Maps and Learning Vector Quantization, pp. 249–257. Springer (2014)

14. Ultsch, A., Siemon, H.P.: Kohonen's self organizing feature maps for exploratory data analysis. vol. 1, pp. 305–308 (1990)
15. Davies, D.L., Bouldin, D.W.: A cluster separation measure. IEEE Trans. Pattern Anal. Mach. Intell. **2**, 224–227 (1979)
16. Taşdemir, K., Merényi, E.: A validity index for prototype based clustering of data sets with complex structures. IEEE Trans. Syst. Man Cybern. Part B **41**, 1039–1053 (2011). August
17. Bauer, H.-U., Villmann, T.: Growing a hypercubical output space in a Self-Organizing Feature Map. IEEE Trans. Neural Netw. **8**(2), 218–226 (1997)

Part V
Learning Vector Quantization Theories and Applications I

Big Data Era Challenges and Opportunities in Astronomy—How SOM/LVQ and Related Learning Methods Can Contribute?

Pablo A. Estévez

Abstract Astronomy is facing a paradigm shift caused by the exponential growth of the sample size, data complexity and data generation rates of new sky surveys. For example, the Large Synoptic Survey Telescope (LSST), which will begin operations in northern Chile in 2022, will generate a nearly 150 Petabyte imaging dataset. The LSST is expected not only to improve our understanding of time varying astrophysical objects, but also to reveal a plethora of yet unknown faint and fast-varying phenomena. In this talk I will present big data era challenges and opportunities in astronomy from the point of view of computational intelligence, machine learning and statistics. In particular, I will address the question of how SOM/LVQ and related learning methods can contribute to cope with these challenges and opportunities.

Keywords Astronomy · Astrophysics · Big data · Self-organizing map · Learning vector quantization

P.A. Estévez (✉)
Department of Electrical Engineering, University of Chile and Millennium Institute of Astrophysics, Santiago, Chile
e-mail: pestevez@cec.uchile.cl

© Springer International Publishing Switzerland 2016
E. Merényi et al. (eds.), *Advances in Self-Organizing Maps and Learning Vector Quantization*, Advances in Intelligent Systems and Computing 428, DOI 10.1007/978-3-319-28518-4_23

Self-Adjusting Reject Options in Prototype Based Classification

T. Villmann, M. Kaden, A. Bohnsack, J.-M. Villmann, T. Drogies, S. Saralajew and B. Hammer

Abstract Reject options in classification play a major role whenever the costs of a misclassification are higher than the costs to postpone the decision, prime examples being safety critical systems, medical diagnosis, or models which rely on user interaction and user acceptance. While optimum reject options can be computed analytically in case of a probabilistic generative classification model, it is not clear how to optimally integrate reject strategies into efficient deterministic counterparts, such as popular learning vector quantization (LVQ). Recently, first techniques propose promising a posteriori strategies for an efficient reject in such cases (Fischer et al. Neurocomputing, 2015 [7]). In this contribution, we take a different point of view and formalize optimum reject via an integrated cost function. We show that an efficient approximation of these costs together with a geometric reject rule leads to an extension of LVQ which not only aligns the classification model along the reject costs but also self-adjusts an optimum reject threshold while training.

T. Villmann (✉) · M. Kaden
Computational Intelligence Group, University Applied Sciences Mittweida,
Mittweida, Germany
e-mail: thomas.villmann@hs-mittweida.de

A. Bohnsack
Staatliche Berufliche Oberschule Kaufbeuren, Kaufbeuren, Germany

J.-M. Villmann · T. Drogies
Institute for Laboratory Medicine, Clinical Chemistry and Molecular
Diagnostic University Leipzig, Leipzig, Germany

T. Drogies
LIFE - Leipzig Research Center for Civilization Diseases, University Leipzig,
Leipzig, Germany

S. Saralajew
Dr. Ing. h.c. F. Porsche AG Weissach, Weissach, Germany

B. Hammer (✉)
CITEC Centre of Excellence, University Bielefeld, Bielefeld, Germany
e-mail: bhammer@techfak.unibielefeld.de

© Springer International Publishing Switzerland 2016
E. Merényi et al. (eds.), *Advances in Self-Organizing Maps and Learning
Vector Quantization*, Advances in Intelligent Systems and Computing 428,
DOI 10.1007/978-3-319-28518-4_24

1 Introduction

Machine learning and automated classification constitute integral parts of technical systems in a variety of application domains such as deep learning for image classification, prototype based models for robust online learning in vision systems, or ensemble classifiers for biomedical data analysis, to name just a few examples [3, 4, 14]. While machine learning models often reach a high accuracy close to 100%, there exist cases where errors are unavoidable. One prominent setting is that the observed data incorporate uncertainty such as randomness in the underlying process (aleatoric uncertainty) or incomplete knowledge e.g. due to limited sensorial equipment (epistemic uncertainty) [24]. Another setting where errors arise is given in online learning in changing environments, where concepts are subject to drift as compared to the learned model [18]. In such cases, classification errors necessarily occur, and the question can be raised whether classification of all data is appropriate, or whether it is better to reject the classification of data where the output is subject to a high degree of uncertainty. In particular in safety critical domains, medical diagnosis, or interactive models which rely on a high degree of user acceptance, it might be better to postpone a decision rather than to provide an output with a high degree of uncertainty.

The notion of a *reject option* refers to the possibility of a classifier to output the symbol *reject* rather than a specific class in case of a high degree of uncertainty. A corresponding mathematical treatment relies on the assumption that misclassification costs are higher than postponing the classification, such that it might be favorable not to classify than to make an error. In the pioneering work [5], an optimum reject rule has been derived in case of known Bayesian risk. This treatment can be extended to plug-in rules where the Bayesian risk is only approximated provided certain regularity assumptions apply [11]. This theoretical framework offers a direct interface to integrate rejects into classification models which estimate the classification probability, such as Gaussian processes, Bayes classifiers, or graphical models. It can also be combined with classifiers which are a-posteriori extended towards a probabilistic output [6, 17].

One alternative to a probabilistic treatment is offered by a combination of classifiers with a deterministic reject option which is based on geometrical rules such as the distance to the decision boundary. Quite a number of approaches extend popular classification schemes in this way, mostly relying on heuristic grounds [7, 19]. Interestingly, deterministic reject rules can surpass the accuracy of probabilistic counterparts if the latter do not optimally approximate the underlying generative characteristic [7]. Further, local reject rules can greatly enhance the quality of a global reject strategy [8, 16]. These approaches rely on posterior reject strategies based on the given classification prescription and they do not optimise the decision boundary itself according to the reject costs while training.

Classification with a reject option which optimises the full system according to the underlying loss function has been investigated extensively in the case of support vector machines [2, 10, 26]. Here the standard two class classifier is equipped with a third option, a reject in case the distance to the decision boundary falls into a specified interval. The loss function is approximated by a convex surrogate, for which consistency can be proved under certain conditions.

In this article, we address prototype-based classification schemes, which have become increasingly popular in recent years in the context of interpretable models e.g. in the biomedical application domain as well as life-long learning with adaptive model complexity due to their intuitive and sparse representation of the model in terms of representative prototypes [1, 4, 9, 14, 22, 25, 27]. We will focus on modern variants which adjust prototypes in a discriminative way based on a suitable cost function, and which can be accompanied by strong learning theoretical guarantees [23]. Recently, first reject options which enrich these methods with posterior reject strategies based on geometrical principles have been proposed [7, 8]. Since the techniques rely on a trained model, they do not optimally adjust the prototype positions itself according to the rejection costs, nor do they provide an optimum threshold provided known rejection costs. In this contribution, we will propose a novel strategy which derives prototype-bases training models from an optimization of the classification costs including a reject option. The resulting algorithm not only adjusts the prototype positions accordingly, but it also provides an optimum reject threshold for the given scenario. This technology will be demonstrated in a number of benchmark examples.

2 LVQ as Approximation of the 0-1-loss

Learning vector quantization (LVQ) as introduced by KOHONEN constitutes one of the most popular approaches to adjust a prototype-based model in a discriminative way [15]. Here we introduce a cost function based variant as approximation of the standard 0-1-loss [23]. We assume training data have the form $\mathbf{v} \in V \subseteq \mathbb{R}^n$ with known class labels $c(\mathbf{v}) \in \mathfrak{C} - \{1, \ldots, K\}$. An LVQ model is characterised by a fixed number of M prototypes $\mathbf{w}_j \in W \subset \mathbb{R}^n$, $j = 1, \ldots, M$ with labels $y_j \in \mathfrak{C}$. These prototypes define a classification based on a winner-takes-all rule, mapping a vector $\mathbf{v} \in \mathbb{R}^n$ via

$$\mathbf{v} \mapsto y_{s(\mathbf{v})} \text{ where } s(\mathbf{v}) = \mathrm{argmin}_{k=1...M} d(\mathbf{v}, \mathbf{w}_k) \tag{1}$$

where the index $s(\mathbf{v})$ is the index of the best matching prototype with respect to a fixed distance measure $d(\mathbf{v}, \mathbf{w})$, usually the squared Euclidean distance or a more general quadratic form. The 0-1-loss on these data equals

$$E = \sum_{\mathbf{v} \in V} l(y_{s(\mathbf{v})}, c(\mathbf{v})) \text{ where } l(y_{s(\mathbf{v})}, c(\mathbf{v})) = \begin{cases} C_e & \text{if } y_{s(\mathbf{v})} \neq c(\mathbf{v}) \\ 0 & \text{otherwise} \end{cases} \tag{2}$$

with error costs $C_e > 0$ (e.g. $C_e = 1$) which can be rephrased using the equality

$$l(y_{s(\mathbf{v})}, c(\mathbf{v})) = C_e \cdot H(\mu(\mathbf{v})) \text{ where } \mu(\mathbf{v}) = \frac{d^+(\mathbf{v}) - d^-(\mathbf{v})}{d^+(\mathbf{v}) + d^-(\mathbf{v})} \qquad (3)$$

with the Heaviside function H; $d^+(\mathbf{v}) = d(\mathbf{v}, \mathbf{w}^+)$ denotes the dissimilarity between $\mathbf{v}$ and the closest prototype $\mathbf{w}^+$ with the same class label $y^+ = c(\mathbf{v})$, and $d^-(\mathbf{v}) = d(\mathbf{v}, \mathbf{w}^-)$ denotes the dissimilarity value for the best matching prototype $\mathbf{w}^-$ with a class label y^- different from $c(\mathbf{v})$. Instead of the real-valued *classifier function* $\mu(\mathbf{v})$, every function which is positive iff $d^+(\mathbf{v}) > d^-(\mathbf{v})$ could be used; $\mu(\mathbf{v})$ has the advantage that its values are scaled to the interval $(-1, 1)$. Note that we can analogously define a classifier function for data $\mathbf{v}$ without known label: in that case we set $\mathbf{w}^+$ and $\mathbf{w}^-$, respectively, as the two closest prototypes which have a different class label, i.e. $\mathbf{w}^+$ determines the output of the classification.

Since H is not differentiable, SATO&YAMADA introduced a smoothed variant

$$E_{\text{LVQ}}(W) = \sum_{\mathbf{v} \in V} C_e \cdot f_\theta(\mu(\mathbf{v})) \text{ with sigmoidal } f_\theta(t) = (1 + \exp(-t/\theta))^{-1} \quad (4)$$

which approximates the 0-1-loss for $\theta \to 0$ [21]. Note that this variant can easily be extended to incorporate non-equal misclassification costs by a suitable weighting scheme [12]. Update rules can be derived thereof by a stochastic gradient technique, the resulting learning scheme is the generalized LVQ (GLVQ).

3 Distance Based Reject Options in LVQ

CHOW investigated how to optimally extend a classifier to incorporate a reject option [5]. We assume rejection costs $0 < C_r < C_e$ provided the classifier rejects a classification. C_r is used if the output is chosen as $s(\mathbf{v}) = $ reject.

Provided the class conditional probabilities $p(\mathbf{v}|k)$ and priors π_k for a class $k \in \mathfrak{C}$ are known, according to the Bayes criterion, a data point has to be rejected if

$$\max_k \pi_k \cdot p(\mathbf{v}|k) < (1 - \tau) \sum_{j=1}^{K} \pi_j \cdot p(\mathbf{v}|j) \qquad (5)$$

with the optimal *reject threshold* $\tau = C_r/C_e$ as shown in [5]. The evaluation of this reject rule (5), however, requires a good estimation of the class conditional probabilities $p(\mathbf{v}|k)$ which is difficult, in particular in high-dimensional spaces; hence this confines the practicability of Chow's approach.

We rely on learning vector quantisation and distance based reject options as an alternative, whereby the reject threshold and prototype locations are derived from an approximation of the empirical loss function as a surrogate for the expected loss. As proposed in [7], we extend the classification rule to a reject based on the classifier function $\mu(\mathbf{v})$ as

$$\mathbf{v} \mapsto \begin{cases} \text{reject} & \text{if } |\mu(\mathbf{v})| \le \varepsilon \\ y_{s(\mathbf{v})} & \text{otherwise} \end{cases} \qquad (6)$$

where ε constitutes the rejection threshold or reject margin. Hence the 0-1-costs are extended towards a third case:

$$l(y_{s(\mathbf{v})}, c(\mathbf{v})) = \begin{cases} C_e & \text{if } y_{s(\mathbf{v})} \ne c(\mathbf{v}) \\ C_r & \text{if } y_{s(\mathbf{v})} = \text{reject} \\ 0 & \text{otherwise} \end{cases} \qquad (7)$$

This gives rise to the expected loss $\mathcal{E}(l(y_{s(\mathbf{v})}, c(\mathbf{v}))) = \int l(y_{s(\mathbf{v})}, c(\mathbf{v}))p(\mathbf{v})d\mathbf{v}$ with its empirical counterpart

$$E(W, \varepsilon) = \sum_{v \in V} l(y_{s(\mathbf{v})}, c(\mathbf{v})) = C_r \cdot \sum_{v \in V} H(\mu(\mathbf{v}) + \epsilon) + (C_e - C_r) \cdot \sum_{v \in V} H(\mu(\mathbf{v}) - \epsilon) \qquad (8)$$

As usual in machine learning, we optimize this empirical counterpart, whereby general learning theoretical guarantees guarantee the validity of this approach. Since these extended costs are not differentiable, we take the sigmoid softening function f_θ and obtain a smooth approximation which serves as cost function for *GLVQ with reject option (GLVQ-r)*

$$E_{\text{GLVQ-r}}(W, \varepsilon) = C_r \cdot \sum_{v \in V} f_\theta(\mu(\mathbf{v}) + \epsilon) + (C_e - C_r) \cdot \sum_{v \in V} f_\theta(\mu(\mathbf{v}) - \epsilon) \qquad (9)$$

Parameter updates can easily be derived thereof by means of a stochastic gradient descent, relying on the derivatives

$$\frac{\partial C_{\text{GLVQ-r}}}{\partial \mathbf{w}^+} = \frac{2 \cdot d^-}{(d^+ + d^-)^2} \left[C_r \cdot f_\theta'(\mu(\mathbf{v}) + \varepsilon)) + (C_e - C_r) \cdot f_\theta'(\mu(\mathbf{v}) - \varepsilon)) \right] \cdot \frac{\partial d^+}{\partial \mathbf{w}^+} \qquad (10)$$

and

$$\frac{\partial C_{\text{GLVQ-r}}}{\partial \mathbf{w}^-} = \frac{-2 \cdot d^+}{(d^+ + d^-)^2} \left[C_r \cdot f_\theta'(\mu(\mathbf{v}) + \varepsilon)) + (C_e - C_r) \cdot f_\theta'(\mu(\mathbf{v}) - \varepsilon)) \right] \cdot \frac{\partial d^-}{\partial \mathbf{w}^-} \qquad (11)$$

An adjustment of the threshold (margin) ε takes place based on the derivative

$$\frac{\partial C_{\text{GLVQ-r}}}{\partial \varepsilon} = \left[C_r \cdot f_\theta'(\mu(\mathbf{v}) + \varepsilon)) - (C_e - C_r) \cdot f_\theta'(\mu(\mathbf{v}) - \varepsilon)) \right] \qquad (12)$$

These rules apply for a reject strategy with one global threshold. Local margins could be derived similarly from corresponding costs. So far, GLVQ is used with the standard Euclidean distance. Generalizations to metric adaptation or more general forms as summarized in [13] are immediate.

4 Simulations

We tested the approach for different datasets. The first one is an artificial one-dimensional data set of overlapping Gaussians (G1). The other two data sets are real world data. The medical data set (MD) is about prediction of myocardial infarct based on blood properties. The real world technical data (TD) set deals with the detection of headlight calibration situations. For all application we scaled the cost such that $C_e = 1$ and the rejection costs C_r are scaled accordingly.

4.1 Artificial Datasets

The first artificial dataset consists of two overlapping one-dimensional Gaussians with centers $\mu_1 = -0.75$ and $\mu_2 = 0.75$ and variances $\sigma_1 = \sigma_2 = 0.3$. The rejection costs for GLVQ-r was set to $C_r = 0.2$. For this setting, the theoretical threshold $\varepsilon_T = 0.277$ is calculated according to [5, formula (26)]. In our simulation, each Gaussian was sampled by 5000 data points and we used one prototype per class. GLVQ-r achieves (one prototype per Gaussian) a value $\varepsilon = 0.28$, which is a good verification of the theoretical results.

The second artificial dataset is the 'Moon' dataset, which is a composition of two non-linear and overlapping subsets, see Fig. 1. We applied GLVQ-r with 5 and 2 prototypes per class. Depending on the increasing relative rejection costs C_r we observe a decreasing rejection rate and an increasing classification error rate, see Fig. 2. Thus, lower rejection costs yield better accuracy values as expected.

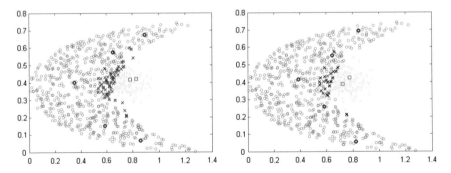

Fig. 1 Visualization of the 'Moon' data. The prototypes for the classes are 'o' and '□'. We observe a clear difference for the prototype distribution in dependence on the reject costs. For higher reject cost the prototypes move to the class borders in critical regions. *left relative* rejection costs $C_r = 0.1$ with 66 rejected data (*red* ×) and two misclassified data (*black* ×). **right** *relative* rejection costs $C_r = 0.45$ with 3 rejected data (*red* ×) and 21 misclassified data (*black* ×)

Fig. 2 Logarithmic classification error and reject rate in dependence on the *relative* rejection costs C_r for the 'Moon dataset'

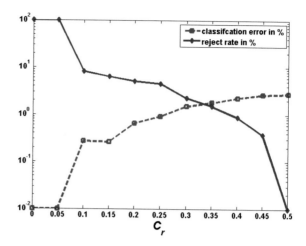

4.2 Medical Dataset

This dataset is a subset from the Leipzig LIFE Heart Study and contains the data of 1126 male patients suffering from a from coronary artery disease with a stenotic lumen reduction of at least 50 % with (460) and without (666) a myocardial infarction. For these patients several blood parameters among others were measured: CRP (C reactive protein), WBC (white blood cells), PLT (thrombocytes), Hkt (hematocrit), MPV (mean platelet volume), FIB C (fibrinogen, based on the Clauss method), vWF (von-Willebrand-factor ristocetin cofactor activity), vWF:RCo (ristocetinvon co-factor activity of the vWF) and sGPVI (soluble glycoprotein VI).

The data were trained with both GLVQ and GLVQ-r to predict a myocardial infarct by three-fold cross validation. All reported results are obtained for the test data. The rejection costs for GLVQ-r varied according to $C_r = 0.275 \ldots 0.475$, see Fig. 3. GLVQ corresponds to $C_r = 0.475$, because here the reject rate was zero yielding an accuracy of 82.8 %. Otherwise, rejection costs of $C_r = 0.3$ yield an accuracy of 93.1 %, which underlines the strong overlap of the classes.

4.3 Technical Dataset

Trends in automobile-headlights show an increasing interest in intelligent lighting-systems. The high-performance LED-headlamps and the growing number of information, delivered by the specific sensors of the car, establish the base of these lighting-systems. For example, the matrix-beam-system, which realizes a glare-free high-beam, is one of the first intelligent lighting-systems, which are available in current cars [20]. The performance of those systems depends particularly on the interaction between the headlamp and the driver-assistance-camera. A correctly

Fig. 3 Logarithmic classification error and reject rate in dependence on the *relative* rejection costs C_r for the medical dataset

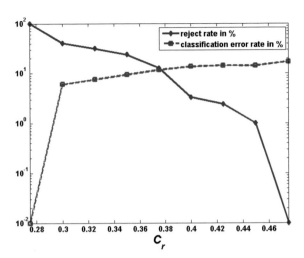

adjusted headlamp is as important as the calibration of the used camera. An online calibration system like the Porsche Automatic Headlamp Setting (PAHS) can provide both demands. The PAHS system combines a movement of the headlamp's light-distribution with a detection of the headlamp's cut-off-line (COL), see Fig. 4, by the driver-assistance-camera.

As a result of this calibration the horizontal and the vertical mis-aiming of the headlamp can be calculated, which finally allows a continuous control of the headlamp. Hence, correct classification of the COL crucially influences precision of the whole system.

For the detection of the upper COL (uCOL) and lower COL (lCOL) 6 parameters were provided by the PAHS to the classifier system, which are easy to measure and to calculate including contrast values for COLs, canny-edge descriptors and characteristic angles [20]. In our example, the resulting *Porsche data* comprises 27 % uCOL- and 60 % lCOL-samples.

Application of GLVQ as classifier yields a test accuracy of 91 % using 1 prototype per class (three-fold cross validation). The *relative* rejection costs for GLVQ-r are chosen as $C_r = 0.0 \ldots 0.5$. The results are depicted in Fig. 5.

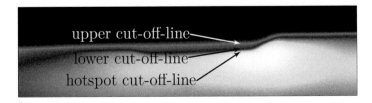

Fig. 4 Logarithmic gray-scaled luminance-measurement of a headlamp and detectable cut-off-lines in PAHS

Fig. 5 Logarithmic classification error and reject rate in dependence on the *relative* rejection costs C_r for the *Porsche dataset* obtained from three-fold cross-validation

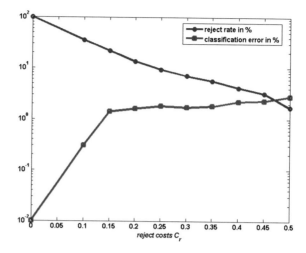

Fig. 6 Comparison of classification errors and reject rates in dependence of the reject margin ϵ for GLVQ-r and GLVQ applied to the *Porsche dataset*. For GLVQ, we used the a-posteriori variant proposed in [7, 8] to generate the curves

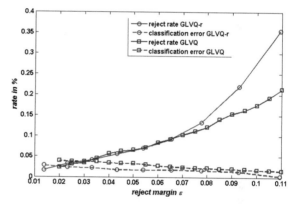

We obtain an approximately constant accuracy for a broad range of rejection costs C_r although the corresponding reject rate is slightly decreasing. Further, comparing GLVQ-r with standard GLVQ and a-posteriori determination of reject and error rates in dependence on the reject margin ϵ according to [7, 8], we observe that GLVQ-r leads to both lower reject rates and lower error rates, see Fig. 6. Thus we can conclude that the prototypes are better adapted to the reject regime for GLVQ-r than the prototypes of standard GLVQ.

5 Conclusions

We have introduced reject options for LVQ schemes which rely on efficient geometrical reject rules, and which optimize the prototype positions and reject threshold according to the chosen objective. The latter constitutes a direct approximation of the 0-1-loss function with reject. We have demonstrated the superior behavior of this approach in a couple of benchmarks.

So far, the algorithmic development and experiments have been restricted to the most simple GLVQ version. There exist powerful alternative schemes such as metric learning for LVQ, functional approaches, kernel schemes, and variants for relational data, to name just a few. The proposed framework can directly be transferred to these settings. This will be the subject of future research.

Acknowledgments This publication was supported by the LIFE—Leipzig Research Center for Civilization Diseases, University Leipzig. LIFE is funded by means of the European Union, by the European Regional Development Fund (ERDF) and by means of the Free State of Saxony within the framework of the excellence initiative.

References

1. Arlt, W., Biehl, M., Taylor, A.E., Hahner, S., Libe, R., Hughes, B.A., Schneider, P., Smith, D.J., Stiekema, H., Krone, N., et al.: Urine steroid metabolomics as a biomarker tool for detecting malignancy in adrenal tumors. J. Clin. Endocrinol. Metab. **96**(12), 3775–3784 (2011)
2. Bartlett, P.L., Wegkamp, M.H.: Classification with a reject option using a hinge loss. J. Mach. Learn. Res. **9**, 1823–1840 (2008)
3. Bengio, Y.: Learning deep architectures for AI. Found. Trends Mach. Learn. **2**(1), 1–127 (2009)
4. Biehl, M., Sadowski, P., Bhanot, G., Bilal, E., Dayarian, A., Meyer, P., Norel, R., Rhrissorrakrai, K., Zeller, M.D., Hormoz, S.: Inter-species prediction of protein phosphorylation in the sbv IMPROVER species translation challenge. Bioinformatics **31**(4), 453–461 (2015)
5. Chow, C.: On optimum recognition error and reject tradeoff. IEEE Trans. Inf. Theory **16**(1), 41–46 (1970)
6. fan Wu, T., Lin, C.-J., Weng, R.C.: Probability estimates for multi-class classification by pairwise coupling. J. Mach. Learn. Res. **5**, 975–1005 (2003)
7. Fischer, L., Hammer, B., Wersing, H.: Efficient rejection strategies for prototype-based classification. Neurocomputing, page to appear (2015)
8. Fischer, L., Hammer, B., Wersing, H.: Optimum reject options for prototype-based classification. CoRR, arXiv:1503.06549 (2015)
9. Giotis, I., Bunte, K., Petkov, N., Biehl, M.: Adaptive matrices and filters for color texture classification. J. Math. Imaging Vis. **47**(1–2), 79–92 (2013)
10. Grandvalet, Y., Rakotomamonjy, A., Keshet, J., Canu, S.: Support vector machines with a reject option. In: Advances in Neural Information Processing Systems 21, Proceedings of the Twenty-Second Annual Conference on Neural Information Processing Systems, Vancouver, British Columbia, Canada, December 8–11, pp. 537–544 (2008)
11. Herbei, R., Wegkamp, M.: Classification with reject option. Can. J. Stat. **34**(4), 709–721 (2006)
12. Kaden, M., Hermann, W., Villmann, T.: Attention based classification learning in GLVQ and asymmetric classification error assessment. In: Villmann, T., Schleif, F.-M., Kaden, M., Lange, M. (eds.) Advances in Self-Organizing Maps and Learning Vector Quantization: Proceedings of 10th International Workshop WSOM 2014. Mittweida, Advances in Intelligent Systems and Computing, vol. 295, pp. 77–88. Springer, Berlin (2014)
13. Kaden, M., Lange, M., Nebel, D., Riedel, M., Geweniger, T., Villmann, T.: Aspects in classification learning—review of recent developments in learning vector quantization. Found. Comput. Decis. Sci. **39**(2), 79–105 (2014)
14. Kirstein, S., Wersing, H., Gross, H., Körner, E.: A life-long learning vector quantization approach for interactive learning of multiple categories. Neural Netw. **28**, 90–105 (2012)

15. Kohonen, T.: Learning vector quantization for pattern recognition. Report TKK-F-A601, Helsinki University of Technology, Espoo, Finland (1986)
16. Pillai, I., Fumera, G., Roli, F.: Multi-label classification with a reject option. Pattern Recognit. **46**(8), 2256–2266 (2013)
17. Platt, J.C.: Probabilistic outputs for support vector machines and comparisons to regularized likelihood methods. In: Advances in Large Margin Classifiers, pp. 61–74. MIT Press (1999)
18. Polikar, R., Alippi, C.: Guest editorial learning in nonstationary and evolving environments. IEEE Trans. Neural Netw. Learn. Syst. **25**(1), 9–11 (2014)
19. de Stefano, C., Sansone, C., Vento, M.: To reject or not to reject: That is the question - an answer in case of neural classifiers. IEEE Trans. Syst. Man Cybern.—Part C: Appl. Rev. **30**(1), 84–94 (2000)
20. Saralajew, S., Villmann, T., Söhner, S.: An application of the generalized matrix learning vector quantization method for cut-off-line classification of automobile-headlights. Machine Learning Reports **9**(MLR-01-2015), 1–16 (2015). ISSN:1865–3960, http://www.techfak.uni-bielefeld.de/~fschleif/mlr/mlr_01_2015.pdf
21. Sato, A.S., Yamada, K.: Generalized learning vector quantization. In: Tesauro, G., Touretzky, D., Leen, T. (eds.) Advances in Neural Information Processing Systems, vol. 7, pp. 423–429. MIT Press (1995)
22. Schleif, F., Villmann, T., Kostrzewa, M., Hammer, B., Gammerman, A.: Cancer informatics by prototype networks in mass spectrometry. Artif. Intell. Med. **45**(2–3), 215–228 (2009)
23. Schneider, P., Hammer, B., Biehl, M.: Adaptive relevance matrices in learning vector quantization. Neural Comput. **21**, 3532–3561 (2009)
24. Senge, R., Bösner, S., Dembczynski, K., Haasenritter, J., Hirsch, O., Donner-Banzhoff, N., Hüllermeier, E.: Learning classifiers that distinguish aleatoric and epistemic uncertainty: reliable classification. Inf. Sci. **255**, 16–29 (2014)
25. de Vries, G., Pauws, S.C., Biehl, M.: Insightful stress detection from physiology modalities using learning vector quantization. Neurocomputing **151**, 873–882 (2015)
26. Yuan, M., Wegkamp, M.H.: Classification methods with reject option based on convex risk minimization. J. Mach. Learn. Res. **11**, 111–130 (2010)
27. Zhu, X., Schleif, F., Hammer, B.: Adaptive conformal semi-supervised vector quantization for dissimilarity data. Pattern Recognit. Lett. **49**, 138–145 (2014)

Optimization of Statistical Evaluation Measures for Classification by Median Learning Vector Quantization

D. Nebel and T. Villmann

Abstract Prototype-based classification is mainly influenced by the family of learning vector quantizers (LVQ) as introduced by KOHONEN. The main goal is to optimize the classification accuracy while the prototypes explore the class distribution in the data space. Recent variants can deal also with dissimilarity data, i.e. only the dissimilarities between the data objects are given. Otherwise, classification accuracy may be not appropriate to judge the classification performance, for example for imbalanced data or in medical applications, where frequently sensitivity and specificity are favored. In this paper we develop a median LVQ-variant optimizing those statistical classification evaluation measures, if only dissimilarity data are available. Thus, the presented approach is the discrete counterpart of a recently proposed LVQ-approach for optimization of statistical measures in case of vectorial data. For this purpose, we make use of a probabilistic description of the classification decision proposed in Robust Soft LVQ.

1 Introduction and Motivation

Learning vector quantization (LVQ) as introduced by TEUVO KOHONEN is a popular approach for classification of vector data [13, 14]. The basic idea of this approach is to represent the data classes by prototype vectors and optimize their positions to optimize the classification accuracy. Many variants of the basic Hebbian learning scheme were developed since the initial work by Kohonen. Gradient based variants

D. Nebel · T. Villmann (✉)
Computational Intelligence Group, University of Applied Sciences Mittweida,
09648 Mittweida, Germany
e-mail: thomas.villmann@hs-mittweida.de

D. Nebel · T. Villmann
Institut fr Computational Intelligence Und Intelligente Datenanalyse e.V.,
09648 Mittweida, Germany

© Springer International Publishing Switzerland 2016
E. Merényi et al. (eds.), *Advances in Self-Organizing Maps and Learning Vector Quantization*, Advances in Intelligent Systems and Computing 428,
DOI 10.1007/978-3-319-28518-4_25

as well as probabilistic approaches were developed [23, 24]. An actual overview can be found in [12, 19]. Yet, the main learning task, the optimization of the classification accuracy, as well as the requirement of a differentiable dissimilarity measure in data space for comparison of prototypes and data were kept most the time.

During the last years the focus was shifted to more advanced classification goals like optimization of sensitivity, specificity or the F_β-measure developed by van Rijsbergen [21], which are based on the evaluation of the confusion matrix for binary classifications. These statistical quality measures are more adequate for class-imbalanced training data [11].

Recently, the topic of LVQ-extensions for classification of dissimilarity or relational data came into the research focus of classification learning, as such variants are already known for unsupervised vector quantization [2, 5, 7, 8]. For relational approaches the prototypes are assumed as linear combination of the data. For general dissimilarity data prototypes are restricted to be data samples. The latter strategy is known as median-learning. First attempts for relational and median LVQ-variants optimizing the classification accuracy were provided in [6, 17, 18]. In the present publication, we extend these ideas to the previously mentioned statistical measures derived from the confusion matrix. For this purpose, we make use of a probabilistic variant of LVQ, which we combine with a technique called *generalized Expectation-Maximization-scheme* (gEM) recently developed for Median-GLVQ [18].

The remainder of the paper is as follows: First we introduce useful notations and give a brief overview about the general gEM-scheme. Thereafter we consider the classifier function of the Robust Soft LVQ (RSLVQ, [24]) in the light of the entries of the confusion matrix. From these considerations we derive mathematical expressions, which can be plugged into the gEM-scheme for optimization of basic statistical classification evaluation measures, i.e. sensitivity, specificity, etc. Equipped with this knowledge we are able to built up gEM-models for optimizing more sophisticated evaluation measures like the previously mentioned F_β-measure or Matthews correlation coefficient. Exemplary simulation accompany the theoretical explanations.

2 Basic Notations and the Generalized Expectation Maximization Approach

2.1 *Notations and Abbreviations*

In the following we clarify notation and abbreviations. We suppose data objects $\mathbb{X} = \{x_i\}_{i=1,...,N}$ and M prototypes $\theta_k \in \Theta$, i.e. the cardinality of Θ is M. We assume a binary classification problem with the classes $C = \{\oplus, \ominus\}$. Let $c(\cdot)$ be the formal class label function, which assigns to each data object the class label $y_i = c(x_i)$. Analogously, $c_j = c(\theta_j)$ returns the predefined class label of the prototype. Further, M^+ denotes the number of prototypes assigned to the class $\oplus$. We introduce prototype dependent Kronecker-symbol abbreviations like

$$\delta_k^+ = \begin{cases} 1 \text{ if } & c_k = \oplus \\ 0 \text{ if } & c_k = \ominus \end{cases} \text{ and } \delta_k^- = \begin{cases} 1 \text{ if } & c_k = \ominus \\ 0 \text{ if } & c_k = \oplus \end{cases}$$

as short-hand notations. Analogously, we define

$$\delta^+(x_i) = \begin{cases} 1 \text{ if } & y_i = \oplus \\ 0 \text{ if } & y_i = \ominus \end{cases} \text{ and } \delta^-(x_i) = \begin{cases} 1 \text{ if } & y_i = \ominus \\ 0 \text{ if } & y_i = \oplus \end{cases}$$

as data dependent Kronecker-symbols.

2.2 The Mathematical Theory of the Generalized Expectation Maximization Approach

In [16], the general mathematical theory for maximization of a cost function $K(\mathbb{X})$ in the form

$$K(\mathbb{X}) = \sum_i g(x_i, \Theta) \tag{1}$$

with positive, bounded real functions $g(x_i, \Theta)$ was proposed. In particular, it was shown that the logarithmic cost function (LCF) $C(\mathbb{X}) = \ln\left(\sum_i g(x_i, \Theta)\right)$ is decomposible into

$$C(\mathbb{X}) = \mathcal{L}(\gamma, \Theta) + \mathcal{K}(\gamma||p) \tag{2}$$

with the loss term

$$\mathcal{L}(\gamma, \Theta) = \sum_i \gamma_i \ln\left(\frac{g(x_i, \Theta)}{\gamma_i}\right) \tag{3}$$

and the a formal Kullback-Leibler-divergence (KLD)

$$\mathcal{K}(\gamma||p) = \sum_i \gamma_i \ln\left(\frac{\gamma_i}{p(x_i)}\right) \tag{4}$$

where

$$p(x_i) = \frac{g(x_i, \Theta)}{\sum_i g(x_i, \Theta)}$$

is the formal probability for a data object x_i. The values $\gamma_i \geq 0$ fulfill the restriction $\sum_{i=1}^{N} \gamma_i = 1$ and, hence, can also be interpreted as formal probability values. Thus $\mathcal{L}(\gamma, \Theta)$ is a lower bound for the LCF $C(\mathbb{X})$ due to the non-negativeness of the KLD $\mathcal{K}(\gamma||p)$. Using this property we obtain the following maximizing strategy for the LCF $C(\mathbb{X})$:

1. **Expectation-step (E-step)**: *set*

$$\gamma_i := p(x_i) \implies \mathcal{K}(\gamma||p) = 0$$

$$\implies C(\mathbb{X}) = \mathcal{L}(\gamma, \Theta)$$

2. *generalized* **Maximization-step (gM-step)**: take the parameters γ_i as fixed and *find new prototypes Θ^{new}*, such that:

$$\mathcal{L}(\gamma, \Theta^{new}) \geq \mathcal{L}(\gamma, \Theta^{old})$$

3. **Convergence criterion**: if $\Theta^{new} = \Theta^{old}$ *stop*. Else goto 1.

Note that the cost function value $C(\mathbb{X})$ does not change in the E-step, because $C(\mathbb{X})$ is independent from the parameters γ_i. Further, the new prototypes Θ^{new} maybe found by an arbitrary search procedure, which allows the avoidance of calculation of a derivative as it is demanded for a stochastic gradient descent learning. Thus, applying a sophisticated discrete search procedure, with new prototypes Θ^{new} restricted to be data objects, a median-like optimization scheme is obtained. Otherwise, because the new prototypes are not required to maximize the function $\mathcal{L}$ in the second step of the algorithm, it is not a precise maximization and therefore, we denote it as a generalized M-step (gM-step) and the overall procedure a *generalized* EM-optimization (gEM).

3   The RSLVQ-Classifier Decision Function and the Confusion Matrix Entries

Robust Soft LVQ (RSLVQ) is a probabilistic variant of LVQ [24], which keeps the idea of prototype based model form LVQ but relaxes the restriction of crisp classification. The RSLVQ-classifier is based on a likelihood ratio cost function realizing a soft-count of misclassification based on a mixture model, where the model parameters $\Theta = \{\theta_1, \ldots, \theta_M\}$ play the role of the prototypes. In particular, the cost function is given as

$$C_{RSLVQ} = \sum_i \ln\left(\mu_{RSLVQ}(y_i|x_i)\right) \tag{5}$$

Table 1 Confusion matrix

		True		
Predicted		$\oplus$	$\ominus$	
	$\oplus$	TP	FP	$\tilde{N}_+$
	$\ominus$	FN	TN	$\tilde{N}_-$
		N_+	N_-	N

with the RSLVQ-classifier function

$$\mu_{RSLVQ}(\kappa|x_i) = p(\kappa|x_i, \Theta) \tag{6}$$

describing the probability that a data object x_i is assigned to class $\kappa \in C$.

In the following we will consider binary classification tasks such that the confusion matrix entries as depicted in Table 1 is relevant to judge the classification performance.

We will relate these entries to the RSLVQ-classifier function, such we obtain convenient estimates in terms of $\mu_{RSLVQ}(\kappa|x_i)$.

For the binary decision problem the conditional mixture models in (6) become

$$p(\oplus|x_i, \Theta) = \frac{\sum_j \delta_j^+ p(x_i|\theta_j)}{\sum_k p(x_i|\theta_k)} \text{ and } p(\ominus|x_i, \Theta) = \frac{\sum_j \delta_j^- p(x_i|\theta_j)}{\sum_k p(x_i|\theta_k)} \tag{7}$$

where the conditional probabilities

$$p(x_i|\theta_j) = \exp\left(-\left(\frac{d(x_i, \theta_j)}{\sigma_j}\right)^2\right) \tag{8}$$

are usually taken as Gaussians with width's $\sigma_j > 0$. Here, $d(x_i, \theta_j)$ is an arbitrary dissimilarity measure[1] between data objects and prototypes.

Remark 1 We make the observation that if the prototypes are restricted to be data objects, i.e. $\Theta \subseteq \mathbb{X}$, only the dissimilarities between the data objects are required to calculate both probabilities $p(\oplus|x_i, \Theta)$ and $p(\ominus|x_i, \Theta)$. Thus the idea of median learning is realized.

To keep the model simple in the following, we assume $\sigma = \sigma_j$ for all prototypes θ_j, $j = 1 \ldots M$ in (8). Then the conditional probabilities $p(\kappa|x_i, \Theta)$ from (7) become crisp in the limit $\sigma \searrow 0$. Particularly we obtain

$$p(\oplus|x_i, \Theta) \xrightarrow[\sigma \to 0]{} \begin{cases} 1 & \text{if } y_i = \oplus \\ 0 & \text{else} \end{cases} \text{ and } p(\ominus|x_i, \Theta) \xrightarrow[\sigma \to 0]{} \begin{cases} 1 & \text{if } y_i = \ominus \\ 0 & \text{else} \end{cases}$$

---

[1]For a mathematical definition of a dissimilarity measure we refer to [20].

and, therefore, the conditional probabilities $p(\kappa|x_i, \Theta)$ play the role of an indicator function for their classes in this limit. In consequence, both quantities $\mu_{RSLVQ}(\oplus|x_i)$ and $\mu_{RSLVQ}(\ominus|x_i)$ can be used to count approximately the correctly classified data objects x_i, which is essential for determination of the confusion matrix. In particular, we have

$$TP = \sum_i \delta^+(x_i) \cdot \mu_{RSLVQ}(\oplus|x_i) \text{ and } FP = \sum_i \delta^-(x_i) \cdot \mu_{RSLVQ}(\oplus|x_i) \quad (9)$$

$$FN = \sum_i \delta^+(x_i) \cdot \mu_{RSLVQ}(\ominus|x_i) \text{ and } TN = \sum_i \delta^-(x_i) \cdot \mu_{RSLVQ}(\ominus|x_i) \quad (10)$$

as RSLVQ-based estimates for these quantities. They are identical in the structure compared to the GLVQ-counterparts for statistical classification evaluation measures as presented for vector data and prototypes in [11].

4 Specification of $g(x_i, \Theta)$ for gEM-Optimization of Statistical Classification Evaluation Measures based on the Confusion Matrix

In this chapter we will describe several statistical quality measures for classification in the form of (1), which allows to apply the gEM-optimization scheme provided in Sect. 2.2. In particular, we will determine suitable functions $g(x_i, \Theta)$ for several statistical classification evaluation measure. For this purpose, we will use the descriptions (9) and (10) of the entries of the confusion matrix introduced in the previous section by means of the RSLVQ-classifier function $\mu_{RSLVQ}(\kappa|x_i)$ from (6). According to this strategy, the optimization of the respective cost function (1) by gEM is, in fact, an optimization of the prototypes $\Theta = \{\theta_1, \ldots, \theta_M\}$ appearing as parameters in this model. Thus, we always preserve the idea of prototype based learning. Otherwise, the median learning methodology is also kept according to the Remark 1.

4.1 Simple Classification Quality Measures

We start considering simple quality measure, which are directly derived from the confusion matrix. We will write them in the form of (1) and specify the respective choice of $g(x_i, \Theta)$.

The numerical stable variants (last column in Table 2) can immediately plugged into the gEM-algorithm for optimization of the respective measure, which then is simply of the form (1).

Table 2 Specification of $g(x_i, \Theta)$ for basic classification quality measures based on the confusion matrix

Statistical measure	Definition	$g(x_i, \Theta)$ in $C(\mathbb{X})$	Stability
Recall/Sensitivity ρ	$\rho = \frac{TP}{N_+}$	$\frac{\delta^+(x_i)\cdot\mu(\oplus|x_i)}{N_+}$	$g(x_i, \Theta) + 1$
Specificity ς	$\varsigma = \frac{TN}{N_-}$	$\frac{\delta^-(x_i)\cdot\mu(\ominus|x_i)}{N_-}$	$g(x_i, \Theta) + 1$
Precision π	$\pi = \frac{TP}{TP+FP}$	$\frac{\delta^+(x_i)\cdot\mu(\oplus|x_i)}{\sum_j \delta^+(x_j)\cdot\mu(\oplus|x_j)+\sum_j \delta^-(x_j)\cdot\mu(\oplus|x_j)}$	$g(x_i, \Theta) + 1$
Neg. prediction value ν	$\nu = \frac{TN}{TN+FN}$	$\frac{\delta^-(x_i)\cdot\mu(\ominus|x_i)}{\sum_j \delta^-(x_j)\cdot\mu(\ominus|x_j)+\sum_j \delta^+(x_j)\cdot\mu(\ominus|x_j)}$	$g(x_i, \Theta) + 1$
Fall-out	$\varphi = 1 - \mu$	$\frac{\delta^-(x_i)\cdot\mu(\oplus|x_i)}{N_-}$	$2 - g(x_i, \Theta)$
False discovery rate	$FDR = \frac{FP}{FP+TP}$	$\frac{\delta^-(x_i)\cdot\mu(\oplus|x_i)}{\sum_j \delta^-(x_j)\cdot\mu(\oplus|x_j)+\sum_j \delta^+(x_j)\cdot\mu(\oplus|x_j)}$	$2 - g(x_i, \Theta)$
False negative rate	$FNR = \frac{FN}{N_+}$	$\frac{\delta^+(x_i)\cdot\mu(\ominus|x_i)}{N_+}$	$2 - g(x_i, \Theta)$

The last column gives the numerically stable variant to be used in gEM

4.2 Complex Measures

Now we consider more complex statistical measures frequently applied in classification evaluation.

The F_β-measure developed by C.J. VAN RIJSBERGEN combines precision π and recall ρ (sensitivity) into a single quantity

$$F_\beta = \frac{(1 + \beta^2)\pi\rho}{\beta^2\pi + \rho} \tag{11}$$

depending on the balancing parameter β and $F_\beta \in [0, 1]$ [21]. Frequently, this balancing parameter is chosen as $\beta = 1$ yielding the measure to be the ratio of the arithmetic and the geometric mean between both quantities precision and recall. Using the quantities from the confusion matrix, we can rewrite (11) as

$$F_\beta = \frac{(1 + \beta^2)TP}{(1 + \beta^2)TP + \beta^2 FN + FP}$$

such that we have $F_\beta = \sum_i g(x_i, \Theta)$ with positive functions

$$g(x_i, \Theta) = \frac{\delta^+(x_i) \cdot (1 + \beta^2)\mu(\oplus|x_i)}{\sum_j \left((1 + \beta^2)\delta^+(x_j) \cdot \mu(\oplus|x_j) + \beta^2\delta^+(x_j) \cdot \mu(\ominus|x_j) + \delta^-(x_j) \cdot \mu(\oplus|x_j)\right)}$$

to be maximized. The functions $g(x_i, \Theta)$ have to be substituted by $g(x_i, \Theta) + 1$ to avoid numerical instabilities.

The Jaccard Index

$$J = \frac{TP}{FP + TP + FN}$$

with range $0 \leq J \leq 1$ is explained in [4, 10]. It can be expressed as $J = \sum_i g(x_i, \Theta)$ using the positive functions

$$g(x_i, \Theta) = \frac{\delta^+(x_i) \cdot \mu(\oplus|x_i)}{\sum_j \left(\delta^-(x_j) \cdot \mu(\oplus|x_j) + \delta^+(x_j) \cdot \mu(\oplus|x_j) + \delta^+(x_j) \cdot \mu(\ominus|x_j)\right)}$$

and has to be maximized. Interestingly, J is closely related to the Tanimoto distances [22]. Again, the numerically stable behavior of gEM the choice $g(x_i, \Theta) + 1$ is recommended.

Matthews correlation coefficient

$$MCC = \frac{TP \cdot TN - FP \cdot FN}{\sqrt{(TP + FP)(TP + FN)(TN + FP)(TN + FN)}} \tag{12}$$

is another popular classification quality measure, which is equivalent to the χ^2-statistics for a 2×2 contingency table [15]. It can be rewritten in the form $MCC = \sum_{i,j} g(x_i, x_j, \Theta)$ with

$$g(x_i, x_j, \Theta) = \frac{\delta^+(x_i) \cdot \delta^-(x_j) \cdot \mu(\oplus|x_i)\mu(\ominus|x_j) - \delta^-(x_i) \cdot \delta^+(x_j) \cdot \mu(\oplus|x_i)\mu(\ominus|x_j)}{\sqrt{(TP + FP)(TP + FN)(TN + FP)(TN + FN)}}$$

and $g(x_i x_j, \Theta) \in [-1, 1]$. To ensure positivity and numerical stability of gEM the substitution $\bar{g}(x_i, x_j, \Theta) = g(x_i, x_j, \Theta) + 2$ is demanded.

Other statistical evaluation measures as well as a median variant based on GLVQ are presented in [18].

5 Exemplary Numerical Simulations

We conducted several experiments to validate the approach. However, we restricted ourselves to the F_β-measure, because it is a complex one comprising also simple measures from Table 2. We denote this median LVQ-scheme as Median-F_β-LVQ. For each experiment we used only one prototype per class.

The reported results were obtained by the twenty repetitions of following 5-fold modified cross validation procedure. For each fold the data are split into five parts. Three parts are taken as training data, one part are the test data to select the best model. The last part remains as validation part. Only the validation results are presented, because these are the results with the lowest model bias. We denote this procedure as *unbiased cross validation* (five-UCV).

We considered three data sets. Two of them are from the UCI-Repository [1]. These are the *Pima Indians Diabetes database* (PIMA) consisting of non-diabetic (500) and diabetic (268) samples and the *Haberman's Survival dataset* (HS) containing long- and short-time breast cancer survivor samples after surgery. The third data set is from *Wilson's disease* (WD) already used in [11]. This dataset is a vectorial set comprising

Table 3 Results of RSLVQ and F_β-Median-LVQ for the WD-dataset after five-UCV

RSLVQ		F_β-Median-LVQ		β^2
Mean	Std	Mean	Std	
F_β				
0,762	0,12	0,77	0,13	0.5
0,783	0,116	0,782	0,118	1.0
0,826	0,12	0,8434	0,098	2.0
Accuracy				
0,774	0,11	0,774	0,115	0.5
0,774	0,11	0,772	0,121	1.0
0,774	0,11	0,752	0,11	2.0
AUROC				
0,886	0,093	0,866	0,105	0.5
0,886	0,093	0,875	0,11	1.0
0,886	0,093	0,87	0,102	2.0

The F_β-Median-LVQ was trained with the β-values depicted in the last column. The performances of the algorithms are compared with respect to the F_β-measure (first block), the accuracy (second block) and the area under the ROC-curve (AUROC, [3])

the glucose consumption profiles according to selected brain regions for volunteers and patients. In [11] this dataset was considered for the vectorial GLVQ-variant optimizing statistical measures. Thus we can compare our results obtained for the present median variant. For a detailed data description we refer to [9, 11].

Starting with the WD-data we obtain the classification results for RSLVQ and Median-F_β-LVQ after five-UCV as depicted in Table 3. As we can see, the F_β-Median-LVQ outperforms RSLVQ when optimizing the F_β-value. Looking at the other evaluation criteria, F_β-Median-LVQ is at least comparable. Further, if we compare the F_β-Median-LVQ results with those obtained by the vectorial counterpart (which are 0.907, 0.910 and 0.926 for $\beta = 0.5/1.0/2.0$, respectively—see [11]) we detect only a small deterioration. This can be dedicated to the restricted prototype variability in case of median learning.

For the PIMA and the HB we achieve similar performance results as for the WD data, see Table 4. Thus we can conclude that the F_β-Median-LVQ performs well for dissimilarity data.

6 Conclusions

In this paper we presented an approach for prototype based classification learning of dissimilarity data to optimize statistical classification evaluation measure based on the confusion matrix. The method adopts ideas from a probabilistic variant of LVQ and combines them with the generalized EM technique. The resulting algorithm can be seen as the median counterpart to the already proposed vectorial variant of LVQ optimizing those statistical measures.

Table 4 Performance results fo the PIMA (top) and HD (bottom) obtained by RSLVQ and F_β-Median-LVQ applying UFC

RSLVQ		F_β-Median-LVQ		β^2
Mean	Std	Mean	Std	
F_β				
0,774	0,026	0,789	0,029	0.5
0,793	0,032	0,818	0,021	1.0
0,815	0,046	0,898	0,01	2.0
Accuracy				
0,719	0,035	0,709	0,039	0.5
0,719	0,035	0,733	0,031	1.0
0,719	0,035	0,6673	0,026	2.0
AUROC				
0,757	0,048	0,778	0,042	0.5
0,757	0,048	0,7734	0,05	1.0
0,757	0,048	0,6352	0,12	2.0
F_β				
0,774	0,026	0,789	0,029	0.5
0,793	0,032	0,818	0,021	1.0
0,815	0,046	0,898	0,01	2.0
Accuracy				
0,719	0,035	0,709	0,039	0.5
0,719	0,035	0,733	0,031	1.0
0,719	0,035	0,6673	0,026	2.0
AUROC				
0,757	0,048	0,778	0,042	0.5
0,757	0,048	0,7734	0,05	1.0
0,757	0,048	0,6352	0,12	2.0

For detailed explanation of the table structure, see Table 3

References

1. Blake, C., Merz, C.: UCI repository of machine learning databases. University of California, Department of Information and Computer Science, Irvine, CA. http://www.ics.edu/mlearn/MLRepository.html (1998)
2. Cottrell, M., Hammer, B., Hasenfu, A., Villmann, T.: Batch and median neural gas. Neural Netw. **19**, 762–771 (2006)
3. Fawcett, T.: An introduction to ROC analysis. Pattern Recogn. Lett. **27**, 861–874 (2006)
4. Geweniger, T., Fischer, L., Kaden, M., Lange, M., Villmann, T.: Clustering by fuzzy neural gas and evaluation of fuzzy clusters. Comput. Intell. Neurosci. **2013**, Article ID 165248 (2013). doi:10.1155/2013/165248
5. Hammer, B., Hasenfuss, A.: Topographic mapping of large dissimilarity data sets. Neural Comput. **22**(9), 2229–2284 (2010)

6. Hammer, B., Hofmann, D., Schleif, F.-M., Zhu, X.: Learning vector quantization for (dis-)similarities. Neurocomputing, page in press (2013)
7. Hathaway, R., Bezdek, J.: NERF c-means: non-Euclidean relational fuzzy clustering. Pattern Recogn. **27**(3), 429–437 (1994)
8. Hathaway, R., Davenport, J., Bezdek, J.: Relational duals of the c-means clustering algorithms. Pattern Recogn. **22**(3), 205–212 (1989)
9. Hermann, W., Barthel, H., Hesse, S., Grahmann, F., Kühn, H.-J., Wagner, A., Villmann, T.: Comparison of clinical types of Wilson's disease and glucose metabolism in extrapyramidal motor brain regions. J. Neurol. **249**(7), 896–901 (2002)
10. Jaccard, P.: The distribution of the flora in the alpine zone. New Phytol. **11**, 37–50 (1912)
11. Kaden, M., Hermann, W., Villmann, T.: Optimization of general statistical accuracy measures for classification based on learning vector quantization. In: Verleysen, M. (ed.) Proceedings of European Symposium on Artificial Neural Networks, Computational Intelligence and Machine Learning (ESANN'2014), pp. 47–52, Louvain-La-Neuve, Belgium (2014). i6doc.com
12. Kaden, M., Lange, M., Nebel, D., Riedel, M., Geweniger, T., Villmann, T.: Aspects in classification learning—Review of recent developments in Learning Vector Quantization. Found. Comput. Decis. Sci. **39**(2), 79–105 (2014)
13. Kohonen, T.: Learning vector quantization for pattern recognition. Report TKK-F-A601, Helsinki University of Technology, Espoo, Finland (1986)
14. Kohonen, T.: Self-Organizing Maps. Springer Series in Information Sciences, vol. 30. Springer, Berlin, Heidelberg (1995)
15. Matthews, B.: Comparison of the predicted and observed secondary structure of T4 phage Iysozyme. Biochimica et Biophysica Acta **405**, 442–451 (1975)
16. Nebel, D., Hammer, B., Frohberg, K., Villmann, T.: Median variants of learning vector quantization for learning of dissimilarity data. Neurocomputing **169**, 295–305 (2015)
17. Nebel, D., Hammer, B., Villmann, T.: Supervised generative models for learning dissimilarity data. In: Verleysen, M. (ed.) Proceedings of European Symposium on Artificial Neural Networks, Computational Intelligence and Machine Learning (ESANN'2014), pp. 35–40, Louvain-La-Neuve, Belgium (2014). i6doc.com
18. Nebel, D., Villmann, T.: A median variant of generalized learning vector quantization. In: Lee, M., Hirose, A., Hou, Z.-G., Kil, R. (eds.) Proceedings of International Conference on Neural Information Processing (ICONIP). LNCS, vol. II, pp. 19–26. Springer-Verlag, Berlin (2013)
19. Nova, D., Estévez, P.: A review of learning vector quantization classifiers. Neural Comput. Appl. (2013)
20. Pekalska, E., Duin, R.: The Dissimilarity Representation for Pattern Recognition: Foundations and Applications. World Scientific (2006)
21. Rijsbergen, C.: Information Retrieval, 2nd edn. Butterworths, London (1979)
22. Rogers, D.J., Tanimoto, T.: A computer program for classifying plants. Science **132**(3434), 1115–1118 (1960)
23. Sato, A., Yamada, K.: Generalized learning vector quantization. In: Touretzky, D.S., Mozer, M.C., Hasselmo, M.E. (eds.) Advances in Neural Information Processing Systems 8. Proceedings of the 1995 Conference, pp. 423–429. MIT Press, Cambridge, MA, USA (1996)
24. Seo, S., Obermayer, K.: Soft learning vector quantization. Neural Comput. **15**, 1589–1604 (2003)

Complex Variants of GLVQ Based on Wirtinger's Calculus

Matthias Gay, Marika Kaden, Michael Biehl,
Alexander Lampe and Thomas Villmann

Abstract This paper addresses the application of gradient descent based machine learning methods to complex-valued data. In particular, the focus is on classification using Learning Vector Quantization and extensions thereof. In order to apply gradient-based methods to complex-valued data we use the mathematical formalism of Wirtinger's calculus to describe the derivatives of the involved dissimilarity measures, which are functions of complex-valued variables. We present a number of examples for those dissimilarity measures, including several complex-valued kernels, together with the derivatives required for the learning procedure. The resulting algorithms are tested on a data set for image recognition using Zernike moments as complex-valued shape descriptors.

1 Introduction

Complex-valued data (cv-data) are available in a variety of classification problems. Applications can be found in the fields of signal processing, telecommunications or image processing, where complex signal transformations like the Fourier transform are used. Moreover, many methods of machine learning can be extended to cv-data. Examples are neural networks with complex-valued weights and thresholds [1], the

M. Gay (✉) · M. Kaden (✉) · A. Lampe · T. Villmann (✉)
Computational Intelligence Group, University of Applied Sciences Mittweida,
Mittweida, Germany
e-mail: mgay@hs-mittweida.de

M. Kaden
e-mail: marikakaestner@googlemail.com

T. Villmann
e-mail: thomas.villmann@hs-mittweida.de

M. Biehl
Johann-Bernoulli-Institute for Mathematics and Computer Science,
University of Groningen, Groningen, The Netherlands

© Springer International Publishing Switzerland 2016
E. Merényi et al. (eds.), *Advances in Self-Organizing Maps and Learning
Vector Quantization*, Advances in Intelligent Systems and Computing 428,
DOI 10.1007/978-3-319-28518-4_26

Support Vector Machine (SVM) with complex-valued kernels [2], or the complex independent component analysis [3].

The Generalized Learning Vector Quantization (GLVQ) and its matrix versions (GMLVQ) are established prototype-based classification methods [4]. The prototypes are adapted in a learning process and, together with a suitable measure of dissimilarity or distance, parametrize the classification scheme. The training process is guided by the minimization of a cost function which, for instance, approximates the classification error [5, 6]. This optimization is frequently done by stochastic gradient descent or related schemes. To this end, the cost function and the distance measure have to be differentiable with respect to the prototypes and, potentially, with respect to further adaptive parameters. For the application to cv-data, a distance measure with appropriate complex derivatives is required. In [7] the problem of cv-data has already been addressed in the context of GLVQ. There, the derivatives of the real-valued costs were chosen intuitively by treating the real and imaginary parts of all adaptive parameters independently. In this paper, we will provide a more solid mathematical foundation based on *Wirtinger's calculus* [8].

For real-valued data, kernel methods like the SVM or the Kernelized GLVQ [9] are very popular and were successfully employed in many practical applications. Complex-valued kernels also exist, but are not very widespread in machine learning. In this contribution, we integrate complex-valued kernels into the GLVQ formalism and compute the derivatives for some examples. We apply the cv-GLVQ with kernels on a real world data set, where the features correspond to complex-valued shape descriptors.

2 Fundamentals

2.1 Generalized Learning Vector Quantization and Variants

We consider a classification problem with training data samples $v \in V \subset \mathbb{R}^D$ and the corresponding class labels $c(v) \in C$. Additionally, an initial prototype set $W = \{w_k \in \mathbb{R}^D, k = 1, \ldots, |W|\}$ is given, together with the class assignments $y(w_k) \in C$ and at least one prototype per class. The prototypes are learned during the training phase of GLVQ. After the training, a new data point v is assigned to the class $y(w_{s(v)})$ of the winner prototype. This prototype is determined by the *winner-takes-all* rule (WTA)

$$s(v) = \underset{k=1,\ldots,|W|}{\operatorname{argmin}} \, d(v, w_k), \tag{1}$$

where $d(v, w)$ is the same dissimilarity measure that was used during the training procedure.

For the training in GLVQ, a *classifier* function

$$\mu_W(v) = \frac{d\left(v, w^+\right) - d\left(v, w^-\right)}{d\left(v, w^+\right) + d\left(v, w^-\right)} = \frac{d^+(v) - d^-(v)}{d^+(v) + d^-(v)} \tag{2}$$

is defined for a training data point v. Here, $d^+(v) = d\left(v, w^+\right)$ is the minimal distance of v to the prototype w^+ belonging to the same class as v. Likewise, $d^-(v) = d\left(v, w^-\right)$ is the distance to the closest prototype w^- with a different class label, i.e. $y(w^-) \neq c(v)$. Thus, a data point is correctly classified if $\mu_W(v)$ is negative, i.e. $d^-(v) > d^+(v)$.

With this, the cost function

$$E_{\text{GLVQ}}(V, W) = \frac{1}{|V|} \sum_{v \in V} f\left(\mu_W(v)\right) \tag{3}$$

has to be minimized. A popular choice for the monotonously increasing transfer function f is the sigmoid function $f_\theta(x) = (1 + \exp(-\theta x))^{-1}, \theta > 0$, which allows (3) to be interpreted as a smooth approximation of the classification error [6].

A common optimization strategy for general, non-convex and non-linear cost functions is stochastic gradient descent [10]. If the data points are real-valued, the prototypes are updated via

$$w^\pm \leftarrow w^\pm - \alpha \Delta w^\pm \tag{4}$$

$$\text{with} \quad \Delta w^\pm = \frac{\partial f(\mu_W(v))}{\partial \mu_W(v)} \cdot \frac{\partial \mu_W(v)}{\partial d^\pm(v)} \cdot \frac{\partial d^\pm(v)}{\partial w^\pm} \tag{5}$$

with the learning rate $0 < \alpha \ll 1$ and the derivative of the classifier function

$$\frac{\partial \mu_W(v)}{\partial d^\pm(v)} = \frac{\pm d^\mp(v)}{(d^+(v) - d^-(v))^2}. \tag{6}$$

Obviously, the dissimilarity measure $d^\pm(v)$ in (5) has to be differentiable with respect to the prototypes. Yet, this is the only restriction on d; symmetry or other metric properties are not required, see [11] for an example.

If the squared Euclidean distance $d_E^2(v, w) = \|v - w\|_2^2$ is chosen, the respective derivative in (5) yields a scaled vector shift

$$\frac{\partial d_E^2(v)}{\partial w} = -2(v - w). \tag{7}$$

A more general choice is the quadratic form

$$d_\Omega^2(v, w) = (v - w)^\mathsf{T} \Lambda (v - w) = \|\Omega(v - w)\|^2 \tag{8}$$

with

$$\frac{\partial d_\Omega^2(v)}{\partial w^\pm} = -2\,\Omega^\top\Omega(v - w^\pm) \tag{9}$$

$$\text{and} \quad \frac{\partial d_\Omega^2(v)}{\partial \Omega} = \Omega(v - w)(v - w)^\top \tag{10}$$

where $\Omega \in \mathbb{R}^{m \times D}$ with $m \leq D$ is a linear mapping matrix. In Generalized Matrix Relevance Learning (GMLVQ) [12, 13], this mapping matrix is also adapted and the resulting matrix $\Lambda = \Omega^\top\Omega$ is termed the relevance matrix of the classifier. Furthermore, if $m \ll D$ the scheme provides a discriminative low-dimensional representation of the considered data set [14].

Beside modifications of the Euclidean distance, other more unconventional dissimilarities have attracted attention, recently. Depending on the characteristics of the data, ℓ_p-norms, divergences or kernel distances can be chosen for $d(v, w)$. In [9] specific differential kernel distances were applied in the GLVQ formalism, including the radial basis function (RBF) or exponential kernel. The authors showed that the topological richness of GLVQ with kernels is equivalent to the SVM applying the same kernel. Thus, the differential kernel GLVQ (DK-GLVQ) is a powerful tool for solving non-linear classification problems.

2.2 A Brief Review of Wirtinger's Calculus

When having gradient-based learning for cv-data in mind, one first has to agree on a proper definition of the involved derivatives. Classical complex analysis imposes strong conditions on a function to be called *differentiable*. In particular it has to satisfy the Cauchy-Riemann differential equations [15]. Unfortunately, almost no function in engineering or optimization is differentiable in that sense—especially none with a real-valued co-domain, such as cost functions or distance measures. However, one can circumvent this issue by relaxing the definition of differentiability in a reasonable way. This can be accomplished by using Wirtinger's calculus, as it is done extensively in engineering.

We want to motivate the idea of Wirtinger's calculus, also known as $\mathbb{C}$-$\mathbb{R}$-calculus, by considering a real-valued cost function

$$f : \mathbb{C} \to \mathbb{R} \quad \text{with} \quad z = x + iy \mapsto f(z) = u(x, y)$$

with $x, y, u \in \mathbb{R}$, as a function of two real-valued arguments. When we consider the minimization of f in this sense, we require

$$\frac{\partial u(x, y)}{\partial x} = 0 \quad \text{and} \quad \frac{\partial u(x, y)}{\partial y} = 0. \tag{11}$$

These necessary conditions can be written very compactly as

$$a \cdot \frac{\partial f}{\partial x} + i \, b \cdot \frac{\partial f}{\partial y} = 0$$

for any nonzero $a, b \in \mathbb{R}$. A smart choice of a and b yields a consistent calculus and a powerful tool for complex-valued optimization.

Definition 1 The *Wirtinger differential operators* for a function of $z = x + i \, y$ are defined as

$$\frac{\partial}{\partial z} := \frac{1}{2} \left(\frac{\partial}{\partial x} - i \frac{\partial}{\partial y} \right) \quad \text{and} \quad \frac{\partial}{\partial z^*} := \frac{1}{2} \left(\frac{\partial}{\partial x} + i \frac{\partial}{\partial y} \right),$$

where $\frac{\partial}{\partial z^*}$ is the *conjugate differential operator* with respect to the complex conjugate z^* of the variable z.

Although introduced for real-valued functions above, it is important to note that these operators can also be applied to $f : \mathbb{C} \to \mathbb{C}$ with $z = x + i \, y \mapsto f(z) = u(x, y) + i \, v(x, y)$. The function f is required to be only real differentiable. Moreover, most rules from real-valued calculus apply directly to this calculus, e.g. the sum, product and quotient rule. However, there are differences that one has to take care of:

(I) The chain rule is different for the general case. It reads

$$\frac{\partial h(g(z))}{\partial z} = \frac{\partial h}{\partial g} \cdot \frac{\partial g}{\partial z} + \frac{\partial h}{\partial g^*} \cdot \frac{\partial g^*}{\partial z} \tag{12}$$

and likewise with z^* for the conjugate derivative. However, in the practically relevant special case where $g : \mathbb{C} \to \mathbb{R}$ and h maps real numbers to real numbers, the term $\frac{\partial h}{\partial \operatorname{Im}(g)}$ vanishes in both $\frac{\partial h}{\partial g}$ and $\frac{\partial h}{\partial g^*}$, reducing the chain rule to the one known from real calculus.

(II) In $\frac{\partial f}{\partial z}$, the term z^* is treated as a constant and vice versa. For instance, consider $f(z) = |z|^2 = z \cdot z^*$. Then $\frac{\partial f}{\partial z} = z^*$ and $\frac{\partial f}{\partial z^*} = z$.

(III) The above example shows an interesting difference between the square and the absolute square: analogous to the real-valued square, $\frac{\partial}{\partial z} z^2 = 2z$ stays valid, whereas above we have $\frac{\partial}{\partial z^*} |z|^2 = z$. However, in gradient-based learning, the constant factor of 2 is of minor concern.

The actual power of Wirtinger's calculus arises in the multi-dimensional case, where a 2-dimensional interpretation of each entry in terms of (11) becomes extremely difficult. Using Wirtinger gradients circumvents this problem in an elegant way.

Definition 2 For $f \colon \mathbb{C}^N \to \mathbb{C}$ or $\mathbb{R}$, the *Wirtinger gradients* are

$$\frac{\partial f}{\partial z} := \left[\frac{\partial f}{\partial z_1}, \ldots, \frac{\partial f}{\partial z_N}\right]^\mathsf{T} \quad \text{and} \quad \frac{\partial f}{\partial z^*} := \left[\frac{\partial f}{\partial z_{1^*}}, \ldots, \frac{\partial f}{\partial z_{N^*}}\right]^\mathsf{T}.$$

There are two important examples that will be of interest later on. First, consider the squared Euclidean norm $\|z\|_2^2 = z^\mathsf{H} z$, where $(\cdot)^\mathsf{H}$ denotes the complex conjugate transpose. Its gradients read

$$\frac{\partial}{\partial z}\|z\|_2^2 = z^* \quad \text{and} \quad \frac{\partial}{\partial z^*}\|z\|_2^2 = z. \tag{13}$$

Secondly, for the quadratic form $\|z\|_\mathsf{A}^2 = z^\mathsf{H} \mathbf{A} z$, we have

$$\frac{\partial}{\partial z}\|z\|_\mathsf{A}^2 = \mathbf{A}^\mathsf{T} z^* \quad \text{and} \quad \frac{\partial}{\partial z^*}\|z\|_\mathsf{A}^2 = \mathbf{A} z. \tag{14}$$

Observe that using the conjugate Wirtinger operator often yields simpler, more readable expressions.

3 Complex-Valued Data and Gradient-Based Learning

In the following we want to specify a few basic distance measures for complex-valued data and their derivatives, especially as generalizations of their real-valued counterparts. We focus on the squared distances, for which the derivatives are easier to write down. For the non-squared distances one simply has to add the factor $\frac{1}{2\sqrt{d}}$ according to the chain rule.

Since even for $v \in \mathbb{C}^D$, $d^\pm(v)$ still remains real-valued (and so does $f \circ \mu_W \circ d$), we can apply the conventional chain rule for calculating $\frac{\partial d^\pm(v)}{\partial w^\pm}$ and any outer derivative, as reasoned in point (I) in Sect. 2.2. This yields the same update rule as in (5), where in the case of cv-data, the innermost derivative $\frac{\partial d^\pm}{\partial w^\pm}$ is to be understood as a (preferably conjugate) Wirtinger derivative. If the distance measure depends on additional parameters that have to be adapted as well, the same arguments regarding the chain rule apply and the derivatives of d with respect to those parameters can be written down accordingly.

For the squared Euclidean distance for cv-data

$$d_\mathrm{E}^2(v, w) = \|v - w\|_2^2 = (v - w)^\mathsf{H}(v - w), \tag{15}$$

we obtain

$$\frac{\partial d_\mathrm{E}^2}{\partial w^*} = -(v - w) \tag{16}$$

directly from (13). In analogy to (8), the matrix Euclidean distance reads

$$d_{\boldsymbol{\Omega}}^2(\boldsymbol{v}, \boldsymbol{w}) = \|\boldsymbol{\Omega}(\boldsymbol{v} - \boldsymbol{w})\|_2^2 = (\boldsymbol{v} - \boldsymbol{w})^H \boldsymbol{\Omega}^H \boldsymbol{\Omega}(\boldsymbol{v} - \boldsymbol{w}) \tag{17}$$

and measures the squared Euclidean norm of the projected vector difference. From (14) we get

$$\frac{\partial d_{\boldsymbol{\Omega}}^2}{\partial \boldsymbol{w}^*} = -\boldsymbol{\Omega}^H \boldsymbol{\Omega}(\boldsymbol{v} - \boldsymbol{w}) \tag{18}$$

$$\text{and} \quad \frac{\partial d_{\boldsymbol{\Omega}}^2}{\partial \boldsymbol{\Omega}^*} = \boldsymbol{\Omega}(\boldsymbol{v} - \boldsymbol{w})(\boldsymbol{v} - \boldsymbol{w})^H, \tag{19}$$

which look very similar to the real-valued versions in (9) and (10).

4 Kernels for Complex-Valued Data

Kernel methods [2] are powerful tools whenever the data are not linearly separable in the original data space. Although the kernel methods are formulated in general Hilbert spaces, the use of complex-valued kernels is not very widespread in machine learning.

When imposing a kernel function $\kappa(\boldsymbol{v}, \boldsymbol{w})$, which corresponds to an inner product $\langle \phi(\boldsymbol{v}), \phi(\boldsymbol{w}) \rangle$ in some functional Hilbert space to which the feature map ϕ maps our data, we have to be aware that κ itself can be complex-valued. Hence, we want to bring the most important properties of a complex-valued kernel function to attention:

1. *Conjugate* symmetry: $\kappa(\boldsymbol{v}, \boldsymbol{w}) = \kappa(\boldsymbol{w}, \boldsymbol{v})^*$
2. Positiveness: $\kappa(\boldsymbol{v}, \boldsymbol{v}) \geq 0$ especially implies $\kappa(\boldsymbol{v}, \boldsymbol{v}) \in \mathbb{R}$

The squared kernel distance measure then reads

$$d_{\kappa}^2(\boldsymbol{v}, \boldsymbol{w}) = \kappa(\boldsymbol{v}, \boldsymbol{v}) - \kappa(\boldsymbol{v}, \boldsymbol{w}) - \kappa(\boldsymbol{w}, \boldsymbol{v}) + \kappa(\boldsymbol{w}, \boldsymbol{w}), \tag{20}$$

which remains real-valued since the conjugate symmetry implies $\kappa(\boldsymbol{v}, \boldsymbol{w}) + \kappa(\boldsymbol{w}, \boldsymbol{v})$ $= 2 \, \mathrm{Re}(\kappa(\boldsymbol{v}, \boldsymbol{w}))$. The modifications include real-valued kernels as a special case. For gradient-based learning we use the derivatives of (20) with respect to the parameter of interest.

To give some examples, we now specify a number of kernels for cv-data as a generalization of real-valued ones. The formulas for the kernels and the derivatives of the corresponding distance measures are summarized in Table 1.

- **Gaussian radial basis function (RBF) kernel**: The definition and the derivatives of the widely used Gaussian kernel are very similar to its real-valued analogue. The derivative is obtained by plugging in (15) and (16) via chain rule.

Table 1 Kernels for complex-valued data

Gaussian RBF	$\kappa(v, w) = \exp\left(-\frac{1}{\sigma^2}\|v - w\|_2^2\right)$
	$\dfrac{\partial d_\kappa^2(v, w)}{\partial w^*} = -\dfrac{2}{\sigma^2}\kappa(v, w) \cdot (v - w)$
RBF with metric adaptation	$\kappa(v, w) = \exp\left(-(v - w)^{\mathsf{H}}\Omega^{\mathsf{H}}\Omega(v - w)\right)$
	$\dfrac{\partial d_\kappa^2(v, w)}{\partial w^*} = -2\kappa(v, w) \cdot \Omega^{\mathsf{H}}\Omega(v - w)$
	$\dfrac{\partial d^2(v, w)}{\partial \Omega^*} = 2\kappa(v, w) \cdot \Omega(v - w)(v - w)^{\mathsf{H}}$
Complex Gaussian	$\kappa(v, w) = \exp\left(-\frac{1}{2\sigma^2}(v - w^*)^{\mathsf{T}}(v - w^*)\right)$
	$\dfrac{\partial d_\kappa^2(v, w)}{\partial w^*} = \dfrac{1}{\sigma^2}\left(\kappa(w, w) \cdot (w - w^*) - \kappa(v, w) \cdot (v - w^*)\right)$
Complex Gaussian with metric adaptation	$\kappa(v, w) = \exp\left(-(v - w^*)^{\mathsf{T}}\Omega^{\mathsf{H}}\Omega(v - w^*)\right)$
	$\dfrac{\partial d_\kappa^2(v, w)}{\partial w^*} = 2\,\mathrm{Re}\left(\Omega^{\mathsf{H}}\Omega\right)\left(\kappa(w, w)(w - w^*) - \kappa(v, w)(v - w^*)\right)$
	$\dfrac{\partial d_\kappa^2(v, w)}{\partial \Omega^*} =$ $-\Omega\Big(\kappa(v, v)(v - v^*)(v - v^*)^{\mathsf{T}} + \kappa(w, w)(w - w^*)(w - w^*)^{\mathsf{T}}$ $- 2\,\mathrm{Re}\left(\kappa(v, w)(v - w^*)(v - w^*)^{\mathsf{T}}\right)\Big)$
Exponential	$\kappa(v, w) = \exp\left(w^{\mathsf{H}}v\right)$
	$\dfrac{\partial d_\kappa^2(v, w)}{\partial w^*} = \kappa(w, w) \cdot w - \kappa(v, w) \cdot v$
Exponential with metric adaptation	$\kappa(v, w) = \exp\left(w^{\mathsf{H}}\Omega^{\mathsf{H}}\Omega v\right)$
	$\dfrac{\partial d_\kappa^2(v, w)}{\partial w^*} = \Omega^{\mathsf{H}}\Omega\left(\kappa(w, w) \cdot w - \kappa(v, w) \cdot v\right)$
	$\dfrac{\partial d_\kappa^2(v, w)}{\partial \Omega^*} = \Omega\left(\kappa(v, v)vv^{\mathsf{H}} + \kappa(w, w)ww^{\mathsf{H}} - 2\,\mathrm{Re}\left(\kappa(v, w)vw^{\mathsf{H}}\right)\right)$

- **Gaussian RBF kernel with metric adaptation**: The introduction of a projection matrix to the RBF kernel is obtained by plugging in (17). The chain rule, and (18) and (19) yield the equations in Table 1. Note that this kernel function still remains real-valued for complex data.
- **Complex Gaussian kernel**: As an actually complex-valued generalization of the Gaussian RBF kernel, the complex Gaussian kernel was introduced in [16],

together with an explicit description of the feature space. It does use phase information, but it is even less interpretable than the RBF kernel. Note that the equivalence
$w - w^* = 2 \, \mathrm{i} \, \mathrm{Im}(w)$ holds and can be applied to the formula in the table.

- **Complex Gaussian kernel with metric adaptation**: As with the RBF kernel, we
  can introduce a projection matrix $\boldsymbol{\Omega}$ in the exponent. Note that we have to choose
  $\boldsymbol{\Omega}^H$ in favor of $\boldsymbol{\Omega}^T$ to maintain the conjugate symmetry of the kernel. However,
  it turns out that $\boldsymbol{\Omega}$ can be chosen to be real-valued anyway, since its update will
  always be real-valued and its imaginary part does not affect the kernel value.
  The derivatives for the prototype update and for the matrix update can be found
  in Table 1. Note again that, e. g. for v, $(v - v^*)(v - v^*)^T = -4 \, \mathrm{Im}(v) \, \mathrm{Im}(v)^T$ and
  analogously in the exponent of $\kappa(v, v)$ itself, yielding a completely real-valued
  expression in the parentheses of the matrix update equation.
- **Exponential kernel**: The trivial kernel is the inner product itself, as long as the
  data lives in a Hilbert space. For any given kernel, its exponential is a kernel as well
  [17], hence another basic example is the exponential kernel, as stated in Table 1.
  However, in its basic version, this kernel is very sensitive to scaling of the data in
  terms of numerical stability.
- **Exponential kernel with metric adaptation**: In the exponential kernel we can
  introduce a projection matrix $\boldsymbol{\Omega}$ as well. With a suitable initialization and scaling
  of $\boldsymbol{\Omega}$ the stability problem of the plain exponential kernel can be handled.

There is a variety of real-valued kernels that can be extended for complex data
as well. Often this can be done by simply replacing the scalar products involved
using the conjugate transpose. The derivatives can then be found by straightforward
application of Wirtinger's calculus.

5 Experiments

Image classification is a major application in machine learning. The images are often
given in a high resolution and direct usage is computationally too expensive. Many
feature extraction schemes have been developed to pre-process high-dimensional
data. Some of these methods yield complex-valued features. One example are the
Zernike-Moments [18], which can be used to describe the shape of an object.

The Flavia-Dataset [19] is a set of images of 32 different leaves. It consists of
1907 images with a resolution of 800×600 pixels. In [7], the Zernike-Moments
from the green-channel of the images are generated up to order 20 resulting in 121
complex-valued features per sample. In addition, the data set is z-score normalized.
We concentrate on the same experimental set-up as in [7], performing 10-fold cross-
validation for LVQ systems with 1 prototype per class. Further, the kernel parameter
σ for the single kernels in GLVQ is chosen by systematic search. The values are $\sigma = 5$
for the RBF kernel and $\sigma = 0.01$ for the complex Gaussian kernel. The advantage
of GMLVQ is the indirect automatic determination of σ by the matrix $\boldsymbol{\Omega}$. Therefore,
no additional normalization of $\boldsymbol{\Omega}$ is required.

Table 2 Mean test error in % for the leaf data set with complex-valued features

Algorithm	Euclid	RBF	Complex Gaussian	Exponential
GLVQ	13.1 (±0.5)	12.2 (±2.1)	14.2 (±3.5)	21.5 (±2.1)
GMLVQ	21.3 (±1.7)	13.6 (±4.4)	22.5 (±3.6)	16.0 (±5.7)

The standard deviation is given in parentheses

The performances (test errors) of the cv-G(M)LVQ with different dissimilarity measures are given in Table 2. The cv-GMLVQ using the Wirtinger calculus to derive the update formulas gives similar accuracy as the heuristic approach in [7]. Interestingly, the GLVQ without relevance learning achieves a better mean test accuracy than the GMLVQ, while the latter yields better training accuracies. Thus, the learning of the mapping matrix does not seem beneficial in this case. Applying a cv-kernel, only the RBF kernel results in a slightly better accuracy. The complex Gaussian as well as the Exponential kernel do not seem suitable for this particular data set.

Note that, here, our aim is to demonstrate the applicability of the framework. Achieving superior performance with more suitable kernel functions should be possible, but is beyond the scope of this contribution.

6 Conclusion and Remarks

In this contribution we extended the matrix and kernel versions of GLVQ to complex-valued data. To achieve this in a unified framework, we utilized the theory of Wirtinger's calculus to derive and express the gradients in an elegant way. Bringing this theory to attention was one of the main objectives of this paper, and we successfully applied it to a number of kernels to exemplify the principles. Although the methods could not develop their full potential on the chosen data set, we are confident that they can improve performance more significantly in a variety of practical applications that involve complex-valued data. Current and future work involves the test of the algorithms on different data sets, as well as the derivation and generalization of further kernels for complex-valued data.

Acknowledgments M.B. thanks the Aspen Center for Physics and the NSF Grant No. PHYS-1066293 for hospitality while the writing of this paper was finalized.

References

1. Nitta, T.: Solving the XOR problem and the detection of symmetry using a single complex-valued neuron. Neural Netw. **16**, 1101–1105 (2003)
2. Schölkopf, B., Smola, A.: Learning with Kernels. MIT Press, Cambridge (2002)
3. Hyvärinen, A., Karhunen, J., Oja, E.: Independent Component Analysis. Wiley, New York (2001)

4. Sato, A., Yamada, K.: Generalized learning vector quantization. In: Hasselmo, M.E., Touretzky, D.S., Mozer, M.C. (eds.) Advances in Neural Information Processing Systems, pp. 423–429. MIT Press, Cambridge (1996)
5. Witoelar, A., Gosh, A., de Vries, J., Hammer, B., Biehl, M.: Window-based example selection in learning vector quantization. Neural Comput. **22**(11), 2924–2961 (2010)
6. Kästner, M., Riedel, M., Stickert, M., Hermann, W., Villmann, T.: Border-sensitive learning in kernelized learning vector quantization. In: International Work Conference on Artificial Neural Networks (IWANN), Tenerife (2013)
7. Bunte, K., Schleif, F., Biehl, M.: Adaptive learning for complex-valued data. In: 20th European Symposium on Artificial Neural Networks, ESANN 2012, Bruges, Belgium, April 25–27 2012
8. Wirtinger, W.: Zur formalen Theorie der Funktionen von mehr komplexen Veränderlichen. Mathematische Annalen **97**(1), 357–375 (1927)
9. Villmann, T., Haase, S., Kaden, M.: Kernelized vector quantization in gradient-descent learning. Neurocomputing, **147**(0), 83–95, Advances in Self-Organizing Maps Subtitle of the special issue: Selected Papers from the Workshop on Self-Organizing Maps 2012 (WSOM 2012) (2015)
10. Robbins, H., Monro, S.: A stochastic Approximation Method. Ann. Math. Stat. 400–407 (1951)
11. Mwebaze, E., Schneider, P., Schleif, F.-M., Aduwo, J.R., Quinn, J.A., Haase, S., Villmann, T., Biehl, M.: Divergence-based classification in learning vector quantization. Neurocomputing **74**(9), 1429–1435 (2011)
12. Schneider, P., Biehl, M., Hammer, B.: Adaptive relevance matrices in learning vector quantization. Neural Comput. **21**(12), 3532–3561 (2009)
13. Schneider, P., Hammer, B., Biehl, M.: Distance learning in discriminative vector quantization. Neural Comput. **21**, 2942–2969 (2009)
14. Bunte, K., Schneider, P., Hammer, B., Schleif, F., Villmann, T., Biehl, M.: Limited rank matrix learning, discriminative dimension reduction and visualization. Neural Netw. **26**, 159–173 (2012)
15. Ahlfors, L.V.: Complex Analysis, 2nd edn. McGraw'Hill Publishing Comp. Ltd, Maidenhead (1966)
16. Steinwart, I., Hush, D., Scovel, C.: An explicit description of the reproducing kernel hilbert spaces of gaussian RBF kernels. IEEE Trans. Inf. Theory **52**, 4635–4643 (2006)
17. Genton, M.G.: Classes of kernels for machine learning: a statistics perspective. J. Mach. Learn. Res. **2**, 299–312 (2002)
18. Teague, M.R.: Image analysis via the general theory of moments. J. Opt. Soc. Am. **70**, 920–930 (1980)
19. Wu, S.G., Bao, F.S., Xu, E.Y., Wang, Y., Chang, Y.-F., Xiang, Q.-L.: A leaf recognition algorithm for plant classification using probabilistic neural network, CoRR (2007). arXiv: 0707.4289

A Study on GMLVQ Convex and Non-convex Regularization

David Nova and Pablo A. Estévez

Abstract In this work we investigate the effect of convex and non-convex regularization on the Generalized Matrix Learning Vector Quantization (GMLVQ) classifier, in order to obtain sparse models that guarantee a better generalization ability. Three experiments are used for evaluating six different sparse models in terms of classification accuracy and qualitative sparseness. The results show that non-convex models outperform traditional convex sparse models and non-regularized GMLVQ.

Keywords Learning vector quantization · Sparse models · Regularization · Generalization ability

1 Introduction

Learning Vector Quantization (LVQ) is a well-known classifier that represents class regions by using prototypes [9]. In [16] a generalization scheme called Generalized Learning Vector Quantization (GLVQ) was proposed. GLVQ introduces a continuous and differentiable cost function that aims at margin maximization. An extension is the Generalized Relevance Learning Vector Quantization (GRLVQ) [7] which introduces weighted factors over the original features with the aim of extracting relevance and increasing class separability. Furthermore, the Generalized Matrix Learning Vector Quantization (GMLVQ) [17] is a generalization of the relevance learning which

D. Nova · P.A. Estévez (✉)
Department of Electrical Engineering,
University of Chile, Casilla 412-3, Santiago, Chile
e-mail: pestevez@ing.uchile.cl

D. Nova
e-mail: dnovai@ug.uchile.cl

P.A. Estévez
Advanced Mining Technology Center, University of Chile, Santiago, Chile

D. Nova · P.A. Estévez
Millennium Institute of Astrophysics, Santiago, Chile

© Springer International Publishing Switzerland 2016
E. Merényi et al. (eds.), *Advances in Self-Organizing Maps and Learning Vector Quantization*, Advances in Intelligent Systems and Computing 428,
DOI 10.1007/978-3-319-28518-4_27

305

uses a full matrix to project linearly the original data space into a subspace. This aims at enhancing the separability of the class regions and increasing the generalization ability of the classifier. The reader can find recent reviews of LVQ classifiers in [8, 14].

In high-dimensional data applications sparsity of the relevance matrix may help to achieve a better generalization ability. With this aim, penalty or regularization terms have been added to the GMVLQ cost function [17]. In our contribution, we compare six different sparse penalty functions on GMLVQ using three benchmark datasets. The research question is how to choose an appropriate sparse penalty function and what is the difference between convex and non-convex regularization.

The remainder of this paper is organized as follows: In Sect. 2, a background on GMLVQ and sparse models is presented. In Sect. 3, six sparse penalty functions are introduced and added to the GMLVQ cost function. In Sect. 4, the results using three datasets are presented. Finally, in Sect. 5 the conclusions are drawn.

2 Background

2.1 Generalized Matrix Learning Vector Quantization

The GLVQ cost function aiming at achieving a margin maximization prototype-based classifier was introduced in [16]. GLVQ defines the following receptive field for the i-th prototype:

$$R^i = \{\mathbf{x} \in X | \forall \mathbf{w}_j (j \neq i) \leftarrow d(\mathbf{w_i}, \mathbf{x}) < d(\mathbf{w_j}, \mathbf{x})\}, \tag{1}$$

where $d(\cdot)$ is the square Euclidean distance, $\mathbf{w_i}$ is the i-th vector prototype and $\mathbf{x}$ is a vector sample from data set $\mathbf{X} \in \mathbb{R}^{d \times n}$.

GMLVQ is a variant of GLVQ which introduces the following cost function:

$$E = \sum_k f\left(\mu(\mathbf{x}_k, \mathbf{W}, \mathbf{\Lambda})\right), \tag{2}$$

where $\mathbf{W} = \{\mathbf{w}_1, \mathbf{w}_2, \ldots, \mathbf{w}_n\}$ is a set of prototypes, $\mathbf{\Lambda} \in \mathbb{R}^{d \times d}$ is the relevance matrix, $f(\mu) = \frac{1}{1+e^{-\mu}}$ is a sigmoid function, and $\mu(\mathbf{x}) = \frac{d^+ - d^-}{d^+ + d^-}$ is the relative distance difference with d^+ as the minimum distance to the right class prototype, and d^- as the minimum distance to the wrong class prototype. The distance is rewritten as:

$$d^\Lambda(\mathbf{x}, \mathbf{w}) = (\mathbf{x} - \mathbf{w})^T \Lambda (\mathbf{x} - \mathbf{w}), \tag{3}$$

where Λ must be positive semi-definite. By substituting

$$\mathbf{\Lambda} = \Omega^T \Omega, \tag{4}$$

the following result holds $\mathbf{u}^T \Lambda \mathbf{u} = \mathbf{u}^T \Omega^T \Omega \mathbf{u} = \left(\Omega^T \mathbf{u} \right)^2 \geq 0$ for all $\mathbf{u}$. As a consequence the receptive field of the i-th prototype is redefined as:

$$R_\Lambda^i = \{\mathbf{x} \in X | \forall \mathbf{w}_j (j \neq i) \leftarrow d^\Lambda(\mathbf{w_i}, \mathbf{x}) \leq d^\Lambda(\mathbf{w_j}, \mathbf{x})\}. \qquad (5)$$

2.2 Sparse Models

Sparse models aim at constraining the cost function by adding a penalty function in order to guarantee sparse solutions. There is a trade-off between model accuracy and sparsity but in classifiers a sparse solution usually gives a better generalization ability [6]. A typical sparse model assumes the following cost function:

$$F^* = \arg \min_{\mathbf{x}} E(\mathbf{x}) + \lambda \sum_n \phi(\mathbf{x}(n)), \qquad (6)$$

where $E(\mathbf{x})$ is the cost function associated with the model (Eq. (2) in our case), $\phi(\cdot)$ is a sparsity function (e.g. ℓ_1 or ℓ_0 norm) and λ is a constant which controls the trade-off between sparsity and reconstruction error.

A popular penalty function is the ℓ_0 norm. However, since this penalty function is discrete and non-convex, the problem becomes NP-hard and hardly tractable when the dimensionality of the data is large. Several works have tried to deal with this issue [4, 15, 18, 19]. In [18] the Least Absolute Shrinkage and Selection Operator (Lasso) or ℓ_1 penalty was introduced and its use has had an increasing popularity. Compared to the ℓ_2-norm, the ℓ_1-norm produces sparser models and it performs feature selection within the learning algorithm. This allows reducing the model complexity, but since the ℓ_1-norm is not differentiable it requires changes to gradient based learning algorithms or using a mathematical approximation. Moreover, ℓ_1 does not guarantee uniqueness of the solution as ℓ_2-norm does. Some authors provide arguments against the Lasso and they advocate that penalty functions should be singular at the origin for achieving sparsity by using a non-smooth, and non-convex penalty function instead of the ℓ_1 regularization [1, 5, 13].

Convex functions are attractive because they are more reliably minimized than non-convex functions. However, non-convex penalty functions can lead to enhanced sparsity of solutions [11, 12]. Smoothly Clipped Absolute Deviation (SCAD) [5] is a non-convex penalty function that overcomes the drawbacks of the ℓ_1-norm at the cost of introducing a new parameter. Furthermore, Zhang's penalty function [21] is a linear approximation of the SCAD penalty function that can be interpreted as a two-stage reweighted ℓ_1 penalized optimization problem. This approximation solves the Lasso problem but large parameters are not penalized anymore leading to an unbiased model. The Bridge penalty function [10], also called ℓ_q pseudo-norm when $0 < q < 1$ provides a quasi-smooth approximation of the ℓ_0 sparsity measure as q tends towards to zero, and it yields sparser solutions than Lasso. Depending on the q value the ℓ_q pseudo-norm can change from a convex to a non-convex function as is illustrated in Fig. 1c. Another popular non-convex penalty function is the logarithmic penalty [20] which can be interpreted as an approximation of the ℓ_0-norm. The penalty function must be singular at the origin to produce sparse solutions, it must be bounded by a constant to produce nearly unbiased estimates for large coefficients, and their derivatives should vanish for large values [5]. Furthermore, a sparse penalty function should be chosen so as to promote the sparsity of $\mathbf{x}$ in Eq. (6).

Figure 1 shows the six penalty functions that are used in this contribution. The first row shows a convex and two pseudo-norm penalty functions, and the second row shows three non-convex penalty functions.

In Table 1 a summary of the convex and non-convex penalty functions and their respective derivatives are shown. The first column contains the name of the penalty functions, the second column shows the mathematical expression of the penalty functions, and the third column shows the derivatives of the penalty functions.

3 Sparsity on GMLVQ

In this section we obtain update rules for GMLVQ equipped with six different penalty functions. The aim is to obtain a sparse matrix $\mathbf{\Lambda}$, which is reflected in the main diagonal having a low number of non-zero values. Additionally the eigenvalues of the relevance matrix $\mathbf{\Lambda}$ are smaller. Let us add to Eq. (2) a penalty function as follows:

$$E = \sum_k f\left(\mu(\mathbf{x}_k, \mathbf{W}, \mathbf{\Lambda})\right) + \lambda \sum_{lm} \phi(\Omega_{lm}), \tag{7}$$

where $\phi\left(\cdot\right)$ is any of the functions in Table 1, and $\Omega \in \mathbb{R}^{d \times k}$ is defined as in Eq. (4).

By using gradient descent the following GMLVQ update rules are obtained for the prototypes:

$$\mathbf{w}^+ = \mathbf{w}^+ + 2\beta f' \mu^+ \Lambda(\mathbf{x} - \mathbf{w}^+), \tag{8}$$

$$\mathbf{w}^- = \mathbf{w}^- - 2\beta f' \mu^- \Lambda(\mathbf{x} - \mathbf{w}^-), \tag{9}$$

where f' is the derivative of the sigmoid function $f(x) = \frac{1}{1+e^{-x}}$, $\mu^+ = \frac{\partial \mu}{\partial \mathbf{w}^+}$ is the derivative of the relative distance difference with respect to the right class prototype $\mathbf{w}^+$, $\mu^- = \frac{\partial \mu}{\partial \mathbf{w}^-}$ is the derivative of the relative distance difference with respect to the wrong class prototype $\mathbf{w}^-$, and β is the learning rate. Besides, the Ω matrix is adapted in order to enhance the relevance of the original data dimensions. The update rule for the (l, m)th element of the Ω matrix is the following:

$$\begin{aligned}
\Delta\Omega_{lm} = &-\beta \cdot 2 \cdot f'(\mu(\mathbf{x})) \\
&\cdot(\mu^+(\mathbf{x})((x_m - w_{J,m})[\Omega(\mathbf{x} - \mathbf{w}_J)]_l) \\
&-\mu^-(\mathbf{x}) \cdot ((x_m - w_{K,m})[\Omega(\mathbf{x} - \mathbf{w}_K)]_l)) \\
&-\nabla\phi_\lambda(\Omega_{lm}).
\end{aligned} \tag{10}$$

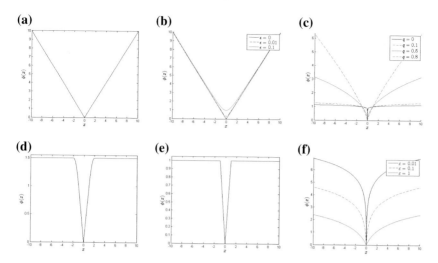

Fig. 1 Plots of six sparse penalty functions used with the GMLVQ cost function: ℓ_1-norm **a**, epsilon-ℓ_1 pseudo-norm **b**, ℓ_q pseudo-norm **c**, SCAD **d**, Zhang's **e**, and log **f**

Table 1 Penalty functions

Name	Formula	Derivative
$\ell_1 - norm$	$\phi_\lambda(\mathbf{x}) = \lambda\|\mathbf{x}\|$	$\frac{\partial\phi_\lambda(\mathbf{x})}{\partial\mathbf{x}} = \lambda\,\mathrm{sgn}(\mathbf{x})$
Epsilon-ℓ_1 pseudo-norm	$\phi_\lambda(\mathbf{x}) = \lambda\left(\mathbf{x}^2 + \epsilon\right)^{\frac{1}{2}}$	$\frac{\partial\phi_\lambda(\mathbf{x})}{\partial\mathbf{x}} = \frac{\lambda\mathbf{x}}{\sqrt{\mathbf{x}^2+\epsilon}}$
ℓ_q pseudo-norm	$\phi_\lambda(\mathbf{x}) = \lambda\|\mathbf{x}\|^q$ with $0 < q < 1$	$\frac{\partial\phi_\lambda(\mathbf{x})}{\partial\mathbf{x}} = q\mathbf{x}\|\mathbf{x}\|^{q-2}$
SCAD	$\phi_\lambda(\mathbf{x}) = \begin{cases} \lambda\|\mathbf{x}\| & \|\mathbf{x}\| \leq \lambda \\ \frac{-\|\mathbf{x}\|^2+2a\lambda\|\mathbf{x}\|-\lambda^2}{2(a-1)} & \lambda < \|\mathbf{x}\| \leq a\lambda \\ \frac{(a+1)\lambda^2}{2} & \|\mathbf{x}\| > a\lambda \end{cases}$	$\frac{\partial\phi_\lambda(\mathbf{x})}{\partial\mathbf{x}} = \begin{cases} \lambda\,\mathrm{sgn}(\mathbf{x}) & \text{if } \|\mathbf{x}\| \leq \lambda \\ \frac{\mathbf{x}-2a\lambda\mathbf{x}}{2\|\mathbf{x}\|-2a\lambda} & \lambda < \|\mathbf{x}\| \leq a\lambda \\ 0 & \|\mathbf{x}\| > a\lambda \end{cases}$
Zhang's	$\phi_\lambda(\mathbf{x}) = \begin{cases} \lambda\|\mathbf{x}\| & \|\mathbf{x}\| < \eta \\ \lambda\eta & \text{otherwise} \end{cases}$	$\frac{\partial\phi_\lambda(\mathbf{x})}{\partial\mathbf{x}} = \begin{cases} \lambda\,\mathrm{sgn}(\mathbf{x}) & \|\mathbf{x}\| < \eta \\ 0 & \text{otherwise} \end{cases}$
Log	$\phi_\lambda(\mathbf{x}) = \lambda\log\left(\|\mathbf{x}\| + \epsilon\right) - \lambda\log(\epsilon)$	$\frac{\partial\phi_\lambda(\mathbf{x})}{\partial\mathbf{x}} = \frac{\lambda\,\mathrm{sgn}(\mathbf{x})}{\mathbf{x}\,\mathrm{sgn}(\mathbf{x})+\epsilon}$

In what follows we use the following notation for the six different GMLVQ sparse models: ℓ_1-norm, epsilon-ℓ_1 pseudo-norm, ℓ_q pseudo-norm, SCAD, Zhang's and log penalty functions.

The sub-gradient approach is used [2, 3] in order to solve the problem of an undefined derivative at $x = 0$ for the functions shown in Table 1. A function $\phi_\lambda(x)$ is called subdifferentiable at x if there exists at least one subgradient at x. The set of subgradients of $\phi_\lambda(x)$ at the point x is called the subdifferential of $\phi_\lambda(x)$ at x, and is denoted $\partial\phi_\lambda(x)$. A function $\phi_\lambda(x)$ is called subdifferentiable if it is subdifferentiable for all x in the domain of $\phi_\lambda(x)$. For example, consider $\phi_\lambda(x) = |x|$; for $x \neq 0$ the subgradient is unique: $\partial\phi_{lambda}(x) = \mathrm{sgn}(x)$. At $x = 0$ the subdifferential is defined by the inequality $|z| \geq gz \; \forall z$, which is satisfied if and only if $g \in [-1, 1]$. Therefore the subgradient is equal to $\partial\phi_\lambda(x) \in [-1, 1]$. In our contribution a gradient descent update approach was used.

4 Simulation Results

In this section three experiments are performed. In all experiments, several values for each parameter are tried in order to tune the different models. The trade-off parameter λ of the penalty function in Eq. (7) was set as $\lambda = \{0.01, 0.05, 0.1, 0.5, 1, 3, 5, 10\}$, the number of prototypes was chosen as a percentage size of the dataset $N_p = \{1, 5, 10, 15, 20\%\}$. In the case of Epsilon-ℓ_1 penalty function the parameter ϵ was varied in the set $\epsilon = \{0.01, 0.05, 0.1, 0.5, 1, 3, 5, 10\}$, for the ℓ_q pseudo-norm penalty function $q = \{0.01, 0.05, 0.1, 0.5, 0.8\}$, for SCAD penalty function $\alpha = \{0.01, 0.05, 0.1, 0.5, 1, 3, 5, 10\}$, for Zhang's penalty function $\eta = \{0.01, 0.05, 0.1, 0.5, 1, 3, 5, 10\}$, and for the log penalty function $\varepsilon = \{0.01, 0.05, 0.1, 0.5, 1, 3, 5, 10\}$. Besides, in every case a 10-fold cross validation, and a one-way analysis of variance (ANOVA) with Bonferroni correction was performed using a significant level of 0.05. The following notation is used to express the results: the mean plus/minus the standard deviation of the mean.

The Pipeline dataset consists of 1000 samples with 3 classes and 12 features. In this dataset the best result is obtained by the log penalty function with $9.9010e-4 \pm 0.0031$ (see Table 2 first row). However, this result is not statistically different from the other two non-convex penalty functions. All non-convex functions outperform the ℓ_1-norm and the difference is statistically significant. Figure 2a shows the effect of the trade-off parameter λ for the SCAD non-convex penalty function. Each curve corresponds to a different value of α. In Fig. 2 it can be observed that in general the higher the λ value the higher the classification error which is an expression of the trade off between accuracy and sparsity representation of the Ω matrix. Notice that for $\alpha = 0.01$ the best regularization effect is obtained since for a wide range of λ values the classification error is kept constant. Figure 2b shows the classification error as a function of λ for the logarithmic penalty function for different ε values. Contrary to SCAD, the logarithmic penalty function does not keep the accuracy constant for different values of λ for any value of ε. Also, in this dataset all non-convex penalty functions outperform the original GMLVQ algorithm (0.0060 ± 0.0085) and GMVLQ with ℓ_1 regularization (0.0030 ± 0.0048) and the differences are statistically significant.

Figures 3a, c illustrate a visualization of the matrix Λ for GMLVQ when using the original GMLVQ algorithm and the logarithmic penalty function with $\lambda = 0.01$, respectively. For didactic purposes, the bar graph illustrates the values of the matrix by using the mode of the 10 fold cross-validation. There is a significant difference between Fig.3a, c due to the sparsity of the solution obtained with the log penalty function. Non-diagonal elements with values different from zero indicate interactions between features for the matrix transformation. The main diagonals are shown in Figs. 3b, d where a high value represents a high relevance for the classification. The boxplots show that for the original GMLVQ there are more non-zero values in

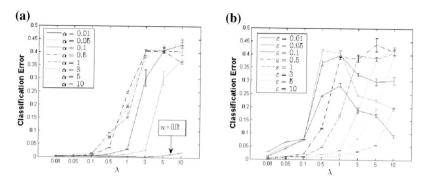

Fig. 2 Plots of the results for different regularization methods over GMLVQ for the Pipeline dataset. SCAD penalty function regularization by using 5% of number of prototypes **a**, and logarithmic penalty function regularization with $N_p = 5\%$ of the size of the dataset as the number of prototypes **b**

the main diagonal compared to the result obtained by using regularization with the logarithmic penalty function. Also, the boxplot indicates the dispersion for 10-runs of the Λ matrix. In the case of the GMLVQ there are many non-zero elements and the distribution shows some outliers marked with red plus sign. However for regularization with the logarithmic penalty function there are fewer non-zero elements and a more compact distribution per feature is obtained.

The Image Segmentation dataset consists of 19 features (where 3 have been eliminated for being constant), 2100 samples and 7 classes. A summary of the results is shown in Table 2, second row. In this dataset all non-convex penalty functions yield similar performances. But they outperform GMLVQ (0.0405 ± 0.0125) and GMLVQ with ℓ_1 regularization (0.0358 ± 0.0099). The logarithmic function obtain a minimum classification error of 0.0305 ± 0.0144 using 5% of the number of prototypes and $\lambda = 0.01$. However, the difference is not statistically significant from those obtained with other non-convex regularization methods. In this case the logarithmic function shows a compact distribution for the 10-runs of the cross validation. An interpretation is that the model is not over-fitting the training data. Figures 4a, b

Table 2 Summary of classification errors for the three datasets by using the original GMLVQ model and adding different penalty functions, mean (standard deviation) for 10-fold cross validation

	GMLVQ	ℓ_1	$Epsilon - \ell_1$	ℓ_q	SCAD	Zhang	log
D1	0.0060	0.0030	0.0010	0.0020	0.0020	0.0020	**9.9010e-4**
	(0.0085)	(0.0048)	(0.0032)	(0.0042)	(0.0133)	(0.0134)	**(0.0031)**
D2	0.0405	0.0358	0.0314	0.0348	0.0310	0.0338	**0.0305**
	(0.0125)	(0.0099)	(0.0154)	(0.0108)	(0.0139)	(0.0102)	**(0.0144)**
D3	0.2537	0.2290	0.2187	0.2214	**0.2147**	0.2175	0.2252
	(0.0574)	(0.0387)	(0.0387)	(0.0260)	**(0.0539)**	(0.0332)	(0.0424)

Pipeline (D1), image segmentation (D2) and pima (D3) datasets

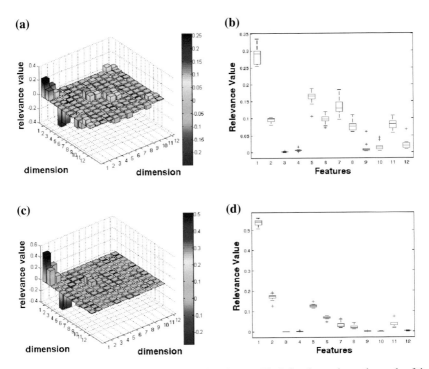

Fig. 3 Visualization of the results for the Pipeline dataset. The left column shows the mode of the matrix relevance Λ for classification error using 10-fold cross-validation with the original GMLVQ **a** and using regularization with the log penalty function with $\lambda = 0.01$ **c**. The right column shows the boxplot of the main *diagonal* for GMLVQ **b** and for the log penalty function with $\lambda = 0.01$ **c**, both for 10 runs

show the eigenvalues when using a low $\lambda = 0.01$ and a high $\lambda = 10$, respectively. These figures show that the higher the sparsity the lower the number of eigenvalues obtained. Figures 4c, d show the main diagonal for relevance matrix Λ by using the original GMLVQ and regularization with a logarithmic penalty function, respectively. It can be observed that there is a significant difference between the number of non-zero values and the distribution of those elements. In the case of logarithmic penalty function there exists a compact distribution of the non-zero elements for 10 runs. Notice that the non-diagonal elements of the relevance matrix Λ add information for improving the classification. A significant non-zero value in a non-diagonal element indicates interactions between features.

The Pima dataset consists of 8 features, 768 samples and 2 classes. A summary of results for this dataset is shown in Table 2, third row. The regularization with the SCAD penalty function obtains the minimum classification error with 0.2147 ± 0.0539 but it is not statistically significant from the errors obtained by the epsilon-ℓ_1 (0.2187 ± 0.0387), and Zhang's (0.2175 ± 0.0332) penalty functions. It is worth to notice that all non-convex functions outperform the ℓ_1 regularization sparse model

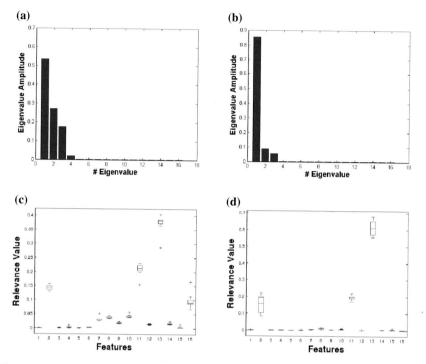

Fig. 4 Plots of the results for different regularization methods using the Image segmentation dataset. Eigenvalues of the relevance matrix Λ for different values of λ: $\lambda = 0.01$ **a**, and $\lambda = 10$ **b**. Boxplots of main diagonal of the relevance matrix for 10-fold cross-validation by using $N_p = 5\%$ of the size of the dataset for SCAD **c**, and logarithmic **d** penalty functions

and the differences are statistically significant. Although, we should take into account that the ℓ_1 norm does not have an extra parameter in contrast to all other methods. Naturally, as in the previous experiments, when increasing the number of prototypes per class a lower classification error is obtained but after 5 % the best performance reaches a saturation level. In all cases small values of the λ parameter allow a better generalization ability without losing accuracy and reaching a sparse representation of the Ω matrix.

5 Conclusions

There is a trade-off between accuracy and sparsity of a solution. However, to a certain point a sparse solution helps to improve the accuracy of the GMLVQ model. Furthermore, when a sparse solution is obtained there are less relevant components of the relevance matrix which improves its interpretability and the regularization adds robustness to the solution. In this contribution we have showed that the non-convex penalty functions outperform the original GMLVQ model and the ℓ_1 regularization in all cases in terms of classification accuracy and sparseness. But the non-convex functions have an extra parameter in order to facilitate the tuning of each penalty function. Sparse models using non-convex functions could be extended to local matrix relevance in the near future.

Acknowledgments This contribution was funded by CONICYT-CHILE under grants Fondecyt 1140816 and DPI20140090.

References

1. Antoniadis, A., Fan, J.: Regularization of wavelet approximations. J. Am. Stat. Assoc. 939–967 (2011)
2. Bertsekas, D.P.: Nonlinear programming. Athena Sci. (1999)
3. Boyd, S., Vandenberghe, L.: Convex Optimization. Cambridge University Press (2004)
4. Candes, E., Tao, T.: The dantzig selector: statistical estimation when p is much larger than n. Ann. Stat. 2313–2351 (2007)
5. Fan, J., Li, R.: Variable selection via nonconcave penalized likelihood and its oracle properties. J. Am. Stat. Assoc. **96**(456), 1348–1360 (2001)
6. Figueiredo, M.A.: Adaptive sparseness for supervised learning. IEEE Trans. Pattern Anal. Mach. Intell. **25**(9), 1150–1159 (2003)
7. Hammer, B., Villmann, T.: Generalized relevance learning vector quantization. Neural Netw. **15**(8), 1059–1068 (2002)
8. Kaden, M., Lange, M., Nebel, D., Riedel, M., Geweniger, T., Villmann, T.: Aspects in classification learning-review of recent developments in learning vector quantization. Found. Comput. Decis. Sci. **39**(2), 79–105 (2014)
9. Kohonen, T.: Self-organizing maps, vol. 30. Springer (2001)
10. Leahy, R.M., Jeffs, B.D.: On the design of maximally sparse beamforming arrays. IEEE Trans. Antennas Propag. **39**(8), 1178–1187 (1991)
11. Lee, H., Battle, A., Raina, R., Ng, A.Y.: Efficient sparse coding algorithms. In: Advances in Neural Information Processing Systems, pp. 801–808 (2006)
12. Ng, A.Y.: Feature selection, l 1 vs. l 2 regularization, and rotational invariance. In: Proceedings of the Twenty-First International Conference on Machine Learning, p. 78. ACM (2004)
13. Nikolova, M.: Local strong homogeneity of a regularized estimator. SIAM J. Appl. Math. **61**(2), 633–658 (2000)
14. Nova, D., Estévez, P.A.: A review of learning vector quantization classifiers. Neural Comput. Appl. **25**(3–4), 511–524 (2014)
15. Osborne, M.R., Presnell, B., Turlach, B.A.: A new approach to variable selection in least squares problems. IMA J. Numer. Anal.-Inst. Math. Appl. **20**(3), 389–404 (2000)
16. Sato, A., Yamada, K.: Generalized learning vector quantization. Adv. Neural Inf. Process. Syst. 423–429 (1996)
17. Schneider, P., Bunte, K., Stiekema, H., Hammer, B., Villmann, T., Biehl, M.: Regularization in matrix relevance learning. IEEE Trans. Neural Netw. **21**(5), 831–840 (2010)
18. Tibshirani, R.: Regression shrinkage and selection via the lasso. J. Royal Stat. Soc. Ser. B (Methodological) 267–288 (1996)
19. Tropp, J., et al.: Just relax: convex programming methods for identifying sparse signals in noise. IEEE Trans. Inf. Theor. **52**(3), 1030–1051 (2006)
20. Weston, J., Elisseeff, A., Schölkopf, B., Tipping, M.: Use of the zero norm with linear models and kernel methods. J. Mach. Learn. Res. **3**, 1439–1461 (2003)
21. Zhang, T.: Some sharp performance bounds for least squares regression with l1 regularization. Ann. Stat. **37**(5A), 2109–2144 (2009)

Part VI
Learning Vector Quantization Theories and Applications II

Functional Representation of Prototypes in LVQ and Relevance Learning

Friedrich Melchert, Udo Seiffert and Michael Biehl

Abstract We present a framework for distance-based classification of functional data. We consider the analysis of labeled spectral data by means of Generalized Matrix Relevance Learning Vector Quantization (GMLVQ) as an example. Feature vectors and prototypes are represented as functional expansions in order to take advantage of the functional nature of the data. Specifically, we employ truncated Chebyshev series in the context of several spectral datasets available in the public domain. GMLVQ is applied in the space of expansion coefficients and its performance is compared with the standard approach in original feature space, which ignores the functional nature of the data. Data smoothing by polynomial expansion alone is also considered for comparison. Computer experiments show that, beyond the reduction of dimensionality and computational effort, the method offers the potential to improve classification performance significantly.

Keywords Classification · Supervised learning · Functional data · Learning vector quantization · Relevance learning · Dimensionality reduction

1 Introduction

A large number of unsupervised and supervised machine learning techniques are based on the use of distances or dissimilarity measures [4, 5]. Such measures can be employed for pairwise comparison of feature vectors, as for instance in the

F. Melchert (✉) · M. Biehl (✉)
University of Groningen, Johann Bernoulli Institute for Mathematics
and Computer Science, P.O. Box 407, 9700 Groningen, AK, The Netherlands
e-mail: Friedrich.Melchert@gmail.com

M. Biehl
e-mail: m.biehl@rug.nl

F. Melchert · U. Seiffert (✉)
Fraunhofer Institute for Factory Operation and Automation IFF,
Sandtorstrasse 22, 39106 Magdeburg, Germany
e-mail: Udo.Seiffert@iff.fraunhofer.de

© Springer International Publishing Switzerland 2016
E. Merényi et al. (eds.), *Advances in Self-Organizing Maps and Learning
Vector Quantization*, Advances in Intelligent Systems and Computing 428,
DOI 10.1007/978-3-319-28518-4_28

317

well-known K-Nearest-Neighbor (KNN) classifier [5, 8, 10]. Prototype-based methods replace the reference data by a number of typical representatives. In unsupervised Vector Quantization (VQ), for instance, they might represent clusters or other structures in the data. In the popular Learning Vector Quantization (LVQ) [12], prototypes serve as characteristic exemplars of the classes and, together with a distance measure, parameterize the classification scheme.

Prototype- and distance-based systems are generally intuitive and straightforward to implement. The selection of appropriate distance measures is a key issue in the design of VQ or LVQ schemes [5]. Frequently, simple Euclidean or other Minkowski measures are employed without taking into account prior knowledge of the problem. Obviously, available insight into the nature of the data should be taken advantage of. So-called relevance learning schemes offer greater flexibility and are particularly suitable for supervised learning due to the use of adaptive distance measures, see [20, 25] for examples in the context of KNN- and LVQ classifiers and [5] for a recent review. In relevance learning, only the basic form of the distance measure is specified in advance. Its parameters are determined in a data-driven training process, which is frequently guided by a suitable cost function, e.g. [18, 20, 25].

Here, we examine the use of prototype-based systems for functional data [17], where feature vectors do not simply comprise a set of more or less independent quantities, but represent a functional relation. This is the case, for instance, in the presence of temporal or other dependencies, which impose a natural ordering of the features. Functional data is found in a large variety of practical application areas [17]. Perhaps, time series and sequences come to mind first in the context of, e.g., bioinformatics, meteorology or economy. Similarly, densities or histograms can be used to represent statistical properties of observations. Another important example is that of spectral data, for instance optical or mass spectra obtained in various fields ranging from remote sensing to chemistry and bioinformatics.

Formally, discretized functional data can be treated by standard prototype-based methods, including relevance learning. However, this naive approach ignores the specific properties of the data, which may result in suboptimal performance due to several problems: Obviously, disregarding the order and intrinsic correlation of neighbored features in functional data can lead to nominally very high-dimensional systems and inappropriately large numbers of adaptive parameters. Thus, the learning problem is rendered unnecessarily complex and may suffer from convergence problems and overfitting effects. In addition, standard measures such as Euclidean metrics are often insensitive to the reordering of features, yielding misleading results when functions are compared [14, 24].

An appropriate functional representation of relevances was proposed and analyzed in [11]. However, there, data samples and prototypes are still considered in the original feature space. Here, we follow a complementary approach and investigate the functional representation of adaptive prototypes and all sample data in terms of appropriate basis functions. We include the implementation of relevance learning in the corresponding space of coefficients, illustrate the method in terms of benchmark datasets and compare with the standard approaches in original feature space.

One previously studied example of this basic idea was considered in [19], where highly specific wavelet representations were employed in the analysis of sharply peaked mass spectra. Here, we discuss the use of supervised LVQ and relevance learning for relatively smooth functional data and restrict the discussion to polynomial expansions. We would like to point out, however, that the framework is quite versatile and can be readily extended to other types of data and basis functions.

In the following section we summarize the mathematical framework and provide details of the classifiers and training prescriptions. In Sect. 3 we introduce benchmark datasets and present the results of computer experiments. We discuss our findings in Sect. 4 before we conclude with a brief Summary and Outlook.

2   The Mathematical Framework

We first review the generic form of the considered supervised learning problem, before detailing the specific approaches for functional data.

2.1   Generalized Matrix Relevance LVQ for Classification

We consider a standard supervised scenario, where d-dimensional feature vectors

$$\boldsymbol{\xi}^{\mu} \in \mathbb{R}^d \text{ and labels } S^{\mu} = S(\boldsymbol{\xi}^{\mu}) \in \{1, 2, \ldots, C\} \text{ for } \mu = 1, 2 \ldots P, \qquad (1)$$

serve as examples for the target classification. For simplicity we restrict the following to binary classification schemes with $C = 2$. The extension to multi-class LVQ is conceptually and technically straightforward.

An LVQ system parameterizes the classifier by means of a set of prototypes

$$\left\{ \boldsymbol{w}_{\xi}^{j} \in \mathbb{R}^d \right\}_{j=1}^{M} \text{ with labels } \sigma^j = \sigma(\boldsymbol{w}_{\xi}^{j}) \in \{1, 2\}, \qquad (2)$$

equipped with an appropriate distance measure $d(\boldsymbol{w}_{\xi}, \boldsymbol{\xi})$. A simple *Nearest Prototype Classifier* (NPC) assigns a vector $\boldsymbol{\xi}$ to the class of the closest prototype, i.e. the $\boldsymbol{w}_{\xi}^{j}$ with $d(\boldsymbol{w}_{\xi}^{j}, \boldsymbol{\xi}) = \min_m \{d(\boldsymbol{w}_{\xi}^{m}, \boldsymbol{\xi})\}_{m=1}^{M}$.

Note that the prototypes are defined in the same space as the considered feature vectors, as indicated by the subscript ξ, here. This fact often facilitates intuitive interpretation of the classifier. Prototypes are usually determined in heuristic iterative procedures like the popular LVQ1 algorithm [12] or by means of cost function based optimization, e.g. [18, 21].

We employ general quadratic distance measures of the form

$$d(\boldsymbol{w}_{\xi}, \boldsymbol{\xi}) = \left(\boldsymbol{w}_{\xi} - \boldsymbol{\xi}\right)^{\top} \Lambda_{\xi} \left(\boldsymbol{w}_{\xi} - \boldsymbol{\xi}\right) = \left[\Omega_{\xi} \left(\boldsymbol{w}_{\xi} - \boldsymbol{\xi}\right)\right]^2. \qquad (3)$$

The parameterization $\Lambda_\xi = \Omega_\xi^\top \Omega_\xi$ ensures that the measure is positive semi-definite [20]. We additionally impose the normalization $\text{Tr}(\Lambda_\xi) = \sum_{i,j} [\Omega_\xi]_{ij}^2 = 1$ in order to avoid numerical problems.

For the sake of clarity, we restrict the formalism to one *global* distance given by a single square matrix $\Omega_\xi \in \mathbb{R}^{d \times d}$. Extensions of this basic framework in terms of local or rectangular matrices are technically straightforward, see [7]. Note that the general form, Eq. (3), reduces to standard (squared) Euclidean distance for $\Lambda_\xi = I/d$ with the d-dim. identity I.

Generalized Matrix Relevance LVQ (GMLVQ) [20] optimizes the prototype positions and the distance measure, i.e. the matrix Ω_ξ, in one and the same data driven training process. It is guided by a cost function of the form

$$E = \sum_{\mu=1}^{P} \Phi \left(\frac{d(\boldsymbol{w}_\xi^J, \xi^\mu) - d(\boldsymbol{w}_\xi^K, \xi^\mu)}{d(\boldsymbol{w}_\xi^J, \xi^\mu) + d(\boldsymbol{w}_\xi^K, \xi^\mu)} \right), \tag{4}$$

which was introduced in [18]. Given a particular feature vector ξ^μ, $\boldsymbol{w}_\xi^J$ denotes the *correct winner*, i.e. the closest prototype representing the class $S^\mu = S(\xi^\mu)$, while $\boldsymbol{w}_\xi^K$ denotes the closest of all prototypes with a class label different from S^μ. Frequently, the function $\Phi(\ldots)$ in Eq. (4) is taken to be a sigmoidal [18]. Here, for simplicity and in order to avoid the introduction of additional hyper-parameters we resort to the simplest setting with $\Phi(x) = x$.

Training can be done in terms of stochastic gradient descent as in [20] or by means of other methods of non-linear optimization; several implemented variants are available online at [6]. Results presented here were obtained by means of a batch gradient descent optimization of E, cf. Eq. (4), equipped with a conceptually simple scheme for automated step size control along the lines of [16].

2.2 Polynomial Expansion of Functional Data

The main idea of this paper is to exploit the characteristics of functional data in terms of representing feature vectors $\boldsymbol{\xi} \in \mathbb{R}^d$ in a parameterized functional form. Such representation can be achieved by expanding the data in terms of suitable basis functions. Interpreting the original features as discretized observations of an underlying continuous function $f(x)$, i.e. $\xi_j = f(x_j)$ for $j = 1, \ldots, d$, we consider expansions of the basic form

$$f_c(x) = \sum_{i=0}^{n} c_i g_i(x). \tag{5}$$

For a given set of basis functions $\{g_i\}_{i=0}^n$, each $\boldsymbol{\xi}$ in original d-dim. feature space can be represented by an $(n+1)$-dim. vector of coefficients denoted as $\boldsymbol{c} = (c_0, c_1, \ldots c_n)^\top$ in the following. In practice, $\boldsymbol{c}$ is determined according to a suitable approximation criterion. A popular choice would be the minimization of the quadratic deviation $\sum_{j=1}^{d} (f_c(x_j) - \xi_j)^2$.

In order to illustrate the basic approach, we consider Chebyshev polynomials of the first kind as a particular functional basis. Hence, the functional representation (5) of the data is given by truncated Chebyshev series, which provide an efficient way to represent smooth non-periodic functions [9]. Specifically, we employed the open source MATLAB™ library *chebfun* [22], which includes a variety of mathematical tools in this context. Here, we make use only of its efficient implementation of Chebyshev approximation of discrete functional data.

2.3 Comparison of Workflows

In Sect. 3 we present results for the application of GMLVQ to a number of benchmark problems. In each of these, d-dimensional spectra—denoted by vectors $\boldsymbol{\xi}$—are assigned to one of two possible classes. We explicitly study and compare the following alternative scenarios:

(A) Training in original feature space

As an obvious baseline, we consider the conventional interpretation of the components of $\boldsymbol{\xi}$ as d individual features, disregarding their functional nature. Hence, the spectra are taken to directly define the feature vectors $\boldsymbol{\xi}$ in Eqs. (1–4) without further processing.

(B) Polynomial representation of data and prototypes

For a given degree n of the approximative expansion (5), each data point $\boldsymbol{\xi}^\mu$ in the dataset can be represented by the vector of coefficients $\boldsymbol{c}^\mu \in \mathbb{R}^{(n+1)}$. Correspondingly, prototypes $\boldsymbol{w}_c^k \in \mathbb{R}^{(n+1)}$ and matrices $\Lambda_c, \Omega_c \in \mathbb{R}^{(n+1)\times(n+1)}$ can be introduced. The GMLVQ formalism outlined above is readily applied in complete formal analogy to Eqs. (1–4). However, data and prototypes are now represented in $(n+1)$-dim. coefficient space. Moreover, the distance measure $d(\boldsymbol{w}_c, \boldsymbol{c}) = (\boldsymbol{w}_c - \boldsymbol{c})^\top \Lambda_c (\boldsymbol{w}_c - \boldsymbol{c})$ cannot be interpreted as a simple generalized Euclidean distance in original feature space anymore.

(C) Polynomial smoothing of the data

In addition to the suggested functional representations (B), we consider the smoothing of d-dim. spectra by applying a Chebyshev expansion with $n < d$ and transforming back to original feature space. After replacing original feature vectors by the resulting smoothed versions $\tilde{\boldsymbol{\xi}}$, the standard GMLVQ approach is applied. As a result we obtain a classifier which is parameterized in terms of prototypes $\boldsymbol{w}_{\tilde{\xi}}^k \in \mathbb{R}^d$ and relevance matrix $\Lambda_{\tilde{\xi}} = \Omega_{\tilde{\xi}}^\top \Omega_{\tilde{\xi}} \in \mathbb{R}^{d\times d}$.

3 Application to Example Datasets

The proposed method is applied and tested in several spectral and, therefore, functional datasets of different sizes and spectral bandwidths.

The wine dataset, available from [15], contains 123 (one outlier removed) samples of wine infrared absorption spectra with 256 values in the range between 4000 and 400 cm^{-1}. The data should be classified according to two assigned alcohol levels (low/high) as specified in [13].

The Tecator dataset comprises 215 reflectance spectra with 100 values each, representing wavelengths from 850 to 1050 nm. The spectra were acquired from meat probes and labeled according to fat content (low/high), see [23] for details.

The orange juice (OJ) dataset is a collection of 218 near infrared spectra with 700 values each. For each spectrum the level of saccharose contained in the orange juice is given. In order to define a two class problem similar to the above mentioned ones, the level is thresholded at its median in the set, defining two classes (low/high saccharose content). The dataset is publicly available at [15].

The fourth example, *the coffee dataset*, was made available in the context of a machine learning challenge by the Fraunhofer Institute for Factory Operation and Automation, Magdeburg/Germany, in 2012 [1]. The full set contains 20000 short wave infrared spectra of coffee beans with 256 values in the range of wavelengths between 970 and 2500 nm. The classification task is to discriminate *immature* and *healthy* coffee beans. Since the dataset is only used for further benchmark of the presented approach it is reduced in size to keep computation time in easy to handle ranges. For this reason 100 samples were selected randomly from each of the two classes.

For each dataset three strategies were evaluated: First, a GMLVQ system is adapted to the labeled set of original spectra, cf. scenario (A). A second set of experiments is guided by (B) in the previous section: Spectra are approximated by polynomials of a certain degree resulting in 5, 10, 15, ..., 100 polynomial coefficients as described above. Subsequently, a GMLVQ classifier is trained in terms of the resulting $(n+1)$-dim. coefficient vectors c^μ. Scenario (C) is considered in order to clarify whether the potential benefit of using a polynomial representation could be only an effect of the smoothing achieved by the approximation. To this end, the system is trained on the basis of smoothed versions of the spectra, obtained by equi-distant discretization of the polynomial approximation.

All experiments were done using the same settings and parameters. GMLVQ systems comprised only one prototype per class. A z-score transformation was performed for each individual training process, achieving zero mean and unit variance features in the actual training set. This transformation balances the varying order of magnitudes observed in different feature dimensions and facilitates the interpretation of the emerging relevance matrices [20]. Relevance matrices were initialized as proportional to the identity, prototypes were placed in the class-conditional means in the training set, initially. A batch gradient descent optimization was performed, employing an automated step size control, essentially along the lines of [16]. MAT-LAB™ demonstration code of the precise implementation used here is available from [3], where a more detailed documentation and description of the modification of the step size adaptation scheme is also provided. We used default values of all relevant parameters specified in [3].

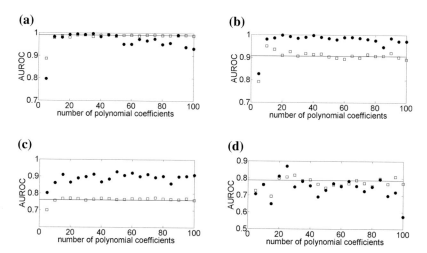

Fig. 1 Comparison of the achieved validation performance, i.e. the area under ROC for different datasets in dependence of the degree of the polynomial curve fitting. The *solid line* represents the value for the classification using the original feature vectors (A). *Filled circles* represent results achieved using polynomial curve fitting, *open squares* correspond to polynomial smoothing of the input data only. All results are displayed as a function of the number of coefficients $(n + 1)$ in the polynomial expansion. **a** Wine dataset. **b** Tecator dataset. **c** OJ dataset. **d** Coffee dataset

A validation scheme was implemented, with each training set containing a random selection of 90 % of the available data. Performance was evaluated in terms of the Area under the ROC (AUROC) with respect to the validation set [10]. The latter is obtained by varying a threshold when comparing the distances of a data point from the prototypes, thus deviating from the NPC scheme. All results were obtained as a threshold average over 10 random splits of the data. Figure 1 displays the performance of systems with full polynomial representation (scenario B), with polynomial smoothing of the data (C), and the baseline of GMLVQ in the original space (scenario A), for comparison.

In order to further illustrate the method, we provide more details for the example of the OJ dataset in Fig. 2. The upper panels display the d-dim. prototypes obtained in scenarios (A) and (C). From the vectors $\boldsymbol{w}_c^{1,2} \in \mathbb{R}^{(n+1)}$ in scenario (B), we also computed their corresponding functional form in original feature space. Diagonal elements of the relevance matrices Λ_ξ, $\Lambda_{\tilde\xi} \in \mathbf{R}^{d \times d}$ and $\Lambda_c \in \mathbb{R}^{(n+1) \times (n+1)}$ are depicted in the lower panels. We show results for the example case of 20 coefficients, which yields typical performance, cf. Fig. 1a.

4 Discussion

Our results demonstrate that the proposed approach has potential to improve classification performance with functional data significantly, as verified by the *Tecator* data (Fig. 1b) and the *OJ* dataset (Fig. 1c). In both cases, the prediction accuracy as

measured by the AUROC is significantly larger when applying GMLVQ in the polynomial representation. For a wide range of degrees n, the AUROC robustly exceeds
that of systems trained from the raw datasets. This is not the case for training according to scenario (C), which shows that the improvement in (B) cannot be explained
as an effect of the smoothing only.

In the other two datasets, cf. Fig. 1a, d, the accuracy of the system with polynomial
representation is comparable or nearly identical to that in original and smoothed
feature space. Also for these datasets, the performance is robust with respect to the
polynomial degree in a wide range of values n.

Clearly, the positive effect of the functional representation on the performance
will depend on the detailed properties of the data and the suitability of the basis
functions. However, even if accuracy does not increase, one advantage of the polynomial representation remains: The dimensionality of the problem can be reduced
significantly in all considered cases without adverse effects on performance. For
approximations with, say, 20 polynomial coefficients, which performs well in all
datasets, the number of input dimensions is reduced by 80 % for *Tecator*, by ca. 92 %
for *Wine* and *Coffee* data and by 97 % for the *OJ* dataset. This leads to a drastically
reduced number of free parameters, which is quadratic in the number of feature
dimensions in GMLVQ, and results in a massive speed-up of the training process.
The gain clearly over-compensates the time needed for performing the polynomial
approximation, generically. The computational effort of the approximation was evaluated experimentally and appeared to increase linearly with the original dimension
d and the polynomial degree n. The observed absolute computation time was on the
order of a few seconds for the representation of 50000 feature vectors.

Another point that deserves attention is the question of interpretability of the prototypes and relevances [2, 5]. One important advantage of LVQ and similar methods
is the intuitive interpretation of prototypes, since they are determined in original feature space. The proposed polynomial representation seems to void this advantage,
by considering data and prototypes in the less intuitive space of coefficients. Note,
however, that the transformation can be inverted after training, hence the prototypes
can be projected back to the original feature space. Figure 2 shows a comparison of
the prototypes achieved from original data, and those achieved in coefficient space,
mapped back to original feature space. Although the prototypes for the polynomially
approximated data are smoother, the comparison shows that they resemble each other
in all three scenarios.

While this provides further evidence for the usefulness of the suggested approach,
it is important to be aware of the significant differences in terms of the applied distance measure. Figure 2 (lower panels) displays the corresponding relevance profiles for the OJ dataset as an example. For original (A) and smoothed data (C),
the diagonal elements of Λ_ξ and $\Lambda_{\tilde{\xi}}$ can be directly interpreted heuristically as the
relevance of the corresponding components in feature space [5, 20]. The diagonal
elements of Λ_c, however, assign an accumulated weight $[\Lambda_c]_{ii} = \sum_j [\Omega_c]_{ij}^2$ to each
dimension in coefficient space. In the example case of the OJ dataset, the first three
coefficients, corresponding to $g_o(x) = 1$, $g_1(x) = x$, $g_2(x) = 2x^2 - 1$, are neglected
almost completely in the GMLVQ system. This reflects the fact that *constant offset*

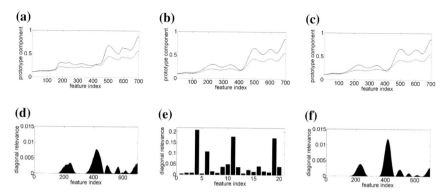

Fig. 2 Prototypes (panels **a**, **b**, **c**) representing low and high saccharose content in the Orange Juice dataset, and diagonal elements of the relevance matrices Λ_ξ, Λ_c and $\Lambda_{\tilde\xi}$ (**d**, **e**, **f**), respectively. *Left panels* (**a**, **d**) Results for using original features (scenario A). *Center panels* (**b**, **e**) Results for polynomial representation (B) with 20 coefficients, prototypes are shown after back-transformation to original feature space. *Right panels* (**c**, **f**): Scenario (C) with 20 coefficients as in (B). **a** OJ original. **b** OJ polynomial fit. **c** OJ smoothened. **d** OJ original. **e** OJ polynomial fit. **f** OJ smoothened

as well as *linear* and *quadratic trend* are not discriminative in the sense of the classification task at hand. Note that relevance learning enables the system to disregard these properties without further pre-processing of the data.

5 Summary and Outlook

We presented a framework for the efficient classification of functional data, based on their appropriate representation in terms of basis functions. As a concrete example, we considered polynomial approximations of smooth spectral data. The framework was applied to and evaluated for four real world datasets. The results show, that for a range of degrees n of polynomial approximations, a comparable or even significantly better classification performance can be achieved. In these cases, the improvement is not due to the smoothing effect of the representation only, but reflects the fact that the suggested approach takes into account the functional nature of the data more appropriately.

Besides its potential effect on the accuracy, the functional representation of the data can lead to a massive reduction of the data dimension. Consequently, a much smaller number of adaptive parameters is employed in the classifier, reducing the risk of over-fitting, avoiding potential convergence problems, and resulting in less computational effort.

Our initial study leaves several interesting questions open for further investigation. In future projects we will study more systematically the properties of the distance measure imposed by the polynomial representation. While our example results display great robustness with respect to n, more generally, the choice of a suitable degree for the polynomial approximation could play an important role in practice. Validation schemes should be devised in order to identify optimal choices systematically according to the complementary criteria of best performance and minimal dimensionality.

The considered polynomial approximation appears to be a suitable approach for reasonably smooth spectral data. The degree to which the classification performance can be improved in comparison with the standard approaches clearly depends on the suitability of the functional basis for the particular dataset. The systematic investigation of alternative representations in the context of other application domains will be in the center of forthcoming studies.

Acknowledgments F. Melchert thanks for support through an Ubbo-Emmius Sandwich Scholarship by the Faculty of Mathematics and Natural Sciences.

References

1. Backhaus, A.: Wettbewerb Detektion von defekten Bohnen im Rohkaffee. Fraunhofer Inst. for Factory Operation and Automation, Biosystems Engineering (2012)
2. Backhaus, A., Seiffert, U.: Classification in high-dimensional spectral data: accuracy vs. interpretability vs. model size. Neurocomputing **131**, 15–22 (2014)
3. Biehl, M.: A no-nonsense beginner's tool for GMLVQ. University of Groningen. http://www.cs.rug.nl/~biehl/gmlvq
4. Biehl, M., Hammer, B., Verleysen, M., Villmann, T. (eds.): Similarity Based Clustering. Springer Lecture Notes in Artificial Intelligence, vol. 5400 (2009)
5. Biehl, M., Hammer, B., Villmann, T.: Distance measures for prototype based classfication. In: Grandinetti, L., Petkov, N., Lippert, T. (eds.) BrainComp 2013, Proceedings of the International Workshop on Brain-Inspired Computing, Cetraro/Italy, 2013. Lecture Notes in Computer Science, vol. 8603, pp. 100–116. Springer (2014)
6. Biehl, M., Schneider, P., Bunte, K.: Relevance and Matrix adaptation in Learning Vector Quantization (GRLVQ, GMLVQ and LiRaM LVQ). University of Gronignen. http://matlabserver.cs.rug.nl/gmlvqweb/web/
7. Bunte, K., Schneider, P., Hammer, B., Schleif, F.M., Villmann, T., Biehl, M.: Limeted rank matrix learning, discriminative dimension reduction and visualization. Neural Netw. **26**, 159–173 (2012)
8. Cover, T., Hart, P.: Nearest neighbor pattern classification. IEEE Trans. Inf. Theory **13**(1), 21–27 (1967)
9. Driscoll, T.A., Hale, N., Trefethen, L.N.: Chebfun Guide. Pafnuty Publications (2014)
10. Duda, R., Hart, P., Stork, D.: Pattern Classification. Wiley (2000)
11. Kästner, M., Hammer, B., Biehl, M., Villmann, T.: Generalized functional relevance learning vector quantization. In: Verleysen, M. (ed.) Proceedings of the European Symposium on Artificial Neural Networks (ESANN), pp. 93–98. d-side (2011)
12. Kohonen, T.: Self-organizing Maps. Springer, Berlin (1995)
13. Krier, C., François, D., Rossi, F., Verleysen, M., et al.: Supervised variable clustering for classification of nir spectra. In: Verleysen, M. (ed.) Proceedings of the European Symposium on Artificial Neural Networks (ESANN), pp. 263–268. d-side (2009)
14. Lee, J.A., Verleysen, M.: Generalization of the Lp norm for time series and its application to self-organizing maps. In: Cottrell, M. (Eds.): Proceedings of the Workshop on Self-Organizing Maps (WSOM) 2005, pp. 733–740. Paris, Sorbonne (2005)
15. Meurens, M.: Orange juice near-infrared spectra dataset and wine mean infrared spectra dataset. University of Louvain. http://mlg.info.ucl.ac.be/index.php?page=DataBases

16. Papari, G., Bunte, K., Biehl, M.: Waypoint averaging and step size control in learning by gradient descent. Machine Learning Reports MLR-06/2011, vol. 16 (2011)
17. Ramsay, J., Silverman, B.: Functional Data Analysis. Springer (2006)
18. Sato, A., Yamada, K.: Generalized learning vector quantization. In: Tesauro, G., Touretzky, D., Leen, T. (eds.) Advances in Neural Information Processing Systems. vol. 7, pp. 423–429. MIT Press (1995)
19. Schneider, P., Biehl, M., Schleif, F.M., Hammer, B.: Advanced metric adaptation in Generalized LVQ for classification of mass spetrometry data. In: Proceedings of the 6th International Workshop on Self-Organizing-Maps (WSOM), 5 pp. Bielefeld University (2007)
20. Schneider, P., Biehl, M., Hammer, B.: Adaptive relevance matrices in Learning Vector Quantization. Neural Comput. 21, 3532–3561 (2009)
21. Seo, S., Bode, M., Obermayer, K.: Soft nearest prototype classification. IEEE Trans. Neural Netw. 14(2), 390–398 (2003)
22. The Chebfun Developers: Chebfun—Numerical Computing with functions. University of Oxford. http://www.chebfun.org
23. Thodberg, H.H.: Tecator meat sample dataset. StatLib Datasets Archive. http://lib.stat.cmu.edu/datasets/tecator
24. Villmann, T., Hammer, B.: Functional principal component learning using Oja's method and Sobolev norms. In: J. Princip, R. Miikulainen (eds.): Advances in Self-Organizing Maps—Proceedings of the Workshop on Self-Organizing Maps (WSOM) 2009. Springer Lecture Notes on Computer Science 5629: 325–333 (2009)
25. Weinberger, K., Saul, L.: Distance metric learning for large margin classification. J. Mach. Learn. Res. 10, 207–244 (2009)

Prototype-Based Classification for Image Analysis and Its Application to Crop Disease Diagnosis

Ernest Mwebaze and Michael Biehl

Abstract In this paper, we provide an application of Learning Vector Quantization (LVQ)-based techniques for solving a real-world problem. We apply LVQ for automated diagnosis of crop disease in cassava plants using features extracted from images of plants' leaves. The problem reduces to a five class problem in which we attempt to distinguish between a leaf from a health plant and leaves representing four different viral and bacterial diseases in cassava. We discuss the problem under additional constraints that the solution must easily be deployable on a mobile device with limited processing power. In this study we explore the right configuration of type of algorithm and type of features extracted from the leaves that optimally solves the problem. We apply different variations of LVQ and compare them with standard classification techniques (Naïve Bayes, SVM and KNN). Results point to a preference of color feature representations and LVQ-based algorithms.

Keywords Prototype-based classification · LVQ · GLVQ · GMLVQ · DLVQ · Multi-class classification · Feature extraction · Image analysis

1 Introduction

Learning Vector Quantization (LVQ) and more general prototype-based classifiers have been used successfully in many applications [1]. Their major attraction is the intuitiveness with which the learned prototypes can be interpreted. A particularly unique advantage of prototype-based methods is the narrow barrier in transitioning the learned classifier to a production system.

E. Mwebaze (✉)
School of Computing & Informatics Technology, Makerere University,
P.O. Box 7062, Kampala, Uganda
e-mail: emwebaze@cit.ac.ug

M. Biehl
Johann Bernoulli Institute for Mathematics and Computer Science, University of Groningen,
P.O. Box 407, 9700 Groningen, AK, The Netherlands
e-mail: m.biehl@rug.nl

© Springer International Publishing Switzerland 2016
E. Merényi et al. (eds.), *Advances in Self-Organizing Maps and Learning Vector Quantization*, Advances in Intelligent Systems and Computing 428,
DOI 10.1007/978-3-319-28518-4_29

329

In this paper, we address a specific practical application; the discrimination of four diseases in cassava plants by analysing images of their leaves. The overarching goal is to be able to do automated diagnosis of the state of health of cassava plants from photos taken using mobile phones.[1] The general context of this problem is that small holder farmers in Uganda and in Africa in general experience heavy yield loss due to viral diseases in cassava. Cassava is a very important staple and cash crop in Africa and has been tagged as one of the key food security crops.

Presently to diagnose disease in gardens, experts travel to the farmers fields and visually inspect their crops. With our work, we can enable farmers in remote places do diagnosis of their crops without need of an expert. This context further constrains the scope of solutions we can deploy. We require a classification algorithm that can be easily deployed on a mobile device which is not too expensive in execution. Prototype-based algorithms are suited for this kind of problem because deployment involves transferring the trained prototypes (which are vectors of numbers) to the device. To classify a new image, features are extracted and the distance from the features to the prototypes is calculated and the class of the *closest* prototype is assigned as the diagnosis of the image. The simplicity in deployment and the efficiency in running the algorithm make this family of methods very attractive for this kind of problem.

In our previous work [2, 3] we highlighted the ease of implementing prototype-based classification techniques on a mobile phone for diagnosing disease from images of crops taken with a camera. There, we mainly looked at a binary classification; trying to distinguish the presence of disease from no disease. One caveat with this particular problem was that it was uncertain whether the trained classifier was predicting presence of disease or it was predicting an anomaly from the health leaf images.

In this work we move further and investigate a multiclass problem where we are classifying between five classes representing healthy leaf images and four viral and bacterial diseases in cassava plants: cassava mosaic disease (CMD), cassava bacterial blight (CBB), cassava green mite (CGM) and cassava brown streak disease (CBSD). We explore the right mix of type of algorithm, type of features extracted from leaf images and type of disease being classified that can optimally solve this problem under the constraints of easy of deployment and economy of processing power. In the sections that follow, we reintroduce prototype-based classification in the context of LVQ, we then explain how we do image processing on the leaf images and finally how we use the data separately and combined in different prototype-based schemes and with different algorithms.

---

[1]http://www.air.ug/mcrops/.

2 Prototype-Based Classification and LVQ

An LVQ system can generally be set up as follows. For a particular classification task, we assume that a set of labelled example data is available:

$$\{\mathbf{x}^\mu, y^\mu\}_{\mu=1}^P ,$$

where the $\mathbf{x}^\mu \in \mathbf{R}^N$ are feature vectors and the labels $y^\mu \in \{1, 2, \ldots C\}$ specify their class membership.

The prototypes of the LVQ system are defined as a set of M prototype vectors $\mathbf{w}^j \in \mathbf{R}^N$ which carry labels $c(\mathbf{w}^j) \in \{1, 2, \ldots C\}$ such that $W = \left\{(\mathbf{w}^j, c(\mathbf{w}^j)\right\}_{j=1}^M$. The system can be set up with one or more prototype vectors per class. The prototype vectors are identified in the feature space and serve as typical representatives of their classes. Together with a given distance measure $d(\mathbf{x}, \mathbf{w})$, they parametrize the classification scheme. To predict the class of a new data point $\mathbf{x}$, the distance between $\mathbf{x}$ and the prototypes of the system is calculated and $\mathbf{x}$ is assigned to the class $c(\mathbf{w}^L)$ of the closest prototype with $d(\mathbf{x}, \mathbf{w}^L) \le d(\mathbf{x}, \mathbf{w}^j)$ for all j.

A variety of modifications to LVQ that have been suggested in the literature, aiming at better convergence or favorable generalization behavior. Generalized LVQ (GLVQ), Generalized Matrix LVQ (GMLVQ) and Divergence-based LVQ (DLVQ) are examples of such extensions that provide better convergence and generalization.

2.1 Generalized LVQ

A key variant of LVQ, the Generalized LVQ (GLVQ) algorithm introduced by Sato and Yamada [4] incorporates an objective (cost) function in the training of the LVQ system. The advantage of an objective function based LVQ system is that one can use gradient methods (online or batch) to optimize it. The GLVQ cost function can be stated in the following form

$$E(W) = \sum_{\mu=1}^P \Phi\left(\frac{d(\mathbf{x}^\mu, \mathbf{w}^J) - d(\mathbf{x}^\mu, \mathbf{w}^K)}{d(\mathbf{x}^\mu, \mathbf{w}^J) + d(\mathbf{x}^\mu, \mathbf{w}^K)}\right), \tag{1}$$

where $\mathbf{w}^J$ denotes the closest correct prototype with $c(\mathbf{w}^J) = y^\mu$ and $\mathbf{w}^K$ is the closest incorrect prototype $(c(\mathbf{w}^K) \ne y^\mu)$. The function Φ generally determines the active region of the algorithm and is restricted to the interval $[-1, +1]$.

In principle, a variety of numerical optimization procedures are available for the minimization of the cost function in Eq. (1) by means of gradient descent techniques. In this work, we employed batch gradient descent.

2.2   Generalized Matrix LVQ

In Generalized Matrix Learning Vector Quantization (GMLVQ) [5], a matrix $\Lambda = \Omega^\top \Omega$ that captures the interplay between different data dimensions in the distance measure is added to the cost-function based GLVQ scheme. The distance measure $d_\Omega(\mathbf{x}, \mathbf{w})$ is defined as

$$d_\Omega(\mathbf{x}, \mathbf{w}) = (\mathbf{x} - \mathbf{w})^\top \Lambda (\mathbf{x} - \mathbf{w}). \tag{2}$$

The cost function in Eq. (1) can be minimized with respect to the prototypes and the matrix Ω for example by means of gradient based methods. The corresponding derivatives yield batch updates of the form

$$\mathbf{w}_{t+1} = \mathbf{w}_t - \alpha_t^{(w)} \frac{\partial E/\partial \mathbf{w}}{|\partial E/\partial \mathbf{w}|} \, , \; \Omega_{t+1} = \Omega_t - \alpha_t^{(\Omega)} \frac{\partial E/\partial \Omega}{|\partial E/\partial \Omega|} \, , \tag{3}$$

where $\alpha^{(\Omega)}$ and $\alpha^{(w)}$ are the step sizes controlling the training of Ω and $\mathbf{w}$, respectively.

2.3   Distance-Based LVQ Variants

GLVQ and GMLVQ form the basis of other improved LVQ algorithms. Many other variants function as extensions of these two algorithms. In previous studies, we have used unconventional dissimilarity measures.

A particular variant of interest is Divergence-based LVQ (DLVQ) which uses divergences as a distance measure. In this work we carry out experiments with the DLVQ algorithm using the Cauchy-Schwarz divergence as the distance measure. In our previous work [3] we have found that the Cauchy-Schwarz is a particularly good divergence distance measure because it is robust to numerical errors during training (for example small values) and it can be used even in the case when the data is non-normalized.

3   Image Processing of Leaves

3.1   Image Collection and Processing

To undertake this work, we collected a number of images of cassava plant leaves manifesting the different diseases by shadowing experts from the Ugandan National Crop Resources Research Institute (NaCRRI), the body in Uganda responsible for cassava disease research, during a country-wide annual survey. Images of 3264×2448

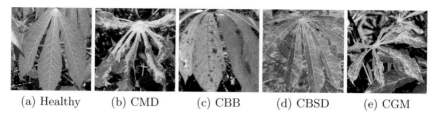

(a) Healthy   (b) CMD   (c) CBB   (d) CBSD   (e) CGM

Fig. 1 Sample images associated with the five (5) classes of the classification problem

resolution were taken using a mobile phone with an 8 MP camera. These images were cropped down to an average size of 500×500 pixels, and then annotated by experts from NaCRRI, who assigned disease classes to the images together with a score of severity. Figure 1 shows examples of cassava leaf images collected and used in this study. For this work, we selected 150 healthy leaf images, 121 leaf images infested with CMD, 98 leaf images infested with CBB, 111 images of crops infested with CBSD and 91 images of crops infested with CGM.

3.2 Feature Processing and Extraction

Different crop diseases manifest differently on the leaves of a cassava plant. Some deform the leaf and decolorize it, some put patches on the leaf, while others manifest as small colored marks on the leaf. Images taken for this work were with noisy backgrounds as depicted in Fig. 1. Three types of features were extracted from the images representing color, shape and oriented gradient features.

In our previous work [2], we extracted color features as a normalized histogram of hues of pixels for each leaf. Shape features were extracted using Speeded Up Robust Features (SURF) and Scale Invariant Feature Transformation (SIFT) techniques that obtain interest point descriptors from the images. We observed that color and shape features were very informative of presence and absence of disease. In this work, we used an opensource MATLAB^{TM} image feature extraction toolbox [6] that standardizes the process of feature extraction. It also provides a more consistent method of obtaining a representative vector per image for a particular feature using a bag-of-words approach.

Extracting Color Features: For each of the images we extract color names. Color names [7, 8] are linguistic labels that humans assign to colors; they represent the color an ordinary human being would assign to an image. These tend to provide a more consistent representation of color than histograms of hue pixels and also provide a more natural fit to the way expert diagnosis is done in the fields presently.

Using the toolbox, each image is described in the form of regions or patches of multiple sizes. These are converted to color names and histograms are calculated from overlapping patches on the image. A bag-of-words plus spatial pyramid pipeline is then applied to obtain a vector of features representing each image. The bag-of-words pipeline as described in the toolbox basically works as follows; using a random sampling of the extracted features from various patches, learn a dictionary using k-means, and apply locality-constrained linear coding (LLC) [9] to soft-encode each patch to some dictionary entries. Then apply max pooling with a spatial pyramid [10] to obtain the final feature vector.

Extracting SIFT Features: We also extract SIFT features [11] to represent shape information in the images. For each image a set of scale-invariant feature vectors corresponding to a set grid of histograms around each key point location on the leaf image are extracted. Different diseases transform the surface of the leaf differently and these differences can be captured by SIFT features. In our previous work [2] we have shown good performance with SIFT features for representing shape distortions in leaf images. With this toolbox, SIFT descriptors are extracted from patches of multiple sizes from each image converted to grayscale and a bag-of-words approach as described for the color features is applied to obtain representative SIFT feature vectors for each of the images.

Extracting HOG Features: HOG represents Histograms of Oriented Gradient features. HOG features define edge and gradient based descriptors that have been shown to be very successful for object recognition particularly human objects [12]. To obtain HOG features, an image is decomposed into small squared cells and a normalized histogram of oriented gradients is computed for each cell. For this case we extract HOG descriptors on a grid of 2×2 cells and concatenate them to obtain a descriptor for each grid location. As before, a bag-of-words plus spatial pyramid pipeline is applied to obtain the final vector representation for each image.

We observe that extraction of SIFT and HOG features is expensive particularly for usage with this specific application on a mobile device. However, we investigate these features here in a bid to understand what levels of accuracy we can theoretically get with these features and how that relates to more inexpensive features like color.

4   Experiments and Results

This section describes the various experiments we undertook applying the different algorithms to the different data representations of the leaf images.

4.1   Experimental Set-Up

Three LVQ-based algorithms were considered: GLVQ, GMLVQ and DLVQ. In our experiments, we average accuracy scores over 25 validation runs of the algorithms with different initializations of the prototypes. In each run we use 10 % of the data for testing. GLVQ and GMLVQ were implemented using a publicly available toolbox.[2] The distance measure used for these algorithms was Euclidean distance and optimization of the algorithms was done using batch gradient descent with heuristic step size adaptation following [13].

For DLVQ we employed stochastic gradient descent with a uniform learning rate throughout the training as described in [3] in greater detail. We obtained the best learning rate of $1e^{-7}$ through experimenting with different learning rates, and results were averaged over 25 runs of the algorithm.

For completeness, we also applied standard algorithms: Naïve Bayes, Support Vector Machine (SVM) with a Radial Basis Kernel and K-Nearest Neighbor (KNN) with $k = 15$. For this we used the *scikit-learn* toolbox[3] which has standard implementations of several machine learning algorithms. For other parameters, we used the defaults from the toolbox.

4.2 Results

We applied these algorithms to the three feature representations of the dataset: color, SIFT and HOG. As a further exploratory experiment we concatenated the three datasets and applied the algorithms again to this concatenated dataset. The idea here was to get some understanding of the effect of manipulating different feature representations on the performance of the classifiers.

Table 1 shows results of our experiments. It shows percentage true positive rates of the three LVQ-based classifiers and the other standard classifiers for the different feature representations and the concatenated feature set. For the LVQ-based classifiers, the table indicates comparable performance amongst the three feature representations. Different feature representations seem to have a slight bias towards a particular algorithm, for example color features offer best performance with GMLVQ and SIFT features with GLVQ. Overall GMLVQ seems to give the best performance across the different feature representations, albeit the difference in performance being slight.

When we consider the non LVQ-based algorithms, we observe comparable performance overall when compared with the LVQ-based algorithms. For the color features however, we observe a marked improvement in performance with the SVM algorithm. From our exploratory experiments with the concatenated features set, we observe significant improvement in performance for all algorithms.

To further investigate what configuration of feature representation, algorithm and disease category works best for practical implementation, we analyzed the class-wise performance of the different LVQ-algorithms. Table 2 shows detailed results of the class-wise performance of the different algorithms. The numbers in the table represent percentage true positive rates in classifying the different classes/diseases. We observe again some reliance of the performance on the class and the type of

---

[2]http://www.cs.rug.nl/~biehl/gmlvq.html.

[3]http://scikit-learn.org/.

Table 1 Overall true positive rates (%) for different algorithms applied to the different leaf image representations

	GLVQ	GMLVQ	DLVQ	Naïve Bayes	SVM	k-NN
Color	80	82	80	85	91	80
HOG	82	85	79	82	83	76
SIFT	86	82	85	85	79	80
Combined dataset	**88**	**95**	**100**	**100**	**100**	**100**

Standard deviation on GLVQ and GMLVQ are approx. 1 %, for DLVQ, 4 % and for Naïve Bayes, SVM and KNN 1 %

Table 2 Classification true positive rates (%) for the different feature representations categorized by disease

	Color			SIFT			HOG		
	GLVQ	GMLVQ	DLVQ	GLVQ	GMLVQ	DLVQ	GLVQ	GMLVQ	DLVQ
Healthy	100	98	100	100	99	100	93	100	90
CMD	**86**	**91**	**83**	83	80	77	73	88	71
CBB	29	68	61	79	72	**79**	**82**	**74**	78
CBSD	72	**93**	65	**83**	71	**85**	81	73	79
CGM	**96**	**90**	**85**	84	80	81	84	87	80

Highlighted numbers show which feature set gives the highest true positive rate for detection of a particular disease. Scores generally have a rounded off standard deviation of 1 %

feature representation of the data for example for CMD and CGM diseases, results indicate greater performance with color feature representations for all algorithms while for other diseases there is even a finer reliance on the type of feature and the type of algorithm.

For practical implementation, the color feature representation still appears to be the most appealing. For this particular problem of diagnosing diseases, we were also interested in understanding what kind of misclassifications occur amongst the diseases. For the color representation we further investigated the two LVQ-based algorithms that were run under exactly the same configuration of parameters; GLVQ and GMLVQ. Table 3 depicts corresponding confusion matrices obtained from our experimentation.

From these two confusion matrices, we observe a pattern of miss-classification between two sets of diseases: CMD and CBSD, and CBB and CGM. It appears for these pairs of diseases, the algorithms are challenged in properly discriminating them. Some implications of this are that prototypes of CMD and CBSD are closer to each other compared to prototypes of CBB and CGM which also appear to be closer to each other.

Table 3 Confusion matrices showing class-wise performance results of applying GLVQ and GMLVQ algorithms on the color feature representation dataset

	Predictions				
	Healthy	CMD	CBB	CBSD	CGM
(a) GLVQ					
Healthy	**100**	0	0	0	0
CMD	0	**86.0**	0	13.2	0
CBB	0.5	7.5	**28.7**	1.3	62.2
CBSD	1.1	26.9	0	**71.54**	0.46
CGM	0	0	4.2	0	**95.8**
(b) GMLVQ					
Healthy	**99.7**	0.3	0	0	0
CMD	0	**90.9**	0	9.1	0
CBB	0.5	0	**68.3**	0	31.5
CBSD	0.3	6.6	0.5	**92.6**	0
CGM	0	0	10.2	0	**89.5**

5 Discussion

In this paper, we have presented a practical application of prototype based algorithms based on LVQ to the problem of predicting the state of health of a cassava plant based on images of the plant's leaves. The goal was to explore different combinations of types of algorithms, image feature representations and diseases and obtain an optimal configuration for deployment of this solution to a mobile device. A sub-goal was also to understand which features to extract from the leaf images that are most informative about the state of disease of the plant.

We investigated color, SIFT and HOG features and applied these to a bank of LVQ-based and non LVQ-based algorithms. Overall, we observe from Table 1 that non LVQ-based algorithms seem to provide comparable performance to LVQ-based algorithms. Particularly we observe SVM offering the best performance for the color feature representation of the images. We note however that from a practical point of view, the advantages gained from applying LVQ-based algorithms may out weight the difference in performance.

Considering the LVQ-based algorithms, Table 1 shows comparable performance for the three feature representations for the three algorithms. We also observe superior performance of all algorithms for the concatenated feature set, however this is highly infeasible given our deployment constraints. Of the other three feature representations, technically it is less costly to extract color features which would make this the choice for practical implementation. Table 2 offers some extra evidence to support this analysis particularly for CMD and CGM diseases which observably, are best classified with color feature representations for all three algorithms.

An interesting result, evident in the confusion matrices in Table 3 is the correlation in the misclassifications between specific pairs of diseases; CMD and CBSD and between CBB and CGM. This is consistent even across different algorithms. One plausible explanation for this result is that these pairs of diseases have similar manifestations on the leaf. Another explanation which is informed from empirical analysis is that in several plants infection can be from a combination of multiple diseases. This is a possible future extension of this work; identifying co-infection in plants.

The overall advantage of using LVQ based algorithms in the practical application of classification techniques in low computational power equipment like mobile phones is the simplicity of scaling the learned algorithm to a live system. This work generates the necessary understanding of the plausible configurations of features, algorithms and disease categories that can form an optimal solution for deployment. A particular advantage with our analysis is that in areas where one disease is suspected or has a high prevalence, we can bias the algorithm to favor certain feature representations that are most accurate for that particular disease. The understanding generated from this research makes this a plausible option for deployment.

Acknowledgments The authors would like to thank the Dr. Titus Alicai and Dr. Chris Omongo of the Uganda National Crop Resources Research Institute (NaCRRI), for granting us permission to access disease and pest surveillance data and for supporting the annotation of the data. This work is carried out with support from the Bill and Melinda Gates Foundation under the *PEARL 1: Automated survey technology and spatial modeling of viral crop disease in cassava* project.

References

1. Neural Networks Research Centre, Bibliography on the self-organizing map (SOM), learning vector quantization (LVQ): University of Technology, Helsinki (2002). http://liinwww.ira.uka.de/bibliography/Neural/SOM.LVQ.html
2. Aduwo, J.R., Mwebaze, E., Quinn, J.A.: Automated vision-based diagnosis of cassava mosaic disease. In: Perner, P. (ed.) Industrial Conference on Data Mining—Workshops, pp. 114–122. IBaI Publishing (2010)
3. Mwebaze, E., Schneider, P., Schleif, F.-M., Aduwo, J.R., Quinn, J.A., Haase, S., Villmann, T., Biehl, M.: Divergence based classification in learning vector quantization. Neural Comput. **74**(9), 1429–1435 (2011)
4. Sato, A.S., Yamada, K.: Generalized learning vector quantization. In: Mozer, M.C., Touretzky, D.S., Hasselmo, M.E. (eds.) NIPS, vol. 8, pp. 423–429. MIT Press, Cambridge (1996)
5. Schneider, Petra, Biehl, Michael, Hammer, Barbara: Adaptive relevance matrices in learning vector quantization. Neural Comput. **21**(12), 3532–3561 (2009)
6. Khosla, A., Xiao, J., Torralba, A., Oliva, A.: Memorability of image regions. In: Pereira, F., Burges, C.J.C., Bottou, L., Weinberger, K.Q. (eds.) Advances in Neural Information Processing Systems 25, pp. 296–304. Curran Associates Inc (2012)
7. van de Weijer, J., Schmid, C., Verbeek, J.: Learning color names from real-world images. In: IEEE Conference on Computer Vision and Pattern Recognition, 2007. CVPR '07, pp. 1–8 (2007)
8. Khan, R., van de Weijer, J., Shahbaz Khan, F., Muselet, D., Ducottet, C., Barat, C.: Discriminative color descriptors. In: 2013 IEEE Conference on Computer Vision and Pattern Recognition (CVPR), pp. 2866–2873 (2013)

9. Wang, J., Yang, J., Yu, K., Lv, F., Huang, T., Gong, Y.: Locality-constrained linear coding for image classification. In: 2010 IEEE Conference on Computer Vision and Pattern Recognition (CVPR), pp. 3360–3367 (2010)
10. Lazebnik, S., Schmid, C., Ponce, J.: Beyond bags of features: spatial pyramid matching for recognizing natural scene categories. In: 2006 IEEE Computer Society Conference on Computer Vision and Pattern Recognition, vol. 2, pp. 2169–2178 (2006)
11. Lowe, D.G.: Distinctive image features from scale-invariant keypoints. Int. J. Comput. Vis. **60**(2), 91–110 (2004)
12. Dalal, N., Triggs, B.: Histograms of oriented gradients for human detection. In: IEEE Computer Society Conference on Computer Vision and Pattern Recognition, 2005. CVPR 2005, vol. 1, pp. 886–893 (2005)
13. Papari, G., Bunte, K., Biehl, M.: Waypoint averaging and step size control in learning by gradient descent (technical report), volume MLR-2011-06 of Machine Learning Reports, pp. 16–26. University of Bielefeld (2011)

Low-Rank Kernel Space Representations in Prototype Learning

Kerstin Bunte, Marika Kaden and Frank-Michael Schleif

Abstract In supervised learning feature vectors are often implicitly mapped to a high-dimensional space using the kernel trick with quadratic costs for the learning algorithm. The recently proposed random Fourier features provide an explicit mapping such that classical algorithms with often linear complexity can be applied. Yet, the random Fourier feature approach remains widely complex techniques which are difficult to interpret. Using Matrix Relevance Learning the linear mapping of the data for a better class separation can be learned by adapting a parametric Euclidean distance. Further, a low-rank representation of the input data can be obtained. We apply this technique to random Fourier feature encoded data to obtain a discriminative mapping of the kernel space. This *explicit* approach is compared with a differentiable kernel vector quantizer on the same but *implicit* kernel representation. Using multiple benchmark problems, we demonstrate that a parametric distance on a RBF encoding yields to better classification results and permits access to interpretable prediction models with visualization abilities.

1 Introduction

Given the increasing amount of large and high-dimensional data sets require a variety of scientific disciplines or application domains and efficient methods for dimension reduction. The feature selection play an essential role in modern data processing.

Besides unsupervised approaches using variants of Principal Component Analysis [23], embedding techniques [17] or random projection strategies [21], supervised

K. Bunte
School of Computer Science, University of Birmingham, Edgbaston, Birmingham, UK

M. Kaden · F.-M. Schleif
Computational Intelligence Group, University of Applied Sciences Mittweida,
Technikumplatz 17, 09648 Mittweida, Germany

F.-M. Schleif (✉)
Université Catholique de Louvain, ICTEAM Institute, Place du Levant 3,
1348 Louvain-la-Neuve, Belgium
e-mail: fmschleif@googlemail.com

© Springer International Publishing Switzerland 2016
E. Merényi et al. (eds.), *Advances in Self-Organizing Maps and Learning Vector Quantization*, Advances in Intelligent Systems and Computing 428,
DOI 10.1007/978-3-319-28518-4_30

341

feature reduction approaches like Recursive Feature Elimination [34] where found to be very efficient in preserving prediction accuracy while reducing the model complexity. However, these approaches are widely *black boxes*, i.e. direct conclusions to the original input data are limited. A very promising alternative is provided by so called Relevance Learning approaches [6, 12]. These techniques identify the so called *relevant* input *features* with respect to a constrained optimization problem, like a supervised learning task. The idea is to adapt the weights of a parametric distance, like the weighted Euclidean distance, during learning such that e.g. the class separation is maximized. In Matrix Relevance Learning (MRL, [27]) this concept was extended to a parametric Euclidean distance, which additionally identify the classification correlations between the features. Standard relevance learning is linear because only the individual input features are weighted whereas for MRL also correlations are considered, leading to quadratic complexity, but with very good results in a variety of applications [2, 7, 19, 28, 30, 32].

Limited Rank Matrix Approximation (LiRaM) [8] is a *random* subspace projection technique where the data are mapped from a originally M dimensional space to m dimensions, with $m \ll M$. LiRaM was formerly mainly considered for standard Euclidean feature representations. In this paper we consider a representation of an explicit kernel feature space, where the data are mapped into a high-dimensional feature space using random Fourier features (RFF). As usual the kernel mapping aims at linear separability of the classes but also makes the approach difficult to interpret. Using LiRaM with its parametric distance learning, we not only make this approach more flexible but are able to obtain interpretable low-rank data representations. An alternative to this explicit strategy is the use of prototype learning with differentiable kernels, called Generalized Learning Vector Quantization with differentiable Kernels (DK-GLVQ, [31]). In the experiments we will use an advanced DK-GLVQ approach [31] where the parameters of the differentiable kernel are optimized during learning as given in Sect. 1.4. This approach is denoted by DK-GMLVQ. The DK-GLVQ can be considered as an implicit strategy where the kernel representation is obtained by a differentiable kernel mapping, as detailed in the following. Combining prototype based classifications with linear low-dimensional representations of the distance parameters provides interpretable models especially for high-dimensional data. For DK-GMLVQ the learned distance parametrization can be interpreted easily with respect to the original input data. In this paper we analyze the classical Euclidean LiRaM, LiRaM with RFF features and DK-GLVQ with a radial basis function (RBF) kernel for low-rank distance matrices. The RBF kernel was chosen due to its flexibility in modeling non-linear separable data. In this paper we show that LiRaM can be effectively kernelized using the proposed approach and that DK-GMLVQ and LiRaM with RFF features permit the low-dimensional inspection of a kernelized data representation from different perspectives. This can be helpful to identify outliers or to get a better understanding of misclassifications given the data can be reasonable embedded into a low-dimensional space. We illustrate these approaches for two benchmark data sets with a rather small number of dimensions and multiple high-dimensional real life data sets taken from the life science domain.

1.1 Limited Rank Matrix Relevance Learning

Dissimilarity based methods play a most important role in, both, unsupervised and supervised machine learning analysis of complicated data sets, see [5] for an overview and further references. In the context of classification problems, Learning Vector Quantization (LVQ) [10, 15, 16, 25] constitutes a particularly intuitive and successful family of algorithms. In LVQ, classes are represented by prototypes which are determined from example data. The prototypes are defined in the original feature space. Together with a suitable dissimilarity or distance measure the prototypes parametrize the classifier, frequently according to a *Nearest Prototype* scheme. Further references also reflecting the impressive variety of application domains in which LVQ has been employed successfully [13, 24].

A key issue in LVQ and other distance based techniques is the choice of an appropriate distance measure. Pre-defined distance measures are, frequently, sensitive to re-scaling of single features or more general linear transformations of the data. Therefore, the Euclidean distance is not always the best choice.

An elegant framework has been developed which can circumvent this difficulty: In so-called Relevance Learning schemes, only the functional form of the distance is fixed, while a set of parameters is determined in the training process [6]. Similar ideas have been formulated for other distance based classifiers, see e. g. [33] for an example in the context of Nearest Neighbor classifiers [9].

A generalized quadratic distance is parametrized by a matrix in Matrix Relevance Learning (GMLVQ, [27]) which is summarized in the following:

GMLVQ employs a distance measure given by the quadratic form

$$d(\vec{y}, \vec{z}) = (\vec{y} - \vec{z})^\top \Lambda (\vec{y} - \vec{z}) \quad \text{for } \vec{y}, \vec{z} \in \mathbb{R}^M. \tag{1}$$

It is required to fulfill the basic conditions $d(\vec{y}, \vec{y}) = 0$ and $d(\vec{y}, \vec{z}) = d(\vec{z}, \vec{y}) > 0$ for all $\vec{y}, \vec{z}$ with $\vec{y} \neq \vec{z}$. These are conveniently satisfied by assuming the parametrization $\Lambda = \Omega \Omega^\top$, i.e.

$$d(\vec{y}, \vec{z}) = (\vec{y} - \vec{z})^\top \Omega \Omega^\top (\vec{y} - \vec{z}) = \left[\Omega^\top (\vec{y} - \vec{z}) \right]^2 = d_\Omega(\vec{y}, \vec{z}) \tag{2}$$

Hence, $\Omega \in \mathbb{R}^{M \times m}$ defines a linear mapping of data and prototypes to a space, in which standard Euclidean distance is applied.

Note that for a meaningful classification and for the LVQ training it is sufficient to assume that Λ is positive semi-definite; the transformation need not be invertible and could even be represented by a rectangular matrix $\in \mathbb{R}^{M \times m}$ [8] with $m < M$. The quadratic complexity of Relevance Matrix Learning is due to the squared number of matrix parameters in Ω for $m \equiv M$ which are adapted during learning. Here we show how Λ can be approximated by restricting Ω to a rectangular matrix $M \times m$ with $m \ll M$, which is basically the idea used in LiRaM [8].

1.2   Limited Rank Matrix Approximation

For the first approach (LiRaM) Ω is defined as a rectangular $M \times m$ matrix, without imposing any symmetry or other constraints on its structure. The elements of Ω can be varied independently. For instance, the derivative of the distance measure with respect to an arbitrary element of Ω is $\frac{\partial d(\vec{y},\vec{z})}{\partial \Omega_{kl}} = 2(y_k - z_k)\left[\Omega^\top (\vec{y} - \vec{z})\right]_l$ or in matrix notation:

$$\frac{\partial d(\vec{y}, \vec{z})}{\partial \Omega} = = 2\,(\vec{y} - \vec{z})\,(\vec{y} - \vec{z})^\top\,\Omega. \tag{3}$$

Initially Ω can be chosen to be a random matrix with entries drawn from a normal distribution. We will use random Fourier features to provide an implicit kernelization of the LiRaM approach.

1.3   Random Fourier Features

Random Fourier features as introduced in [1], projects the data points onto a randomly chosen line, and then pass the resulting scalar through a sinusoid. The random lines are drawn from a distribution so as to guarantee that the inner product of two transformed points approximates the desired shift-invariant kernel. The motivation for this approach is given by Bochners theorem:

Theorem 1 *A continuous kernel $k(\vec{x}, \vec{y}) = k(\vec{x} - \vec{y})$ on $\mathbb{R}^d$ is positive definite if and only if $k(\delta)$ is the Fourier transform of a non-negative measure.*

If the kernel $k(\delta)$ is properly scaled, Bochners theorem guarantees that its Fourier transform $p(\omega)$ is a proper probability distribution. The idea in [1] is to approximate the kernel as

$$k(\vec{x} - \vec{y}) = \int_{\mathbb{R}^d} p(\omega) e^{j\omega^\top (\vec{x} - \vec{y})} d\omega$$

with some extra normalizations and simplifications one can sample the features for k using the mapping $\mathbf{z}_\omega(\vec{x}) = [\cos(\vec{x})\,\sin(\vec{x})]^\top$. In [1] a proof is given for the uniform convergence of Fourier features to the kernel $k(\vec{x} - \vec{y})$. To generate the random Fourier features one eventually needs a kernel matrix $k(\vec{x}, \vec{y}) = k(\vec{x} - \vec{y})$ and a random feature map $\mathbf{z}(\vec{x}) : \mathbb{R}^d \to \mathbb{R}^{2D}$ s.t. $\mathbf{z}(\vec{x})^\top \mathbf{z}(\vec{y}) \approx k(\vec{x} - \vec{y})$. One draws D i.i.d. samples $\{\omega_1 \ldots, \omega_D\} \in \mathbb{R}^d$ from $p(\omega)$ and generates $\mathbf{z}(\vec{x}) = \sqrt{1/D}[\cos(\omega_1^\top \vec{x}) \ldots \cos(\omega_D^\top \vec{x})\,\sin(\omega_1^\top \vec{x}) \ldots \sin(\omega_D^\top \vec{x})]^\top$

This formulation leads to an explicit mapping of an input vector into a high-dimensional features space of the RFF features. The obtained feature representation can be fed into LiRaM leading to an implicit kernelization of LiRaM. Accordingly, the Ω matrix is defined on this rather high-dimensional feature mapping with $M = D$. In the following we analyze the usage of this implicit kernelization of LiRaM with an explicit one using differentiable kernels.

1.4 Matrix Relevance Learning in Kernelized Vector Quantization

A further idea to integrate the kernel concept into prototype learning is to use differentiable kernels [31], referred to as DK-GLVQ. Thereby, the distance measure $d(\vec{v}, \vec{w})$ in a prototype learner is replaced by the distance measure deduced from the kernel $\kappa(\vec{v}, \vec{w})$;

$$d_\kappa(\vec{v}, \vec{w}) = \sqrt{\kappa(\vec{v}, \vec{v}) - 2\,\kappa(\vec{v}, \vec{w}) + \kappa(\vec{w}, \vec{w})} \qquad (4)$$

For symmetric and positive semi-definite kernels κ, d_κ is only a semi-metric, but if the kernel is positive definite, $d_\kappa(\vec{v}, \vec{w})$ becomes a metric and it yields:

$$d_\kappa(\vec{v}, \vec{w}) = ||\Phi(\vec{v}) - \Phi(\vec{w})||_{\mathcal{H}} \,. \qquad \mathcal{H} \text{ indicating a functional Hilbert space} \quad (5)$$

Now, we further assume that the kernel $\kappa(\vec{v}, \vec{w})$ is differentiable with respect to the prototype parameter $\vec{w}$. In this case one obtains:

$$\frac{\partial d_\kappa^2(\vec{v}, \vec{w})}{\partial \vec{w}} = \frac{\kappa(\vec{w}, \vec{w})}{\partial \vec{w}} - 2\frac{\kappa(\vec{v}, \vec{w})}{\partial \vec{w}}$$

which can immediately be plugged into gradient based prototype adaptation, replacing any other metric. According to [31] the DK-GLVQ provide the same topological richness like other kernelized classifiers like the SVM if the applied kernel is universal. One of the most famous and well-known examples of universal positive kernels is the Gaussian kernel

$$\Gamma_\kappa(\vec{v}, \vec{w}) = \exp\left(-\left(\frac{\vec{v} - \vec{w}}{\sqrt{2\sigma}}\right)^2\right) \qquad (6)$$

with the width $\sigma > 0$. This kernel is differentiable and can be used in the former context. The term $\left(\frac{\vec{v} - \vec{w}}{\sqrt{2\sigma}}\right)^2$ is a scaled quadratic Euclidean distance. It turns out that the induced distance remains a metric if this term is replaced by any other quadratic metric [20]. Hence one can combine the idea of matrix learning discussed in the former section with differentiable kernels. In particular one can consider the kernel

$$\Gamma_\kappa(\vec{v}, \vec{w}, \Omega) = \exp\left(-d_\Omega(\vec{v}, \vec{w})\right) \qquad (7)$$

with an arbitrary matrix $\Omega \in \mathbb{R}^{M \times m}$, i.e. the data and the prototypes are mapped into the $\mathbb{R}^m$ and afterwards the quadratic Euclidean norm is calculated. The resulting derivative for a prototype update is obtained as: $\frac{\partial \Gamma_\kappa(\vec{v}, \vec{w}, \Omega)}{\partial \vec{w}} = \Gamma_\kappa(\vec{v}, \vec{w}, \Omega) \cdot 2\Omega(\vec{v} - \vec{w})$. Further the metric parameters can be updated by using the respective derivative: $\frac{\partial \Gamma_\kappa(\vec{v}, \vec{w}, \Omega)}{\partial \Omega_{[i, j]}} = -2\Gamma_\kappa(\vec{v}, \vec{w}, \Omega) \cdot [\Omega(\vec{v} - \vec{w})]_i [\vec{v} - \vec{w}]_j$. A more detailed derivation including an implementation within a prototype learning framework can be found in [14]. If the matrix Ω is rectangular with $m < M$ inherent regularization and discriminative visualization into a lower dimensional space comes possible similar as with LiRaM. Further, the matrix Λ can be interpreted as a correlation matrix determining the correlations between the data dimensions, which are useful for classification [27]. Even for $M = m$ the algorithm shows inherent regularization because the Ω-adjustments can be related to class dependent principal component analysis, such that the learned matrices tend to be generated by the class eigenvectors [3, 4].

2 Experiments

In the experiments we analyze LiRaM with random Fourier features and the differentiable kernelized (Generalized) Learning Vector Quantization (DK-GMLVQ). For both approaches we use a RBF-kernel for the feature encoding.[1] The prototype positions are updated using the cost function of LiRaM or DK-GMLVQ, respectively. We expect that the RBF encoding improves the discrimination power of the underlying classifier, hence LiRaM should lead to higher test accuracies by implicitly using the RBF kernel instead of being restricted on a standard parametrized Euclidean distance. The prediction accuracy is expected to be similar to the one of DK-GMLVQ on the same kernel mapping. The approaches differ in the way how feature vectors can be inspected in the discriminative visualization and also in the computational costs.

For LiRaM with random Fourier features the data points and prototypes are represented in a rather high-dimensional (e.g. 300 dimensional) feature space, spanned by the random Fourier feature induced explicit RBF-kernel *expansion*. For the DK-GMLVQ the data points and prototypes still live in the *original* feature space of the input data and are implicitly mapped into the RBF-kernel space during learning using the kernel trick for distances [29] as shown before. Using $m < M$ we can obtain discriminative visualizations for both approaches as shown in Fig. 1. If m is larger than 2 a discriminative visualization can not directly be shown, but one can use e.g. t-SNE [18] from the already reduced discriminative m dimensional space.

First, we consider a checkerboard of size 3×3 with consecutive labeling 0/1 (see Fig. 2). This data set is not linear separable. Using a classical Euclidean distance measure in LiRaM and only one prototype per class the obtained classification model provides around 65 % percent prediction accuracy. If we consider a projection of the checkerboard data using the learned Ω matrix learned with LiRaM we see that the classes are overlapping (see Fig. 3 left). However, by using a RFF feature encoding the data become linear separable which can be shown in the respective projection using the learned Ω matrix. As shown in Fig. 3 (right) LiRaM was capable to find a discriminative two dimensional projection also in the rather high-dimensional RFF feature space. For the RFF encoded checkerboard data the reconstructed RBF kernel

---

[1]The random Fourier features are generated such that the respective RBF kernel is approximated.

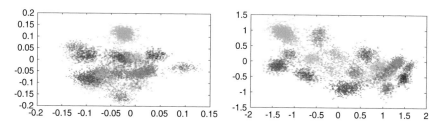

Fig. 1 Two dimensional plot obtained by mapping the points of DS1 using the learned Ω matrix of DK-GMLVQ with RBF kernel (*left*) and LiRaM with random Fourier features (*right*). If the mapping dimension $m > 2$ we additionally applied t-SNE. Varying (*colors*) shades indicate different classes. Axes labeling is arbitrary due to the combination of the provided features

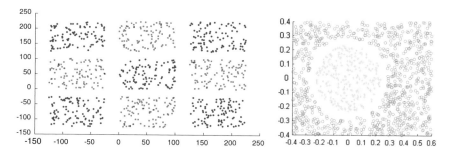

Fig. 2 *Left* checkerboard data set with the two classes $\square$ and $\circ$; *Right* simulated data set (SIM) with the two classes $\circ$ and   illustrating the flag of palau

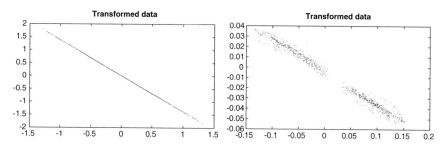

Fig. 3 *Left* checkerboard data mapped by the learning Ω matrix using Euclidean distance; *Right* checkerboard data mapped by using a RFF feature encoding

has roughly an intrinsic dimensionality of 13 by analyzing the eigenvalue spectrum. Moreover, the classification accuracy reaches almost 100 % for this model. For the DK-GMLVQ approach the accuracy is comparable using an implicit mapping in the RBF space. The projection based on the DK-GMLVQ Ω matrix is however not helpful because it can only provide a linear mapping from $2D$ to $2D$, which is often insufficient to get a linear separation.[2] However for the classification model the

---

[2]Note that all vectors in DK-GMLVQ still live in the same D-dimensional space.

implicit RBF space is available, such that the accuracy of DK-GMLVQ is as good as the one of LiRaM with RFF encoding.

Next we consider multiple benchmark data sets as listed below. Due to the assumption that the features are independent, all data sets have been normalized by z-score, i. e. the data vectors are normalized to get zero mean value and a standard deviation of one for each input dimension Experiments where done with a 5-fold cross validation and parameters have been identified on an independent test set (first fold). We used the same σ parameters of the RBF kernel for both approaches. For LiRaM with RFF features we have chosen 300 features. We always used one prototype per class. The following data sets has been considered (pts—number of data points, dim—number of features, ncl—number of classes):

- (DS1) NGS metagenomics data sets (7905 pts, 136 dim, 21 ncl (slightly imbalanced)) [11]
- (DS2) Thyroid data set (2991 pts, 60 dim, 2 ncl) [22]
- (DS3) Bowel cancer mass spectra (95 pts, 1408 dim, 2 ncl) [26]
- (DS4) Breast cancer (263 pts, 9 dim, 2 ncl) [22]
- (DS5) Pima Indians diabetis (768 pts, 8 dim, 2 ncl) [22]
- (DS6) Heart disease (statlog) (270 pts, 13 dim, 2 ncl) [22]
- (DS7) Segmentation data set (2310 pts, 19 dim, 7 ncl) [22]
- SIM—a simulated data set, with points distributed in 2D along a ring enclosed by a rectangle as shown in Fig. 2 (right)

The prediction results of the cross validation are reported in Table 1. In our experiments we found that LiRaM with random Fourier features could *not* be effectively used with $m = 2$. This is directly caused by the RFF encoding, which indeed leads to an intrinsically more high-dimensional data representation. In fact m had to be chosen ≥ 5 to be competitive to the other methods on the training data. On the other hand for DK-GMLVQ the parameter m could be chosen to be $m = 2$ without having a negative effect on the prediction accuracy. We found that a larger m for DK-GMLVQ only improved the training accuracy but had in general no negative impact on the test set. This can potentially be explained by remembering that the prototypes of

Table 1 Test set accuracy (% ± std) of the various data sets

	LiRaM (euclidean)	LiRaM (RFF)	DK-GMLVQ (RBF)
DS1	98.4 ± 0.4 (10)	98.6 ± 0.5 (100)	89.2 ± 1.1 (2)
DS2	84.1 ± 0.7 (2)	84.2 ± 1.3 (100)	73.5 ± 4.0 (2)
DS3	90.5 ± 6.9 (10)	82.1 ± 6.0 (100)	97.7 ± 2.9 (2)
DS4	70.4 ± 2.2 (2)	74.5 ± 5.8 (5)	78.8 ± 3.1 (2)
DS5	76.6 ± 4.2 (2)	70.7 ± 2.3 (5)	78.9 ± 2.9 (2)
DS6	83.7 ± 3.6 (2)	85.2 ± 5.4 (100)	87.0 ± 7.1 (2)
DS7	88.6 ± 2.6 (5)	90.0 ± 3.5 (15)	88.3 ± 3.6 (2)
SIM	67.3 ± 2.4 (2)	100.0 ± 0.0 (2)	100.0 ± 0.0 (2)

Mapping dimension m in brackets

DK-GMLVQ as well as the Ω matrix are still in the original, rather low-dimensional data space.

From the experimental results shown in Table 1 we see that the RBF kernel or RFF encoding is not always helpful. For many data sets LiRaM (with Euclidean distance) is already very effective. For the simulated data, which are not linear separable, we see a clear effect. Accordingly the implicit kernelization of LiRaM can be useful for some data sets. Overall the RBF/RFF encoding does in general not reduce the prediction accuracy (with the exception of DS3), but is often beneficial and as expected most helpful if the data require a non-linear decision function.

Both approaches permit the inspection of a high-dimensional data space containing the data and the prototypes. For LiRaM-RFF the data space is the expanded RBF kernel space whereas for DK-GMLVQ we still have the original data space but the metric and prototype adaptation was substantially influenced by the used RBF kernel. For DS1 and DS3 we show plots of the Λ matrix and the corresponding relevance profile (main diagonal of the Λ matrix) in Figs. 4 and 5. For DS1 one can clearly see the highlighting of individual features by large relevance values in Fig. 5 and correlations to other features in Fig. 4 (left).

If one is interested in better understanding the RBF space one should therefore focus on the (approximated) LiRaM visualizations. If the focus is more on understanding the original data the DK-GMLVQ visualization will be preferable. The RBF encoding should make the data linear separable in the high-dimensional feature space. This is in fact often visible in the obtained low-dimensional plots (see Figs. 6 and 3). For DK-GMLVQ the data still life in the original space and the mapping Ω is adapted to reflect this linear separation. For the simulation data in two dimensions the linear mapping is obviously not able to separate the data, whereas for data with a higher number of given input features it may still be possible to find a linear separation from the mapping. On the other hand LiRaM-RFF works naturally in a high-dimensional space and hence also for the intrinsically two dimensional simulation data set it can provide a (new) two dimensional mapping separating the original data, by exploring the RFF (RBF) feature space. It may however not always be possible to obtain a reliable low-dimensional embedding from LiRaM (RFF) due to the more complicated

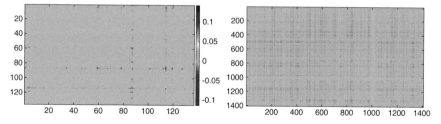

Fig. 4 Classification correlation Λ matrix of DK-GMLVQ with RBF kernel for DS1 and DS3. For DS1, the features 87 and 114 seem to be important for class separation and they are also (negative-) correlated to the other features (*dark blue-negative correlated*). For the DS3 no feature is significantly highlighted in the matrix Λ

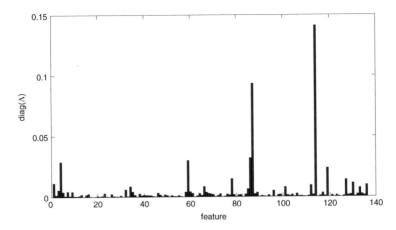

Fig. 5 Relevance profile of DK-GMLVQ with RBF kernel for DS1. Like in Fig. 4, the features 87 and 114 are weighted high, which indicate a high relevance for class separation

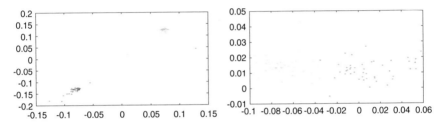

Fig. 6 Two dimensional plot obtained by mapping the points of DS3 using the learned Ω matrix of DK-GMLVQ with RBF kernel (*left*) and LiRaM with random Fourier features (*right*). If the mapping dimension $m > 2$ we additionally applied t-SNE. Axes labeling is arbitrary due to the combination of the provided features

high-dimensional feature space, also complicating the matrix learning. In these cases the mapping dimension m of the learned matrix has to be kept as $m > 2$ which makes a direct inspection impossible. As a simple solution we decided to apply t-SNE as a second embedding strategy, in later work it would be interesting to see if this can be avoided by an improved regularization scheme within LiRaM. Within these constraints the visualizations of LiRaM (RFF) and DK-GMLVQ are helpful to identify overlapping regions in the data space causing misclassifications of unsafe decisions. Especially for classification problems with more than two classes the embeddings help to identify the reasons for these misclassification in more detail than with e.g. a fusion matrix only. Within an interactive framework (as part of future work) it would be possible to analyze different visualizations of the kernelized encoding by modifying the kernel parameter or the originally provided input features.[3]

---

[3]Often the RBF encoding is considered as a silver bullet, but if it fails a controlled inspection framework can be very useful.

3 Conclusions

We have shown a way to kernelize the Limited Rank Matrix Learning using RFF features and compared it with DK-GMLVQ using a RBF kernel. We found that the RFF encoding typically leads to a more complicated data representation such that (very) low-dimensional representation with LiRaM is often not directly available. The RFF encoding will often lead to an *intrinsically* higher dimensional data space (beside the high-dimensional RFF feature space) such that direct low-dimensional embeddings can lead to substantial performance degradation. However it is possible to obtain a discriminative projection into a substantially lower dimensional space e.g. with 5 dimensions which subsequently can be processed by classical embedding techniques with good performance. Another possibility would be to calculate the full RBF kernel (gram matrix) and map it to an Euclidean embedding, which however would be much more costly then using LiRaM-RFF. For the simulated data set it was shown that LiRaM has the potential to identify two or three dimensional embeddings from the kernel space also in cases where DK-GMLVQ fails.

For some data sets the RFF or RBF encoding is beneficial in contrast to the standard Euclidean distance, providing non-linear decision boundaries. We found that DK-GMLVQ is less sensitive to the RBF encoding and still allows for $m = 2$ and low-dimensional visualizations with good quality given the *input* feature space is sufficiently expressive.

For DK-GMLVQ the prototypes are still in the original (rather low-dimensional) data space and one may consider this as a regularization, whereas for LiRaM-RFF such a regularization is not available and the learning of the Ω matrix is more complicated.

In conclusion: one may still prefer DK-GMLVQ as a prototype model to obtain low-dimensional mappings of kernelized data representations as long as LiRaM can not be sufficiently well regularized.

Acknowledgments Marie Curie Intra-European Fellowship (IEF): FP7-PEOPLE-2012-IEF (FP7-327791-ProMoS) is greatly acknowledged. This work has been partially funded by the Belgian FRS-FNRS project 7.0175.13 DRedVis.

References

1. A.Rahimi, Recht, B.: Random features for large-scale kernel machines. In: Platt, J.C., Koller, D., Singer, Y., Roweis, S.T. (eds.) Advances in neural information processing systems 20. In: Proceedings of the Twenty-First Annual Conference on Neural Information Processing Systems. Curran Associates, Inc. (2007). http://books.nips.cc/papers/files/nips20/NIPS2007_0833.pdf
2. Arlt, W., Biehl, M., Taylor, A.: Urine steroid metabolomics as a biomarker tool for detecting malignancy in adrenal tumors. J. Clin. Endocrinol. Metab. **96**, 3775–3784 (2011)
3. Biehl, M., Bunte, K., Schleif, F.M., Schneider, P., Villmann, T.: Large margin linear discriminative visualization by matrix relevance learning. In: Proceedings of IJCNN 2012, 1873–1880 (2012)

4. Biehl, M., Hammer, B., Schleif, F.M., Schneider, P., Villmann, T.: Stationarity of matrix relevance LVQ. In: Proceedings of IJCNN 2015. p. to appear (2015)
5. Biehl, M., Hammer, B., Verleysen, M., Villmann, T. (eds.): Similarity based clustering - recent developments and biomedical applications. In: Lecture Notes in Artificial Intelligence, vol. 5400. Springer (2009)
6. Bojer, T., Hammer, B., Schunk, D., von Toschanowitz, K.T.: Relevance determination in Learning Vector Quantization. In: Verleysen, M. (ed.) European Symposium on Artificial Neural Networks, pp. 271–276 (2001)
7. Bunte, K., Schleif, F.M., Biehl, M.: Adaptive learning for complex-valued data. Proceedings of ESANN 2012, 387–392 (2012)
8. Bunte, K., Schneider, P., Hammer, B., Schleif, F.M., Villmann, T., Biehl, M.: Limited rank matrix learning, discriminative dimension reduction and visualization. Neural Netw. **26**, 159–173 (2012)
9. Cover, T., Hart, P.: Nearest neighbor pattern classification. IEEE Trans. Inf. Theory **13**(1), 21–27 (1967)
10. Crammer, K., Gilad-Bachrach, R., Navot, A., Tishby, A.: Margin analysis of the LVQ algorithm. In: Advances in Neural Information Processing Systems, vol. 15, pp. 462–469. MIT Press, Cambridge, MA (2003)
11. Gisbrecht, A., Hammer, B., Mokbel, B., Sczyrba, A.: Nonlinear dimensionality reduction for cluster identification in metagenomic samples. In: IV, pp. 174–179 (2013)
12. Hammer, B., Villmann, T.: Generalized relevance learning vector quantization. Neural Netw. **15**(8–9), 1059–1068 (2002)
13. Kaden, M., Lange, M., Nebel, D., Riedel, M., Geweniger, T., Villmann, T.: Aspects in classification learning—review of recent developments in learning vector quantization. Found. Comput. Decision Sci. **39**, 79–105 (2014)
14. Kästner, M., Nebel, D., Riedel, M., Biehl, M., Villmann, T.: Differentiable kernels in generalized matrix learning vector quantization. In: 11th International Conference on Machine Learning and Applications, ICMLA, pp. 132–137. IEEE (2012). http://dx.doi.org/10.1109/ICMLA.2012.231
15. Kohonen, T.: Learning Vector Quantization for pattern recognition. Technical report TKK-F-A601, Helsinki Univeristy of Technology, Espoo, Finland (1986)
16. Kohonen, T.: Self-Organizing Maps. Springer, Berlin, Heidelberg (1997)
17. Lee, J.A., Verleysen, M.: Nonlinear Dimensionality Reduction. Springer, Berlin (2007)
18. van der Maaten, L.J.P., Hinton, G.: Visualizing High-Dimensional Data Using t-SNE (2008)
19. Mendenhall, M.J., Merényi, E.: Relevance-based feature extraction for hyperspectral images. IEEE Transactions on Neural Networks. **19**(4), 658–672 (2008)
20. Micchelli, C.A., Xu, Y., Zhang, H.: Universal kernels. J. Mach. Learn. Res. **6**, 2651–2667 (2006). http://www.jmlr.org/papers/v7/micchelli06a.html
21. Mylavarapu, S., Kaban, A.: Random projections versus random selection of features for classification of high dimensional data. In: UKCI, pp. 305–312. IEEE (2013)
22. Newman, D.J., Hettich, S., Blake, C.L., Merz, C.J.: UCI repository of machine learning databases. http://archive.ics.uci.edu/ml/ (1998)
23. Oja, E.: Neural networks, principal components, and subspaces. J. Neural Syst. **1**, 61–68 (1989)
24. Pöllä, M., Honkela, T., Kohonen, T.: Bibliography of self-organizing map (som) papers: 2002–2005 addendum. TKK Reports in Information and Computer Science, Helsinki University of Technology Report TKK-ICS-R23 (2009)
25. Sato, A., Yamada, K.: Generalized learning vector quantization. In: Touretzky, D.S., Mozer, M.C., Hasselmo, M.E. (eds.) Advances in Neural Information Processing Systems, vol. 8. Proceedings of the 1995 Conference, pp. 423–429. MIT Press, Cambridge, MA, USA (1996)
26. Schleif, F.M., Villmann, T., Hammer, B.: Prototype based fuzzy classification in clinical proteomics. Int. J. Approx. Reasoning **47**(1), 4–16 (2008)
27. Schneider, P., Biehl, M., Hammer, B.: Adaptive relevance matrices in learning vector quantization. Neural Comput. **21**(12), 3532–3561 (2009)

28. Schneider, P., Schleif, F.M., Villmann, T., Biehl, M.: Generalized matrix learning vector quantizer for the analysis of spectral data. In: Verleysen, M. (ed.) European Symposium on Artifiacal Neural Networks. Bruges, Belgium, Apr 2008
29. Schölkopf, B.: The kernel trick for distances. In: Advances in Neural Information Processing Systems, vol. 13, Papers from Neural Information Processing Systems (NIPS), pp. 301–307 (2000)
30. Strickert, M., Witzel, K., Mock, H.P., Schleif, F.M., Villmann, T.: Supervised attribute relevance determination for protein identification in stress experiments. In: Proceedings of Machine Learning in Systems Biology (2007)
31. Villmann, T., Haase, S., Kaden, M.: Kernelized vector quantization in gradient-descent learning. Neurocomputing **147**, 83–95 (2015). http://dx.doi.org/10.1016/j.neucom.2013.11.048
32. Villmann, T., Schleif, F.M., Hammer, B.: Comparison of relevance learning vector quantization with other metric adaptive classification methods. Neural Netw. **19**, 610–622 (2006)
33. Weinberger, K., Blitzer, J., Saul, L.: Distance metric learning for large margin nearest neighbor classification. In: Weiss, Y., Schölkopf, B., Platt, J. (eds.) Advances in Neural Information Processing Systems 18, pp. 1473–1480. MIT Press, Cambridge, MA (2006)
34. Weston, J., Mukherjee, S., Chapelle, O., Pontil, M., Poggio, T., Vapnik, V.: Feature selection for SVMs. In: Leen, T.K., Dietterich, T.G., Tresp, V. (eds.) NIPS, pp. 668–674. MIT Press (2000)

Dynamic Prototype Addition in Generalized Learning Vector Quantization

Jonathon Climer and Michael J. Mendenhall

Abstract Learning Vector Quantization (LVQ) is a powerful supervised learning method for classification that uses a network of prototype vectors to form a decision surface. Generalization theory shows there is a non-trivial number of prototype vectors that yield the best generalization. Although it is typical to assign the same number of prototype vectors for each class, other LVQ methods add prototypes dynamically (incrementally) during training. This work offers an extension to the existing dynamic LVQs that minimizes the cost function of Generalized LVQ by focusing on the set of misclassified samples. This cost minimization occurs between the largest cost-contributing class and its nearest "confuser class". A comparison is made between other prototype insertion methods and compares their classification performance, the number of prototype resources required to obtain that accuracy, and the impact on the cost function.

Keywords Dynamic/incremental learning vector quantization · Large margin classifier · Cost minimization

1 Introduction

The family of Learning Vector Quantization (LVQ) [1] methods are supervised learners for statistical pattern recognition. They belong to a class of simple competitive learners and have gained popularity due to their efficiency, ease of implementation, and clear interpretability during training and classification. These algorithms are capable of classifying very high dimensional data and are applied in a variety of fields including machine vision [2, 3], analysis of medical imagery [4], and the classification of hyperspectral data [5].

J. Climer (✉) · M.J. Mendenhall
Department of Electrical and Computer Engineering, Air Force Institute of Technology,
2950 Hobson Way, Wright-Patterson AFB, 45433, USA
e-mail: jonathon.climer@afit.edu; jonclimer@gmail.com

© Springer International Publishing Switzerland 2016
E. Merényi et al. (eds.), *Advances in Self-Organizing Maps and Learning
Vector Quantization*, Advances in Intelligent Systems and Computing 428,
DOI 10.1007/978-3-319-28518-4_31

LVQs use representative "prototype" or "code-book" vectors whose positions in the data space are updated during rounds of "learning". Multiple prototypes per class are typically used in order to achieve accurate classifications. However, Crammer et. al. [6] shows that the generalization error of LVQ-based classifiers (that lead to a "winner-takes-all" classification) are a function of the number of prototype vectors used. Since the goal of any classifier is to generalize the decision surface, it follows that too many prototypes can lead to over-fitting and that there is a non-trivial number of prototypes among the classes for a given problem.

Solutions that do not make *a-priori* assumptions on the prototype distribution between classes add them *dynamically* or *incrementally* as part of the learning process (hereinafter *dynamic*). This concept is supported by [2, 3, 7, 8], based on LVQ2, LVQ3, GRLVQ, and GLVQ respectively. This paper considers a prototype insertion strategy to Generalized LVQ (GLVQ) that minimizes the cost function directly. Our method, in some cases, shows faster convergence due to larger accuracy gains early in the training process, and in some cases requires fewer prototype vectors than other methods in the same class [2, 3, 7, 8].

2   Learning Vector Quantization (LVQ) Background

2.1   LVQ Taxonomy

The taxonomy of LVQ is represented as three phases: competition, winner selection, and synaptic adaptation. Each LVQ defines the set of prototype vectors it allows to **compete**, commonly selecting one or more from the set of "in-class" prototypes (belonging to the same class as the input sample x), "out-of-class" prototypes (belonging to any class other than that of the input sample x), or a "net" prototype (chosen from all prototype vectors, regardless of class label). A **winning** prototype is one that results in the minimum distortion between it and the current sample. When using Euclidean distance as the distortion measure, the winner w_i represents the prototype from the set of competitors, closest to x. (Frequently in LVQ, a second competition selects one more winning prototype, w_j.) After winner selection, many LVQs impose additional conditions in order to apply updates to the prototypes based upon the influence of x, such as windowing functions [9]. Where the additional conditions are satisfied, the authors in [10] show the **synaptic adaptation** rule for LVQ algorithms can be generalized as:

$$w_i \leftarrow w_i - \alpha \frac{\partial S}{\partial w_i}; \qquad w_j \leftarrow w_j - \alpha \frac{\partial S}{\partial w_j}, \tag{1}$$

where α is the (potentially time varying) learn rate and S is the cost function. The completion (or termination) of these three phases for a sample x constitutes one training step. This process is then repeated for all training samples $\{x_1, x_2, \ldots, x_N\}$ to constitute one epoch (where N is the total number of training samples).

2.2 Dynamic LVQ Ancestry

After the introduction of LVQ in [1], several variants have arisen to better estimate the decision surface and overcome divergence and long-term stability challenges [9]. Poirier [7] introduced a new method based on Kohonen's LVQ2 to *dynamically* add prototype vectors as needed to better represent the class distributions and form better decision surfaces. Later works built dynamic LVQ (DLVQ) methods based on LVQ2.1 [11], LVQ3 [8, 12], GLVQ [2], and Relevance GLVQ (GRLVQ) [13]. Of these, the GLVQ variants appear to offer the strongest performance due to their use of a cost function that guides synaptic adaptation. Consequently, GLVQ forms the basis of the methods described in this paper. Following the taxonomy in Sect. 2.1, the competition and winner selection phase of GLVQ uses both the nearest in-class and nearest out-of-class prototypes (w_i and w_j respectively). Instead of explicitly restricting winner selection by a window about the midpoint between w_i and w_j, GLVQ implicitly does this [5, 10] by employing the cost [10]:

$$S = \sum_{n=1}^{N} f(\mu(x_n)), \mu(x) = \frac{d_j - d_i}{d_j + d_i}, \tag{2}$$

where N is the number of training samples and d_i and d_j are squared Euclidean distances between the input sample x_n and the prototype vectors w_i and w_j, respectively. Consistent with [10], $f(\mu)$ is is the sigmoid function $1/(1 + e^{-\mu})$. When $\partial S/\partial w_i$ and $\partial S/\partial w_j$ are substituted into Eq. (1), the resulting synaptic adaptation equations minimize the cost function via gradient descent [10]:

$$w_i \leftarrow w_i - \alpha \frac{\partial f}{\partial \mu} \frac{d_j}{(d_j + d_i)^2} [x - w_i]; \quad w_j \leftarrow w_j + \alpha \frac{\partial f}{\partial \mu} \frac{d_i}{(d_j + d_i)^2} [x - w_j]. \tag{3}$$

2.3 LVQ Taxonomy Addition: Network Structure Modification

In order to characterize the addition and/or deletion of prototype vectors in DLVQs, a new element is incorporated into the standard LVQ taxonomy. **Network structure modification** (NSM) captures the second dynamic and adaptive component that distinguishes DLVQs from its ancestors. This part of the taxonomy is responsible for identifying which class receives the new prototype and its initial location in the data space. While this paper does not exhaustively explore all DLVQ NSM methods in the literature, several are used in our comparative analysis.

Mean: In Zell et al. [14], new prototypes are initialized as the mean of the misclassified samples and is done so across all classes. Although [14] adds, a prototype to each class, we insert a single prototype in the class with the largest classification error.

Closest: Kirsten et al. [3] insert new prototypes into classes where the classification rate exceeds a threshold. This enables the addition of several prototypes at once. New prototypes are placed near class boundaries by initializing them at the same position as the misclassified sample closest to an out-of-class prototype vector. In our use of this method, we insert a single prototype in order to normalize the comparison between insertion methods.

Sampling Cost: Losing et al. [2] represents the latest in DLVQs and has the same overarching goal described in this paper. That is, they desire to minimize the cost function directly in order to maximize the classification accuracy. Conceptually, this is achieved by selecting a random subset of the training samples (e.g., a fixed percentage) as candidate positions for the insertion of a new prototype. One-by-one, candidate positions are tested by calculating the total cost among the subset after the candidate has been added. The position (and class label) of the candidate resulting in the lowest total cost is chosen as the initialization of the new prototype.

Principal Components (PC): Stefano et al. [8] tracks the number of times in-class and out-of-class prototypes are referenced in order to calculate a *split metric* for each prototype. Prototypes with split metrics exceeding a threshold are *split* by replacing the current prototype with two new prototypes placed equidistant from the original along the principal component (eigenvector) direction of the target class's misclassified samples. The distance along the principal component axis is a function of the associated variance (eigenvalue for the corresponding eigenvector).

2.4 Two Proposed NSM Methods

Our *Near-Mean* method is similar to the *Mean* method in [14], but restricts the new prototype to the misclassified sample with the smallest Euclidean Distance to the mean of the misclassified samples. By assuming a "known valid" position closest to the average, we reduce the potential of poorly interpolating the initialization of the new prototype in a sparse area that may represent irregularities in the class distribution, or even represent another class.

Our *Misclassified Cost* method seeks to minimize the cost function directly by focusing on misclassified samples. Conceptually, our method accumulates the cost contribution from each misclassified sample according to Eq. 2 in a confusion matrix according to "true class" and "nearest class". The two classes whose interaction results in the highest total cost define the pool of candidate locations for the new prototype: specifically, their misclassified samples. Each of the candidate solutions are tested and the one with the lowest cost per Eq. 2 is chosen.

2.5   Qualitative Comparison of NSM Methods

The presented NSM methods vary in how they try to improve the configuration of the LVQ network in hopes of improving classification accuracy and generalization. The *Misclassified Cost, Mean, Near-Mean,* and *Closest* methods utilize the set of misclassified training samples to select a recipient class for a new prototype, restricting prototype placement to the class with largest classification error. The *Sampling Cost* and *PC* methods select a candidate location globally (across all samples) where *Sampling Cost* selects from a random subset of the training data. *PC* relies upon misclassified data from each training sample to compute its split metric and assess prototype utilization.

The methods also vary in candidate locations for the new prototype and the resource burden associated with it. *Sampling Cost* and *Misclassified Cost* both perform direct minimization of the cost function. *Sampling Cost* compares placement at $\tilde{N}$ potential locations calculating the cost over all $\tilde{N}$ training samples (in our implementation, $\tilde{N} = N/10$). *Misclassified Cost* calculates the cost over all misclassified samples and once the two classes from the most expensive two-class boundary are identified, candidate locations are evaluated only over the misclassified samples for those two classes. *Misclassified Cost* down selects candidate positions in a way that offers a reduction in total operations. As classification accuracy improves, NSM using the misclassified samples as the candidate locations will reduce in computational complexity as the pool (typically) reduces over continued training. This is in contrast to *Sampling Cost* where the number of samples evaluated as candidate locations remains fixed.

In both cost motivated methods, as well as *Closest* and *Near-Mean*, new prototypes are initialized in positions of known training samples. However, *Mean* and *PC* allow the initialized position to be anywhere. This less restrictive initialization may be beneficial, however, it may also lead to formation of prototypes in sparse (or poorly defined) regions of the *pdf*. Additionally, the necessary computations can be complex. In the case of *Mean*, it is simply the average location of the misclassified samples within the class. *PC* however, requires the additional computation of the covariance matrix and its eigenvectors and eigenvalues.

3   Experimental Process and Results

The NSM methods are compared within the framework of a dynamic GLVQ over three data sets: the Mice Protein Expression data set [15], the USPS Handwritten image data set [16], and the Lunar Crater Volcanic Field hyperspectral data set [17]. In order to promote a fair comparison of NSM methods and allow networks to converge, we restrict the potential for NSM to occur after a fixed number of epochs, allowing at most one new prototype per fixed interval. Consequently, the *Closest* and *Mean* methods add a prototype (when appropriate) to the class with the largest classification error. Training and testing partitions for each data set are preserved between NSM methods, and prototypes in each are initialized to the same values. Additionally, the number of samples used in *Sampling Cost* is restricted to 10 % of the training set, selected randomly.

3.1 Data Sets and Experiment Setup

The Mice Protein Expression data set [15] reports protein expression levels in the brains of eight classes of mice with varying biology and treatments. While the report sought to identify individual proteins linked to learning and Down Syndrome [18], it presents an interesting classification problem. Due to empty fields in the data, 71 of 77 features are selected for 1047 samples. K-Fold cross-validation with $K = 5$ [19] is used to train and validate classification performance. For consistency in training, the protein expression features are linearly scaled on [0, 1]. A learn rate of $\alpha = 0.005$ is used and prototype insertion occurs after 100 epochs.

The USPS Handwritten data set [16] has 9,298 samples where each sample is a 16×16 pixel scan of a handwritten digit $\{0, 1, \ldots, 9\}$. Each 16×16 scan is "linearized" creating a vector with 256 features and each feature is linearly scaled on [0, 1]. K-fold cross-validation is used with $K = 5$ to train and validate performance. An $\alpha = 0.1$ is used and prototype insertion occurs after 150 epochs.

The Lunar Crater Volcanic Field (LCVF) hyperspectral data set [17] contains 1464 samples drawn from 35 classes, each with 194 spectral dimensions. As recommended in [20], each feature vector is normalized with its ℓ_2-norm to compensate for the effect of shadowing due to sensor geometry. Due to the sparsity of several classes, K-fold cross-validation is used with $K = 3$. The learn rate is $\alpha = 0.00001$ and prototype insertion occurs after 100 epochs.

3.2 Results and Discussion

This section introduces the graphs and tables used to draw specific discussion on each data set in Sects. 3.3, 3.4, and 3.5 for the Mice, USPS, and LCVF data sets respectively. The classification and cost-minimization performance on the three data sets previously described is shown in Figs. 1 and 2. They show the performance of *PC, Sampling Cost, Misclassified Cost, Mean*, and *Closest*. In order to improve readability of the figures, *Near-Mean* is omitted. For a baseline, GLVQ as described in [10] is used and is initialized with the maximum number of prototypes listed on the plots. Table 1 shows the training and number of prototypes required to exceed baseline GLVQ accuracy. Locations with a '−' identify NSM methods that did not meet the baseline performance. The strongest overall performers with the peak classification accuracy of each NSM method is listed in Table 2. These numbers are reported along with the total cost and required number of prototypes for the reported configurations.

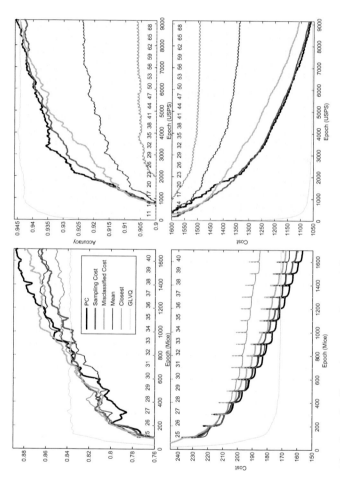

Fig. 1 Accuracy (*top*) and cost function value (*bottom*) versus testing epoch for DGLVQ with the following NSM methods: *PC, Sampling Cost, Misclassified Cost, Mean, Closest,* and GLVQ for the Mice (*left*) and USPS (*right*) data sets. GLVQ is evaluated using the maximum number of allocated prototypes and is included for reference. Plots are the average of the $K = 5$ validation results

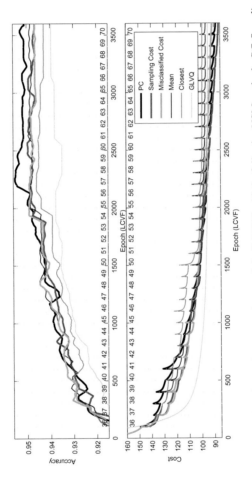

Fig. 2 Accuracy (*top*) and cost function value (*bottom*) versus testing epoch for DGLVQ with the following NSM methods: *PC, Sampling Cost, Misclassified Cost, Mean, Closest,* and GLVQ for the Mice (*left*) and USPS (*right*) data sets. GLVQ is evaluated using the maximum number of allocated prototypes and is included for reference. Plots are the average of the $K = 5$ validation results

Table 1 The number of epochs required to reach peak GLVQ baseline classification performance (value reported next to the data set name) with the resulting number of prototypes and cost reported for each data set and NSM method

Approach	Mice 84.24 %			USPS 94.47 %			LCVF 93.95 %		
	Epochs	Ptypes	Cost	Epochs	Ptypes	Cost	Epochs	Ptypes	Cost
PC	865	32	180.8	8414	66	1076	1036	45	110.2
Sampling cost	832	32	184.8	–	–	–	1349	48	103.0
Misclassified cost	731	31	190.8	9146	70	1107	1118	46	105.8
Mean	918	33	178.5	–	–	–	1108	46	107.3
Near-mean	704	31	195.4	–	–	–	1304	48	110.8
Boundary	822	32	196.6	–	–	–	1605	50.67	107.3

Table 2 Peak classification accuracy, number of prototypes, and total cost averaged over K-folds for each data set and NSM method

Approach	Mice			USPS			LCVF		
	Acc. (%)	Ptypes	Cost	Acc. (%)	Ptypes	Cost	Acc. (%)	Ptypes	Cost
PC	89.11	40	173.9	94.58	70	1065	95.57	57	96.77
Sampling cost	86.72	40	157.0	94.34	67	1080	94.97	66	89.94
Misclassified cost	88.06	40	168.1	94.47	70	1107	94.98	55	101.4
Mean	87.01	40	165.9	92.42	69	1337	94.97	61	99.29
Near-mean	88.06	39	168.1	94.09	70	1185	94.84	57	98.92
Boundary	85.40	40	186.5	90.70	69	1493	94.62	54	101.3
Full GLVQ	84.24	40	173.9	94.47	70	1057	93.95	70	94.88

3.3 Mice Protein Expression Results

As shown in Fig. 1 (left, top), all NSM methods surpass the classification accuracy of the baseline GLVQ and do so with fewer prototypes. Table 1 shows that the *Misclassified Cost* and *Near-Mean* are the first to reach this benchmark, utilizing the fewest prototypes (31 for each method). *Misclassified Cost* appears to dominate classification accuracy until approximately epoch 1200 and the insertion of the 36th prototype. From Fig. 1 (left, top) and seen in Table 2, beyond 1200 epochs *PC* offers the highest accuracy of 89.11 % with *Misclassified Cost* and *Near-Mean* finishing at 88.06 %. Table 2 further shows that *PC* achieves the highest cost, matching that of the baseline GLVQ.

The cost performance of the NSM methods is closely related to classification accuracy as is shown in Fig. 1 (left, bottom). This result is anticipated due to the formulation of GLVQ as a gradient descent algorithm. While all of the NSM methods show decreasing cost, *PC* shows the largest reduction followed by *Sampling Cost*. It is interesting to note that *Misclassified Cost* achieves higher classification accuracy even though several other NSM methods have lower cost curves. This difference is likely due to the fact that *Misclassified Cost* targets cost reduction by evaluating only misclassified samples for the placement of new prototypes. Addressing those misclassifications early on has the potential to strongly shape classification accuracy. Table 2 further shows that each method requires nearly the same number of prototype vectors to achieve their top accuracy, and that no clear trend exists between peak accuracy and associated cost.

3.4 USPS Handwritten Results and Discussion

Figure 1 (right, top) shows that *Misclassified Cost* has superior classification performance for the first 1500 epochs. Beyond 1500 epochs, the classification accuracy achieved by *PC* surpasses the other NSM method. In comparing the two direct cost minimization methods, the classification accuracy of *Sampling Cost* surpasses that of *Misclassified Cost* between 1500 and 7500 epoch. After 7500 epochs, the accuracy of *Misclassified Cost* is slightly better. Unlike the Mice Protein Expression data set, the baseline GLVQ performs on par or better than many of the NSM methods evaluated. This is further supported in Table 1 where we see that *PC* and *Misclassified Cost* are the only NSM methods that meet or exceed GLVQ for the USPS data set. The baseline GLVQ's strong performance might be attributed to the training time it enjoys with the full number of prototypes and that the number of prototypes used adequately represents the classification complexity of the data set. It is also possible that a non-dynamic GLVQ method is appropriate for relatively simple and "well balanced" classes (each class with approximately the same number of samples).

The two NSM methods that directly minimize cost offer some of the best classification performance. It interesting that *PC's* focus on the principal variance direction of the misclassified samples has the effect of indirectly minimizing the cost function yet leads to the lowest overall cost curve and highest classification accuracy as seen in Fig. 1 (right, bottom) and Table 2 respectively. The contextual information from clustering the misclassified samples with *PC* coupled with the "split metric" seems to provide superior initial prototype placement. Table 2 again shows that peak classification accuracy is achieved with approximately the same number of prototypes (*Sampling Cost* doing slightly better), and that the costs associated with those accuracies shows no clear trend.

3.5   LCVF Results and Discussion

Misclassified Cost offers very strong classification (Fig. 2 (top)) for the first half of the results (until approx epoch 1400), which is consistent with the Mice and USPS results. In Table 1, only *PC* achieves the baseline classification accuracy before *Misclassified Cost* with any noticeable lead. Up to 1200 epochs, *Misclassified Cost* also offers the strongest cost reduction as shown in Fig. 2 (bottom). After epoch 1400, the NSM methods (with exception of *PC* and *Boundary*) seem to converge, resulting in maximum classification accuracies in the range of 94.84–94.98 % (a difference of 0.14 %), which is also seen in Table 2. While the convergent result in the second half of Fig. 2 (top and bottom) may not be surprising due to the small sample size and disparate number of elements per class in the LCVF data set, *PC* does seem to offer marked performance gains, peaking at 95.57 % and resulting in the second lowest total cost. We see that *Sampling Cost* obtains a slightly lower cost than *PC*. Table 2 shows that widely varying numbers of prototypes are associated with the peak accuracies achieved by different NSM methods, while again there is no clear trend in resulting costs.

4   Summary

In this paper we promote the dynamic addition of prototype vectors to achieve superior performance and efficiency for GLVQ. We introduce the concept of network structure modification (NSM) into the standard LVQ taxonomy to describe individual methods for dynamic addition/deletion of prototypes within the network. This paper presents two new NSM methods, *Misclassified Cost* and *Near-Mean*, to achieve improved classification accuracy. The former explicitly minimizes the cost function by placing prototypes in way that minimizes the cost due to misclassified samples. *Near-Mean* selects the misclassified sample nearest the mean of the misclassified samples from the class with the largest classification error.

Several NSM methods are evaluated based on training time and prototypes required to meet a baseline classification performance of GLVQ. We find overwhelming evidence of improved classification accuracy with fewer prototype when considering a DGLVQ. We also find that over all three data sets, *Misclassified Cost* is consistently one of the better performing methods based on classification accuracy at the earliest opportunity. This fast convergence with fewer prototypes is a benefit in terms of overall performance and may support a reduced generalization error [6]. Faster convergence coupled with *Misclassified Cost's* diminishing computational complexity as training continues could be beneficial in real-time continuous learning applications.

We examine the trends of NSM methods as prototype networks expand and reach their peak performance configurations. Overall, we see strong classification accuracy from methods that effectively control cost, with some of the best performance from methods that minimize the cost directly. While *Sampling Cost* offers good classification and cost minimization performance by selecting candidate positions from randomly selected samples, *Misclassified Cost* offers promise as an alternative, with better classification accuracy shown all data sets (including a full 2 % gain in the Mice Protein Expression data).

While we anticipated a clear distinct advantage of direct cost minimization NSM methods, *PC* indirectly minimizes cost and consistently results in the best accuracy and cost performance. Our adaptation and implementation of the *PC* method to dynamically add prototypes in GLVQ showed the best results across all data sets. *PC's* use of the misclassified sample variance allowed for a more informed prototype placement. This suggests future work related to cluster metrics to aid prototype placement is warranted. Using the same cluster metrics could improve prototype initialization, to include the number per class and their specific locations, which may result in improved accuracy and cost minimization performance, while reducing training requirements.

Acknowledgments The authors would like to thank Dr. E. Merényi for making the LCVF data set available for this work. The views expressed in this paper are those of the authors and do not reflect the official policy or position of the United States Air Force, the U.S. Department of Defense, or the U.S. Government.

References

1. Kohonen, T.: The self-organizing map. In: Proceedings of the IEEE, vol. 78, no. 9, pp. 1464–1480, Sept 1990
2. Losing, V., Hammer, B., Wersing, H.: Interactive Online Learning for Obstacle Classification on a Mobile Robot. IEEE (2015)
3. Kirstein, S., Wersing, H., Körner, E.: Rapid online learning of objects in a biologically motivated recognition architecture. In: German Pattern Recognition Conference DAGM, pp. 301–308 (2005)
4. Schleif, F.M., Villmann, T., Hammer, B.: Local metric adaptation for soft nearest prototype classification to classify proteomic data. In: International Workshop on Fuzzy Logic and Applications, Lecture Notes in Computer Science, vol. 3849, pp. 290–296. Springer (2006)
5. Mendenhall, M.J., Merényi, E.: Relevance-based feature extraction for hyperspectral images. IEEE Trans. Neural Netw. (2008)

6. Crammer, K., Gilad-bachrach, R., Navot, A., Tishby, N.: Margin analysis of the lvq algorithm. In: Advances in Neural Information Processing Systems, pp. 462–469. MIT press (2002)

7. Poirier, F.: DVQ: dynamic vector quantization application to speech processing. In: Second European Conference on Speech Communication and Technology, EUROSPEECH 1991, Genova, Italy, 24–26 Sept 1991

8. De Stefano, C., D'Elia, C., Marcelli, A., di Frecac, A.: Improving dynamic learning vector quantization. In: 18th International Conference on Pattern Recognition ICPR 2006, vol. 2, pp. 804–807 (2006)

9. Kohonen, T.: Self-organizing Maps, 3rd edn. Springer (2000)

10. Sato, A., Yamada, K.: Generalized learning vector quantization. In: Advances in Neural Information Processing Systems, pp. 423–429. The MIT Press (1996)

11. Grbovic, M., Vucetic, S.: Learning vector quantization with adaptive prototype addition and removal. In: Proceedings of the 2009 International Joint Conference on Neural Networks IJCNN'09, pp. 911–918. IEEE Press, Piscataway, NJ, USA (2009). http://dl.acm.org/citation.cfm?id=1704175.1704308

12. Bermejo, S., Cabestany, J., Payeras, M.: A new dynamic lvq-based classifier and its application to handwritten character recognition. In: ESANN, pp. 203–208 (1998)

13. Kietzmann, T.C., Lange, S., Riedmiller, M.: Incremental GRLVQ: learning relevant features for 3D object recognition. Neurocomput. 71(13–15), 2868–2879 (2008). http://dx.doi.org/10.1016/j.neucom.2007.08.018

14. Zell, A., Mamier, G., Vogt, M., Mache, N., Hübner, R., Döring, S., Herrmann, K., Soyez, T., Schmalzl, M., Sommer, T., Hatzigeorgiou, A., Posselt, D., Schreiner, T., Kett, B., Clemente G., Wieland J.: Stuttgart Neural Network Simulator (SNNS): User Manual, Version 4.1 (1995)

15. Lichman, M.: UCI Machine Learning Repository (2015). http://archive.ics.uci.edu/ml

16. Hull, J.J.: A database for handwritten text recognition research. IEEE Trans. Pattern Anal. Mach. Intell. 16(5), 550–554 (1994)

17. Merényi, E., Farrand, W., Taranik, J., Minor, T.: Classification of hyperspectral imagery with neural networks: comparison to conventional tools. EURASIP J. Adv. in Signal Process. 2014(1), 71 (2014). http://asp.eurasipjournals.com/content/2014/1/71

18. Higuera, C., Gardiner, K.J., Cios, K.J.: Self-organizing feature maps identify proteins critical to learning in a mouse model of down syndrome. In: PLoS ONE 10(6): e0129126 (2015)

19. Hastie, T., Tibshirani, R., Friedman, J.: The elements of statistical learning. In: Springer Series in Statistics. Springer New York Inc., New York, NY (2009)

20. Merényi, E., Singer, R.B., Miller, J.S.: Mapping of spectral variations on the surface of mars from high spectral resolution telescopic images. Icarus 124, 280–295 (1996)

Author Index

© Springer International Publishing Switzerland 2016
E. Merényi et al. (eds.), *Advances in Self-Organizing Maps and Learning
Vector Quantization*, Advances in Intelligent Systems and Computing 428,
DOI 10.1007/978-3-319-28518-4

369

Printed in the United States
By Bookmasters